This book is to be returned on
or before the date stamped below

- 9 JUN 2004

3 0 SEP 2004

14 - 3 - 05

Understanding Wetlands

Understanding Wetlands

Fen, bog and marsh

S. M. Haslam

Taylor & Francis
Taylor & Francis Group

LONDON AND NEW YORK

First published 2003 by Taylor & Francis
11 New Fetter Lane, London EC4P 4EE

Simultaneously published in the USA and Canada
by Taylor & Francis Inc,

29 West 35th Street, New York, NY 10001

Taylor & Francis is an imprint of the Taylor & Francis Group

© 2003 S. M. Haslam (text)

© 2003 Y Bower (illustrations)

Typeset in $9\frac{1}{2}$/12pt Galliard by Graphicraft Limited, Hong Kong
Printed and bound in Great Britain by TJ International Ltd, Padstow,
Cornwall

British Library Cataloguing in Publication Data
A catalogue record for this book is available from the British Library

Library of Congress Cataloging in Publication Data
Haslam, S. M. (Sylvia Mary), 1934–
    Understanding wetlands : fen, bog, and marsh / S. M. Haslam.
        p.   cm.
    Includes bibliographical references and index.
    ISBN 0-415-25794-8 (hc.)
    1. Wetlands.   2. Wetland ecology.   3. Wetland conservation.
    I. Title.
    QH87.3 .H38 2003
    577.68—dc21   2002044380

ISBN 0-415-25794-8

To Peter Brockett, who worked so valuably on reed (*Phragmites*) and who alone saw its unity and brought together its experts on:

growing

reed bed management

harvesting

storage and selling

thatching

fire risk

buildings

thatching history from Anglo-Saxon times

chemical composition

microbial breakdown

reedswamps

conservation

ecology

autecology.

Contents

Preface   xi
Acknowledgements   xiii

1 Introduction   1

Wetland basics 6
Ideas on wetlands: fact and fable 8
The creations of the waters 15

2 Wetlands matter   22

Introduction, and the World Charter for Nature 22
Values 24
Considerations of values 27
Wetland to dryland: the changing from fish and fowl to grain
* and vegetable 28*
Wetland products 31

3 How wetlands work   42

Introduction 42
Integrating wetland processes 43
Wetland landscapes 50
Biodiversity 52

4 In wetland wilds   57

Continuities and discontinuities 57
Geomorphological, hydromorphic, and similar classifications 59
Vegetation classification 60
Wetlands in the landscape 76
Bog 77
Marsh 86
Reedswamp 92
Fens 93

Tall herb and short herb communities 98
Grassland 102
Woodland 102

5   The animals   103

Introduction 103
Invertebrates 104
Fish 110
Birds 110
Mammals 115
Reptiles 118
Amphibia 118
Microorganisms and fungi 118

6   The waters of the wetlands   121

Water in the landscape 121
Soils and drainage 124
Vertical fluctuations 125
Sideways movement 132
Vegetation as an indicator of water regime 132
Case studies 133
Discussion 149

7   Chemical types and vegetation types   150

Introduction 150
Chemical types in the landscape 152
Plants as indicators of nutrient regime 155
Case studies 156
Chemical impact 164

8   The power to purify   166

Principles and definitions 166
How fens, marshes and reedswamps (natural and constructed wetland)
  act chemically 168
Purification 169
Buffer strips 172

9   *Phragmites*: a study in plant behaviour and human use   178

The plant unit 182
The seasonal cycle 184
The seedling and young plant 192
The advancing plant 194

Chemistry and competition 194
The water 196
The thatching reed 197
The maintaining of the stand 199
Conclusion 202

10 The silent battlefield: vegetation changes   203

Vegetation develops 203
*Sallow (*Salix cinerea*) carr invasion 204*
Carex paniculata *and* Phragmites australis *205*
Galium aparine *(goosegrass) in tall-herb fen vegetation 209*
The Schoenus nigricans *community 211*
Schoenus nigricans *and* Molinia caerulea *215*
Schoenus nigricans *and* Cladium mariscus *215*
Phragmites *made sparse in three other vegetation types 217*
Reedswamp invasion of open water 220
A native and an introduced grass in the Camargue, France 221
Phalaris arundinacea *and* Urtica dioica *in flood meadows,*
  River Luznice, Czech Republic 222
Typha *spp. in North America 222*
Cladium mariscoides *and* Typha jamaicense *in the Everglades, Florida 222*
Myrica gale, Cladium mariscus *and the Keeper 222*
Combined ills 223
Conclusions 223

11 Threats and losses, past and present   226

The major dangers 226
Management and loss of Broadland over time in East Anglia 236
Deterioration of waterfowl and wet grassland 241
Reedswamp dieback 241

12 Conservation   243

Introduction 243
Principles of conservation 246

References   248
Basic Figures   259
Basic Charts   272
Index   285

Preface

This is a book about understanding fens, bogs and marshes. It concentrates on habitat, vegetation and animals and human impact, and the interactions between them. It is intended as an introduction for universities and colleges, conservation organisations and the Environment Agency, and indeed for sixth-form and further education work. It is comprehensive for beginners, and should give a 'jumping-off' point for more advanced students and researchers.

Those who find the book valuable should thank Mr T. R. Graham of the (then) Her Majesty's Inspectorate of Pollution. Without his contract, I – who had seen other wetlands but properly studied only East Anglia and reeds – would not have realised there was no such book. I wanted it for that contract! So, several years later, here it is, for others who also want, in one book, a holistic view, and later than Tansley (1939).

All readers of this book probably know wetlands are a threatened habitat, but most probably do not know how little is being done to meet that threat. There is more lip-service than for, say, chalk grasslands, but where is the actual service? Barely half a dozen publishing on British conservation research; a score or two on any British wetland research; money for stopping abstraction, reversing drainage, preventing pollution and necessary management is noticeable by its absence (with few exceptions).

The need is great. Can any reader respond to that need?

S. M. H.
May 2001

Acknowledgements

First, may I thank, most sincerely, Mrs Yvette Bower for her new illustrations (and for leave to use ones drawn for earlier books) and Mrs Tina Bone for word processing.

I am much obliged, for checking parts of the script, to Dr L. A. Boorman, Professor R. S. Clymo, Dr L. Friday, Dr F. Hughes, Dr P. José and Mr D. F. Westlake.

My grateful thanks go to all those who most kindly supplied me with literature or permissions, especially the Aquatic Plant Information and Retrieval System, Florida, Dr R. Buisson, Mr S. Daly (Science Periodicals Library, Cambridge), English Nature Library, Dr P. Grillas, Dr D. Harper, Dr N. Haycock, Dr P. Kirby, Scottish Natural Heritage and Dr M. Wallstein.

1 Introduction

Can the rush grow without mire? Can the flag grow without water?

(Job 8.11)

'When I use a word,' Humpty Dumpty said in rather a scornful tone, 'it means just what I choose it to mean – neither more nor less.'
'The question is,' said Alice, 'whether you can make words mean different things.'
'The question is,' said Humpty Dumpty, 'which is to be master – that's all.'

(Lewis Carroll, *Alice Through the Looking Glass*)

What is a wetland?
A wetland is whatever a competent expert says it is

This, unfortunately, is the only generally agreed definition of a wetland. It is also one that shifts the controversy on to who is a competent expert! In any case, the word wetland is American, and *Webster* defines it as land containing much soil moisture, such as swamp or bog, the word commonly being used in the plural, 'wetlands'. It is only in recent decades that it has crossed the Atlantic as a common word – or, indeed, has been used as a legal term within North America. This is how confusions and indeed unnecessary controversies abound. 'Unnecessary', as all would agree that if it is too wet to go dryshod, and too dry or too covered in vegetation to go by boat, it is a wetland (Figure 1.1). Trouble comes at the fringes – Is this? Is that? – and scientists, quite properly, produce definitions that cover the perspective from which they are working (Florida? Australia? Soil processes? Bird habitat?). Table 1.1 lists a few of the many definitions, all by competent experts, and all valid within their parameters, and containing contradictions – should saltwater be included? Should open water be included? Are plants the best character to use?

Why, then, use the term 'wetlands' at all? It does seem advisable, as it has so quickly come into general use after its ocean crossing, and is easily understood as wet land, 'Gumboot Country' in the New Zealand phrase. In non-ecological English, it means bogs, fens, marsh, swamps, sloughs, morasses, flood meadows, water meadows and many more.

Figure 1.1 What is a wetland?

Table 1.1 Some definitions of wetlands

1   Land containing much soil moisture, as swamp or bog, usually used in the plural (Webster's Dictionary)
2   Areas of saturated ground which may have associated free water surfaces or may be intermittently dry (International Biological Programme)
3   With hydric soils, formed by inundation or saturation for very long periods of time (Florida law)
4   Land having the water table at, near or above the land surface or which is saturated for a long enough period to promote wetland or aquatic processes as indicated by hydric soils, hydrophytic vegetation and various kinds of biological activity which are adapted to the wet environment (Wetlands Registry, Canada)
5   Areas inundated or saturated by surface or ground water at a frequency and duration sufficient to support, and that under normal conditions do support, a prevalence of vegetation typically adapted for life in saturated soil conditions (United States Army Corps of Engineers)
6   Areas dominated by specific herbaceous macrophytes, production of which takes place predominantly in the aerial environment above the water level while plants are supplied with amounts of water that would be excessive for most other higher plants bearing aerial shoots (International Biological Programme)
7   Areas of marsh, fen, peatland or water, whether natural or artificial, permanent or temporary, with water that is static or flowing, fresh, brackish or salt, including areas of marine water the depth of which at low tide does not exceed six metres (Ramsar Convention)
8   Lands transitional between terrestrial and aquatic systems where the water table is usually at or near the surface or the land is covered by shallow water. Wetlands must have one or more of: (1) at least periodically, the land supports predominantly hydrophytes, (2) the substrate is predominantly undrained hydric soil and (3) the substrate is non-soil and is saturated with water or covered by shallow water at some time during the growing season of the year (United States Fisheries and Wildlife Service)
9   Land permanently flooded to saturated, but standing water is rare. Sited at the interface between terrestrial and aquatic habitats, different to each but dependent on both (Mitsch 1994)
10 Lands where saturation with water is the dominant factor determining the nature of soil development and the types of plant and animal communities living on the surface (Novotny and Olem 1994)
11 Areas with: (1) no maritime influence, (2) hydrologic characters, the substrate waterlogged for at least part of the growing season, the water level above, at or below but near the substrate surface, (3) macrovegetation, where present, has morphological, anatomical or physiological adaptations to the aquatic environment, (4) nutrient supplies range from ombrotrophic to mineratrophic, and (5) peat-forming, alluvial or sedimentary (SCOPE)
12 Area of ground that is normally permanently saturated within 10 cm or so of ground surface. Surface water may or may not be associated (Lloyd and Tellam, in Hughes and Heathwaite 1995)
13 Area with its water table near, at or above ground surface (Ratcliffe, in Burnett 1964)
14 Saturated ground which may have an associated saturated free-water surface, normally perennial but may be non-perennial for limited periods
15 Areas where water is at or slightly above or below ground level for most of the year. Also water with a floating raft of vegetation (Löfroth 1991)

Trying to define the subdivisions is as difficult, as seen in Tables 1.2–1.4. The difficulties arise because of the following four reasons.

1   The English words started with wide, often interchangeable, meanings or meanings varying locally, or used only locally (and see the Fact and Fable section, later). Anglo-Saxon Fenn, for instance, described any waterlogged land.

Table 1.2 Wetland names used in Dr Johnson's Dictionary (S. Johnson 1755. *A Dictionary of the English Language*. W. Straham, London)

BOG	A marsh, a morass, a ground too soft to bear the weight of the body
FEN	A marsh; low flat and moist ground; a moor; a bog
MARSH	A fen; a bog; a swamp; a watery tract of land
MEADOW	Ground somewhat watery, not plowed, but covered with grass and flowers
MIRE	Mud; dirt at the bottom of water
MOOR	A marsh; a fen; a bog; a tract of low and watery grounds
MOORLAND	Marsh; fen; watery ground
MORASS	Fen; bog; moor
PASTURE	Ground on which cattle feed
PEAT	A species of turf used for fire
QUAGMIRE	A shaking marsh; a bog that trembles under the feet
SLOUGH	A deep miry place; a hole full of dirt
SWAMP	A marsh; a bog; a fen
SWARD	The surface of the ground, whence green sward or green sword
TURF	A clod covered with grass
BULRUSH	A large rush, such as grows in rivers, without knots
FLAG	A water plant with a broad bladed leaf and yellow flower, so called from its motion in the wind
REED	A hollow knotted [noded] stalk, which grows in wet grounds
(RUSH – description of Genus. RUSHY abounding with rushes)	
SEDGE	A growth of narrow flags; a narrow flag (provincial word)

Table 1.3 Wetland names used outside ecology: recent dictionary definitions

BOG	A piece of wet spongy ground, consisting chiefly of decayed moss and other vegetable matter, too soft to bear the weight of any heavy body upon it
FEN	Lowland covered wholly or partly with shallow water or frequently inundated, a marsh. The 'Fens' are certain low-lying districts inland from the Wash
MARSH	A tract of low-lying land, usually flooded in winter and more or less watery at all times
MEADOW	Permanent grassland mown for hay. Later extended to any grassland especially locally to a tract of low well-watered ground, usually near a river *Water meadow* A meadow periodically overflowed by a stream, 1733. (The true water meadows, penned water systems, became widespread in the seventeenth century)
MIRE	A piece of wet swampy ground, a boggy place; wet or soft mud, slush, dirt. A north-country word for a marsh or boggy place
MOOR	A tract of unenclosed ground, usually a heath; a tract preserved for shooting; a marsh; waste land on which tin is found
MOORLAND	Uncultivated land, modern use as land abounding in heather, a moor
MORASS	A wet swampy tract, a bog, marsh, occasionally boggy land

MOSS	A bog, swamp or morass, a peat bog; wet spongy soil; bog
	    Moss-hag A pit or slough in a bog (Scots)
	    Moss-flow A watery bog (Scots)

(PASTURE A piece of land covered with growing herbage eaten by cattle, grassland)	
QUAGMIRE	Wet boggy ground that yields or quakes under the feet
SLOUGH	A piece of soft, miry or muddy ground, especially a place or hole in a road or way filled with wet mud or mere and impassable by horses, etc. (US: a marsh or reedy pool, small lake, etc.)
SWAMP	A tract of low-lying ground in which water collects; a piece of wet spongy ground, a marsh or bog. Originated in North America, land too moist for cultivation, covered by trees and other vegetation, rich soil

Table 1.4 Wetland names used ecologically

BOG

1   Peat-accumulating wetland that has no significant inflow or outflow and supports acidophilous mosses (Novotny and Olem 1994)
2   Wetland supported by rain (Gilvear *et al.* 1990)
3   Acid nutrient-poor peat area, developed under the influence of precipitation (traditional British)
4   Areas with a prominent ground-layer of peat-building sphagna (*Sphagnum papillosum, S. magellicum, S. capillifolium*); mainly with a low canopy of shrubs (heather, etc.) and some sedges and rushes; damp to waterlogged acid nutrient-poor peat; grades to less nutrient-poor mires with non-peat-forming sphagna (e.g. *S. recurvum, S. squarrosum*) and then grade to nutrient-poor fens (Rodwell 1995)

FEN

1   Peatland fed by minerals from the earth's surface (Heathwaite and Hughes, in Heathwaite and Hughes 1995)
2   Peat-accumulating wetland that receives some drainage (Novotny and Olem 1994)
3   Alkaline or neutral peat, developed mainly or partly under calcium-rich groundwater or run-off water, of medium or high nutrient status (traditional British). (The geographic Fenland is in fact partly silt)
4   Characterised by mixtures of certain of the important swamp emergents and a variety of often tall perennial herbaceous dicotyledons. Not restricted to open-water transitions and flood plains, nor confined to organic substrates (Rodwell 1995)

MARSH

1   Areas of mineral soil watered by river or groundwater (traditional British)
2   Included in fen, an informal term, describing any wet ground and/or its vegetation (Rodwell 1995)
3   Predominantly mineral substrates, not accumulating peat, having regular inundation with surface water, and sedge or herb communities (Heathwaite and Hughes, in Heathwaite and Hughes 1995)
4   Drained fen (in Broadland, East Anglia) neither waterlogged nor necessarily on mineral soil (George 1992)

SWAMP

1   Species-poor vegetation types, generally dominated by bulky emergent monocotyledons (grasses, sedges, etc.) characteristic of open-water transitions with permanently or seasonally submerged substrates (Rodwell 1995)
2   Wetlands dominated by shrubs and trees, with or without peat (Gopal, in Patten 1990)
3   Areas with summer water level usually above soil level, usually dominated by *Phragmites* or other tall, thin species (Tansley 1949)

MIRE

1   Peat-forming wetland (Löfroth 1991)
2   Bryophytes, herbaceous plants and sub-shrubs forming bogs, wet heaths, low-nutrient fens, flushes, springs and soakaways where the ground is permanent or periodically waterlogged by high humidity, high groundwater table (or lateral flow), soils mineral or peat. Peat-building sphagna (and *Eriophorum angustifolium*) are dystrophic, small sedge and rush communities, oligotrophic. Where calcium increases, *Sphagnum* carpets may still occur, although fen, fen meadow and marsh vegetation come with higher calcium and other nutrients (Rodwell 1994)
3   North-country word for a marsh or boggy place (Jackson 1928)
4   Accumulating soils which, in bogs are acid and almost entirely organic, and in fens may be either or both organic and inorganic (Heathwaite & Hughes in Heathwaite & Hughes 1995)

Table 1.4 (cont'd)

MOOR, MOORLAND

1   With peat and ericoid plants as the chief vegetation (Jackson 1928)
2   MOR, Danish. Acid humus of cold wet soils that inhibits soil activity and may form peat. (Opposite: Mull, humus of well-aerated moist soils, formed by the action of soil organisms on plant debris and favouring plant growth) (Jackson 1928)

CARR

1   Bush-level woodland in fens, most often sallow, sometimes alder buckthorn or other. Alderwood may be termed 'alder carr'
2   Extended to any woodland in fen or marsh
3   Open marsh (Yorkshire)

Gradually, as English became more standardised, and, later, as interest in types of countryside increased, the words became more specialised and more 'ecological' – Scottish bogs were usually called mosses until very recently. They were not, however, called fens, a more southern term.

2   English has much in common with various European languages, particularly the Norse and the Saxon. With the passing of centuries, fen (English) and venn (Dutch), moor (English) and mor (Norse) have diverged, but those writing in English with a background of another language are likely to choose the English word looking closest to their own.

3   American has also, of course, diverged considerably from its English ancestry.
      And, if this was not enough:

4   No ecologists agree exactly on what habitats comprise any type of wetland: are valley bogs bog or poor fen? Where does wet grassland become marsh? They also disagree on whether types should be defined by water, soil, vegetation or location. As pointed out above, all are valid within their own terms. No classification can comprise all. A typical dilemma comes in The Fenland of eastern England. This is black peat to inland and south, but silt fens to seawards and north. Are the silt fens 'fens'? In the black peat, most is now fen peat (accumulated under water) but part has some original bog peat (sphagnum and brown-moss) above. Is Holme Fen, with bog peat, fen or bog? Wicken Fen, known world-wide as the exemplar of The Fen, has a little sphagnum – fen, poor fen: bog? The Fenland is drained throughout, and most is drained enough to grow high-yielding crops. Is such arable land fen?

For the purposes of this book, looking mainly at the British Isles, the following can be recognised, freshwater habitats.

1   Reedswamp: winter-flooded or continuously flooded land with thin (linear) dominants, most widely reed (*Phragmites*) but also other grasses and Cyperaceae.
2   Fen: peatland developed under surface water, generally but not always calcium-rich.
    i   Rich fen: water and peat rich in calcium and other nutrients.
    ii   Calcium-dominated fen: water rich in calcium, but not other nutrients. Calcium dominates the peat and habitat, reducing the availability of other nutrients.
    iii Poor fen: water nutrient-poor, but richer than in bog.

3   Bog: peatland developed solely under rain, or resembling this.
    i   Blanket bog (found in the north and west of the British Isles and in Norway):
         rain-fed, growing on soil or directly on rock.
    ii   Raised bog: rain-fed, growing above another wetland type, e.g. infilled lake, flood
         plain or blanket bog.
    iii Valley bog: receiving downslope water from a nutrient-poor catchment as well as
         rain.
4   Marsh: mineral to organic-rich habitat developed under surface water, merging into
    fen.
5   Wet grassland: grassland with soil flooded or at least waterlogged for enough of
    the year to influence species composition. May be developed on any of the other
    habitats.
6   Wet woodland: woodland with soil flooded or at least waterlogged for enough of the
    year to influence species composition; or for established woods, ones that were that
    wet when the woody plants were invading. May be developed on any of the habitats
    listed above.
       These are not very controversial. However, another group is not usually included.
7   Wetland habitats, now drained and protected from flood, and bearing arable, dry
    grassland, dry woodland and settlements. These are on land that was thoroughly wet
    before human impact, and would rapidly become so again if that human effort was
    removed. The superficial water and all biota dependent on it have gone or been
    relegated to drainage dykes, but the underlying features of the wetland remain: posi-
    tion in catchment, elevation in relation to catchment, ability – if given the opportun-
    ity – to catch and store water, to form wetland soils, etc.

Excluded from this group are those peatlands above ordinary surface water level, the
blanket and raised bogs, where all peat has been removed. Here the return of the wet,
growing bog is not rapid, and may require centuries or a change in climate. This denuded
habitat is dry land.

Wet grassland is still very common. Flood meadow is the commonest – on flood plains,
flooded regularly in the past from river (or lake). River floods brought fertilising sediment,
so these could be the most fertile (or, most easily fertile!) fields in the catchment. River
flooding is now infrequent except in highland areas with unembanked streams and undrained
land, although in some of the wetter fields the water table does rise above the surface in
wet winters.

Water meadows, a type of penned water system, were fully developed in the seventeenth century. These were mainly on chalk, where flow was most constant. By means of
a complex system of irrigating ditches, and sluices (hatches) on the river, the meadows
could be flooded with silt-carrying water – and drained – to give the optimum yield from
the earliest grazing to valuable hay. A few are now being restored.

Wetland basics

Through this book are references to displays called 'Basic', both figures and charts, and
starting on p. 259. These give the basics of wetland pattern, habitat, impact, etc. The first,
Basic Figures 1–4, show the typical positions and kinds of various wetlands. There is the

fundamental division between flood plain and hill wetlands (although in practice, as with every ecological division, this is not exact, as flood plains may occur in hills!). There is the reedswamp by open water in the plains, rich fen, calcium-dominated fen and poor fen habitats, some fen carr (wood and shrub), bog-woodland, and rush-pasture, and above the flood plain three types of bog, raised bog raised above the plain, valley bog sloping up from it and blanket bog and moor on the hills above.

While fen occurs where peat can accumulate, where coarser sediment accumulates in the flood plain, fen is absent. As the higher ground dries better in between floods, the communities differ more. Thanks to the value for construction of river gravel and sand, the fastest growing wetland type is the reedswamp, marsh and carr on shallow edges of abandoned gravel and sand pits. Alluvial sand and gravel may be at the surface, or may have been deposited much earlier and been covered by finer substrates. Soon after the last Ice Age, melting ice caused fierce flows capable of carrying gravel. These flows have not since been repeated.

Basic Figures 8, 15 and 16 show farming practices – draining, fertilising to convert the more fertile land to arable, and the less fertile to pasture or conifers.

Water is the driving force of wetland: no water (past or present); no wetland. Quality as well as quantity is relevant. If the water is rain only, bog forms. If it is silt-rich, rich marsh or rich fen forms. Calcium-rich and other-nutrient-poor water lead to calcium-dominated fen. Water quality depends on where it comes from, which may be clouds, springs or rivers. And on what happens to it between there and the wetland, that is, what chemical substances are added (or filtered out). Next in importance is human impact. In fact, in lowland wetlands the habitat and vegetation as it would be without human impact is unknown (except insofar as there is a record in the peat or soil). The present habitats, the water, soil, flora and fauna have developed under management, and can only be preserved with management. There is no such thing, now, as 'natural' fen vegetation, only 'semi-natural' or 'wild' communities. While some bogs date back to, almost, the Ice Age, and so are natural in origin, there is now doubt about whether blanket bogs have been started by management, intentional or otherwise. Bogs, in general, have less management and less varieties of management than fens and marshes.

This intensity of impact is not widely found in the Americas, Africa or Australasia. Here many wetlands can be in a state that, while not truly natural (i.e. untouched by mankind), is what Americans call 'pristine'. The types include more woodland, forested swamps being very typical of the Americas. They may be on peat or more riverine habitats. There are Tropical Rain Forest peatlands and other tropical, savannah and Mediterranean ones. Centred in Asia, at latitudes 10–40°N, are the most ancient intensively managed wetland habitats: rice paddies.

Wetland area does not vary much down the latitudes, from arctic snow-melt hollows to tropical (brackish) mangrove swamps, but rice is a crop of such value, and such yield even before agrochemicals, that where it can be grown well, it will be – extending over, to varying extents, to Europe (Po plain, Italy; Rhone delta, France, etc.). The rice paddies are also the only wetland type that has been greatly extended on to dry land, by people. Rice paddies occur on (prepared, terraced) hillsides where the watery habitat can occur only with management.

Apart from the rice paddies, wetlands have been shockingly reduced in the past few millennia. Wetlands are estimated to cover about 6 per cent of the earth's surface, about 10 per cent in Britain. In the Netherlands, barely 1 per cent of the original peatland remains, about half the British peatland is drained or altered. In East Anglia, the relatively undamaged fenland was about 10 km² in 1983 compared to 3400 km² in 1637. Between 1910 and 1978 about 60 per cent of Welsh lowland peatland was lost. Even in the United States, a newly developed and, now, fairly environmentally aware country, almost half (45 000 km²) the wetland has been lost in the past two centuries (Hughes and Heathwaite, in Hughes and Heathwaite 1995).

Ideas on wetlands: fact and fable

Why should wetlands have attracted so much destruction? In rice-growing regions the reason is plain. The paddies are the best way of feeding an always vast population, and so are superimposed on natural wetlands as well as stretching beyond these. Elsewhere, people in wetlands may drown, catch malaria, other animal-borne illnesses (e.g. bilharzia, amoebic dysentery), and bacterial and viral illnesses (e.g. typhoid, cholera, many abdominal disorders), and prolonged sojourning (with ignorance) can bring rheumatism, chronic digestive and liver disorders, and scrofula (pulmonary and glandular modifications), as recorded in the nineteenth century. In the Fenland, 'Fen ague', considered a form of malaria, was common in both the quotidian and, less frequently, the tertian form (ill every day and every three days, respectively).

And, of course, there was the Will o' the Wisp, Jack o' Lantern, the *ignis fatuus* of the Romans. These were self-lighting balls of marsh gas, presumably mainly methane with traces of spontaneously combustible phosphorus gases. These looked, at a distance, like lanterns, and acquired a bad reputation for leading homecoming travellers astray into the wetland.

In the Broads, Will o' the Wisps were frequent in the nineteenth (and presumably earlier) centuries, until the turf ponds were largely infilled. But by 1900 they were seen but rarely (George 1992). George queries whether this is partly because of less open water, but, as enormous amounts of methane are released when dyke and pond sediment is stirred, is it partly because fewer people are now on the open fen until dark? This author was in the Breck Fens until dark regularly in the 1950s, however, and never saw the lantern-man.

What had the wetlands to compensate? Fowling, fishing (not bog) and, on drier parts, grazing. This was good for a breed of solitary people, inbred, fiercely independent. Not the squirarchy or aristocracy from whom most leaders came who were leaders in appreciation of the country as well as in politics.

As little as 200 years ago, Edward Ferrars, in Jane Austen's *Sense and Sensibility*, could remark:

> 'I am convinced,' said Edward, 'that you [Marianne] really feel all the delight in a fine prospect that you profess to feel . . . I like a fine prospect, but not on picturesque principles. I do not like crooked, twisted, blasted trees. I admire them much more if they are tall and straight and flourishing. I do not like ruined, tattered cottages. I am not fond of nettles, or thistles or heath blossoms. I have more pleasure in a snug farm-house than a watch-tower – and a troop of tidy, happy villagers please me better than the finest banditti in the world.'

Marianne looked with amazement at Edward.

This shows what educated folk, not directly concerned with land management, were saying. It may have had the same influence on the land as, say, Wildlife Trust members do today. The lovers of the picturesque were concerned with form, for the view. The unfashionable were concerned with function, the function of giving a good life to those dependent on the land. Neither is concerned with plants or animals for their own sake. Edward would have disapproved of 'wasted' wetlands, and it is doubtful Marianne would have approved of a landscape so featureless.

Ideas of 'England's green and pleasant land' are traditionally green grass, with garden or lowland countryside behind. Scottish beauty is a heather moor and mountain, not an impassable bog. Wetlands have only very recently become associated with the pleasant – and are still not the typical pleasant. It is quite reasonable to get rid of that which is displeasing. And got rid of, the wetlands were.

Public esteem for, say, bogs in the distance improved with the eighteenth-century interest in the wild, and no doubt also with the drying of displeasing wetlands – particularly, perhaps, those blocking roads? Even in the late nineteenth century and, rarely, even today, some small peatlands or morasses are dangerous to cross. Wetlands improved in estimation as they became less dangerous, less unhealthy and more like 'dry land'. This was also when their main crops, and their characteristic population, were squeezed out.

Memories linger long, and the extracts here are of the awfulness rather than the beauty and value of the English wetland.

St Guthlac lived from 674 to 716. *Felix's Life* was written in the 730s (Colgrave 1956). Guthlac spent his last fifteen years on the Isle of Crowland, six miles into the Fenland from Peterborough: in the desert. Access was by boat. The Fens were known, but sparsely inhabited. Another hermit followed Guthlac, then a monastery was built, which was destroyed by Danes about 870, and refounded in the mid-tenth century.

Guthlac drank only muddy water but at least the absence of people left it unpolluted. Around him were the muddy waters of the black marsh, and dense thickets (so intermittently dry ground). Foes could arrive on foot. There were dense clumps of reed (harundinum, probably *Phragmites*, as a document could be placed in the top of a broken reed), and marshy pools large enough to have waves. Jackdaws lived there. Birds of the untamed wilderness and wandering fishes of the muddy marshes would come swiftly to Guthlac's call, and take food from his hand. Swallows nested where he pointed out.

Guthlac had visions of demons, of evil spirits running amid the black caverns (torturing wicked souls), of British (Celtic) hosts burning his buildings, etc. Fortunately, he also saw St Bartholomew and the radiance of heaven! As horrors also afflicted a helper, malarial-type fever can reasonably be considered a cause: inferring mosquitoes.

This is the earliest graphic description: isles capable of tillage, bearing land birds, set in reedswamp and pools, reeds being both thick and isolated. Various Anglo-Saxons started Christian settlements in wetlands. For instance, in 670 St Etheldreda founded a double monastery (nuns and monks) in the Isle of Ely, becoming its first abbess.

Beowulf is an eighth-century English poem. It is set in Denmark, with no indication whether the author was describing a known Denmark, or placing England there.

Night after night Grendel (a monster) patrolled the fog-bound moors, the mist-ridden fells . . . A lake of water demons, boiling with blood, its terrible waves laced with hot gas, in this fen refuge . . .

They live in an unvisited land among wolf-haunted hills, windswept crags and perilous fen-tracks. The lake . . . overhung with groves of rime-coated trees whose thick roots darken the water . . . Frothing waves rise blackly to the clouds when the wind provokes terrifying storms.

Now, fog is considered a pollution indicator, but this is new. Fogs have cloaked wetlands down the ages, and added to their unpleasantness. In the early twentieth century, it was said that if the idea of hell had originated in England, it would have been to be lost in the fens (then very damp, and with roads impassable in wet), in a fog and a strong east wind. Drainage much reduced wetland fogs. The Fenland became far clearer in the nineteenth century, although low fogs can still occur. Farmers can still be lost in fog in the Shropshire Moors (Purseglove 1989). By 1800, Fenland fogs occurred with north and east winds being blown away by south and southwest ones. Oddly, Borland was not foggy (Miller and Skertchley 1878).

The rich and isolated convents and monasteries on the fens were an easy target for Danish raiders from 787. In 870, the Great Army sacked the Isles of Crowland, Ely, Bardney and (just-upland) Peterborough (Helm 1963). In Crowland, the abbot sent away the able-bodied men, with their relics (including St Guthlac's body). All these they hid *in the nearest fen* [my italics]. The Danes slaughtered all (except one) left behind and fired the buildings. All over NW Europe a new clause was added to the Litany: From the fury of the Northmen, good Lord deliver us.

'Alfred's name will live as long as mankind shall respect the past.'
[Alfred the Great's monument, Wantage]

The *Anglo Saxon Chronicle* records that in 878 the Danish host went secretly in midwinter after Twelfth Night (the reiteration showing the unusualness of this winter manoeuvre),

and rode over Wessex and occupied it, and drove a great part of the inhabitants oversea, and reduced the greater part of the rest, except Alfred the King; and he with a small company moved under difficulties through woods and into fen fastnesse (in the Somerset levels) . . .

And the Easter after, King Alfred with a small company built a fortification at Athelney, and from that fortification, with the men of that part of Somerset nearest to it, he continued fighting against the host. Then . . . he rode . . . to the east of Selwood, and came to meet him there all the men of Somerset and Wiltshire and . . . part . . . of Hampshire, and they received him warmly. And . . . he went . . . to Edington, and there he fought against the entire host, and put it to flight . . . And three weeks later the Danish King Guthrum came to him . . . at Aller which is near Athelney, where the King stood sponsor to him in baptism.

The Danes sued for terms to the man who a short month earlier had been an apparently throneless king. It is this startling reverse of fortune that has caught the imagination of men ever since (Helm 1963). This is one of the great stories of the early English, Alfred defeated, retreating to the wetland, in legend scolded by a peasant's wife who, giving him shelter, asked him to watch the cakes: which he let burn. Then clutching complete victory out of total defeat.

Athelney is only a little rise, but a 1 m flood of the area would make it an island even now, so it was surely near-inaccessible earlier. Glastonbury, on a nearby island, has magical legends of King Arthur (who may have been King, but whose myths are many). With constant confusion between Kings Alfred and Arthur (e.g. in 1066 And All That. 'Then slowly answered Alfred from the marsh, by Arthur, Lord Tennyson'), could it be that Athelney's legends led to Glastonbury's magic?

In 1678, John Bunyan published *The Pilgrim's Progress*, an allegorical tale of the Christian life 'Here is the famous Slough of Despond' (sited, in real life, near Bedford).

> They drew near to a very miry slough, that was in the midst of the plain; and they, being heedless, did both fall suddenly into the bog. The name of the Slough was Despond. Here, therefore, they wallowed for a time, being grievously bedaubed with the dirt; and Christian, because of the burden that was on his back, began to sink into the mire.
>
> At this, Pliable . . . gave a desperate struggle or two, and got out of the mire in that side of the slough which was next to his own house.
>
> Wherefore Christian was left to tumble in the Slough of Despond alone: but still he endeavoured to struggle to that side of the slough that was still further from his own house, and next to the Wicket-gate; the which he did, but could not get out because of the burden that was upon his back.
>
> Then said [Help], 'Give me thy hand.' So he gave him his hand, and he drew him out, and set him upon sound ground, and bid him go on his way.
>
> 'Sir, wherefore . . . is it that this plat is not mended, that poor travellers might go thither with security;' and he said to me, 'This mirey slough is such a place as cannot be mended . . . Labourers also have, by the direction of His Majesty's surveyors been . . . employed about this patch of ground . . . Here have been swallowed up at least . . . thousand cart-loads. There are . . . certain good and substantial steps, placed even through the very midst of this slough: but at such time as this place doth much spue out its filth, as it doth against change of weather, these steps are hardly seen.'

And from Part Two (1692):

> Many there be that pretend to be the King's labourers, and that say they are for mending the King's highway, that bring dirt and dung instead of stones, and so mar instead of mending . . . Then they looked well to the steps, and made a shift to get staggeringly over.

Bunyan uses 'slough', 'mire', and 'bog' almost interchangeably – and there are no bogs in the present sense in the fertile lowlands by Bedford. 'Mire' is the muck in slough or bog, or the slough itself. 'Dirt' is used in a sense intermediate between current American and British usages (Bunyan had had a good plain education).

The Slough is narrow, so, when in, both banks seem close. It has steep edges, to fall off, and to be pulled out from. The mud is deep enough to wallow in, but not to drown in, and in wet weather the mud runs over. Steps can be put in, so the bottom and sides are firm. Government surveyors try to 'mend' the hole by tipping in cartloads of stone. Softer material is useless, but is cheaper, so used by corrupt workmen.

Despite the drainage since Bunyan's time, the slough could still be detected in the 1940s, when it was wet enough for a teenager to call for help to get out (and dry enough for her not to get it). There has been much drainage since!

Nineteenth-century pictures showing Bunyan's house and Bedford jail (where he was imprisoned) put the slough at about 4 m wide across the path, widening in the distance.

Daniel Defoe published his *Tour of Great Britain* in 1724–7. He was very impressed with the yield of waterfowl from the Fenland (e.g. 6000 a week just from St Ives to London), but by this time art (decoy ducks and duck decoys) was needed to crop these: showing overexploitation. The north (silt) fens, in particular, raised fat, large sheep and oxen. The Fenland was so flat that Ely Minster could be seen even from the north.

Especially late in the year, however, fogs blanketed the area, though now and then the Lantern of the Minster stood out (i.e. a low-lying fog). Defoe considered this a horrid air for a stranger to breathe, and also objected to the 'muddy' (peaty) water, the colour of brewed ale, like that of the (peaty) Peak district (south Pennines). He is surprised the population is healthy, apart from the ague (malaria). Flooding in winter was frequent.

The south Essex marshes also yielded innumerable waterfowl – and Essex ague. Romney Marsh has rich fertile grazing. The Somerset levels are also liable to flood, and provide rich grazing for cattle. The River Severn flood plain also has fruitful meadows.

Defoe has much to say about the midland roads (like that with the Slough of Despond). These had constant sloughs, sometimes able to bury both man and horse, impassable in winter. Travellers are prepared to pay either for the new turnpike roads, or to cross farmland instead. Herds of fat bullocks being sent to London were a prime cause of the trouble. The remedy is bridges, and opening drains and watercourses.

Three famous nineteenth-century novels describe lowland wetlands. R. D. Blackmore (1869), in *Lorna Doone* (set in the seventeenth century), describes the Wizard's Slough and the death of the villain Carver Doone on Exmoor:

> There was the Wizard's Slough itself, as black as death, and bubbling, with a few scant yellow reeds in a ring around it. Outside there, bright watergrass of the liveliest green was creeping, tempting any unwary foot to step, and plunge, and founder. And on the marge were blue campanula, sundew and forget-me-not, such as no child could resist. On either side, the hill fell back, and the ground was broken with tufts of rush, and flag, and marestail, and a few rough alder-trees overclogged with water. And not a bird was seen, or heard, neither rail nor water-hen, wag-tail nor reed-warbler.
>
> . . . this horrible quagmire, the worst upon all Exmoor.
>
> Carver Doone turned the corner suddenly. In the black and bottomless bog . . . The black bog had him by the feet; the sucking of the ground drew on him, like the thirsty lips of death . . . I myself might scarcely leap, with the last spring of o'er-laboured legs, from the engulfing grave of slime. He fell back, with his swarthy breast (from which my gripe had rent all clothing), like a hummock of bog-oak, standing out the quagmire; and then he tossed his arms to heaven, and they were black to the elbow . . . Scarcely could I turn away, while, joint by joint, he sank from sight . . . the only sign left . . . was a dark brown bubble, upon a new-formed patch of blackness. But to the centre of its pulpy gorge, the greedy slough was heaving, and sullenly grinding its weltering jaws, among the flags, and the sedges.

Exmoor in general is:

> The mist and the willows . . . All day long the mists were rolling upon the hills and down them, as if the whole land were a wash-house. The moorland was full of snipes and teal, and curlews flying and crying, and lapwings flapping heavily, and ravens hovering round dead sheep . . . It was dismal, as well as dangerous now, for any man to go fowling, . . . because the fog would come down so thick . . . But the danger was . . . in losing of the track, and falling into the mire, or over the brim of a precipice.
>
> [This was an unusually foggy winter, with fog lasting around two months. These occurred every 10 to 50 years.]

John Rudd also went to the Somerset Moors:

> . . . into the open marshes. And thus I might have found my road, in spite of all the spread of water, and the glaze of moonshine; but that . . . fog (like a chestnut tree in blossom, touched with moonlight) met me . . . It was nothing to our Exmoor fogs; not to be compared with them; and all the time one could see the moon; which we cannot do in deep fogs; nor even the sun, for a week together. Yet the gleam of water always makes a fog more difficult: like a curtain on a mirror; none can tell the boundaries. And here we had broad-water patches, in and out, inlaid on land, like a mother-of-pearl in brown Shiltim wood. To a wild duck, born and bred there, it would almost be a puzzle to find her own nest amongst them . . . Through the vapour of the earth, and white breath of the water, and beneath the pale round moon . . . this tangle of spongey banks and of hazy creeks, and reed fringe . . . we came upon a broad open moor, striped with sullen water-courses, shagged with sedge, and yellow iris, and in the drier part with bilberries . . . Broad daylight and upstanding sun, winnowing fog from the eastern hills, and spreading the moors with freshness; all along the dykes they shone, glistened in the willow-trunks, and touched the banks with a hoary grey.

There is much here apart from the drama. Fogs and mists are common. Bird life is abundant – so much that its absence by the Wizard's Slough is noteworthy. The Slough is in a dry land area. It probably has black anaerobic peat, bubbling with methane and hydrogen sulphide. It was screened by flags and (of) reeds, so by tall plants, perhaps *Phragmites* or *Carex rostrata*. If it is *Phragmites*, the yellowing would indicate an un-healthy organic composition. The other species mentioned are:

bright water grass	*Glyceria fluitans*?
blue campanula	*Campanula rotundifolia*?
forget-me-not	*Myosotis* sp.
rush	in tufts, so *Juncus effusus* or other tall species
flag	*Iris pseudacorus*? (*Sparganium erectum*?)
marestail	*Hippuris vulgaris*? *Equisetum* sp. perhaps more likely.

This assemblage is neither *Sphagnum* bog nor particularly nutrient-deficient (nor, of course, nutrient-rich).

While the name of the pit is Slough, the terms 'slough', 'morass' and 'quagmire' are used interchangeably.

Charles Kingsley, who had the opportunity to see partly undrained fens, described what he deduced as eleventh-century ones, in *Hereward The Wake*, 1866.

> From the foot of the wolds, the green flat stretched away, illimitable, to an horizon where, from the roundness of the earth, the distant trees and islands were hulled down like ships at sea. The firm horse-fen lay, bright green, along the foot of the wold; beyond it, the browner peat, or deep fen; and among that, dark velvet alder beds, long lines of reed-rond, emerald in spring and golden under the autumn sun; shining 'eas', or river reaches; broad meres dotted with a million fowl, while the cattle waded along their edges after the side sedge grass [*Cladium*], or wallowed in the mire through the hot summer's day. Here and there, too, upon the far horizon, rose a tall line of ashen trees, marking some island of firm rich soil.
>
> That fair land, like all things on earth, had its darker aspect. The foul exhalations of autumn called up fever and ague, crippling and enervating, and tempting, almost compelling, to that wild and disparate drinking which was the Scandinavian's special sin. Dark and sad were those short autumn days, when all the distances were shut off, and the air choked with foul brown fog and drenching rains from off the eastern sea; and pleasant the bursting forth of the keen north-east wind, with all its whirling snow-storms. For though it sent men hurrying out into the storm, to drive the cattle in from the fen, and lift the sheep out of the snow-wreaths, and now and then never to return, lost in mist and mire, in ice and snow; yet all knew that after the snow would come the keen frost and bright sun and cloudless blue sky, and the fenmen's yearly holiday, when, work being impossible, all gave themselves up to play, and swarmed upon the ice on skates and sledges.

Kingsley sees the beauty! But the beauty of that which he saw was only the remnants. Conan Doyle, in the *Hound of the Baskervilles* (1901), conjured an unforgettable death for the villain, Mr Stapleton (Rodger Baskerville) in Grimpen Mire, on Dartmoor.

> The fog-bank lay like white wool . . . The path zig-zagged from tuft to tuft of rushes among those green-scummed pits and foul quagmires which barred the way to the stranger. Rank reeds and lush, slimy water plants sent an odour of decay and a heavy miasmic vapour into our faces, while a false step plunged us more than once thigh deep into a dark, quivering mire, which shook for yards in soft undulations around our feet. Its tenacious grip plucked at our heels as we walked, and when we sank into it it was as if some malignant hand was tugging us down into those obscene depths, so grim and purposeful was the clutch in which it held us . . . From amid a tuft of cotton grass which bore it up out of the slime some dark thing was projecting . . . the rising mud oozed swiftly in . . . Somewhere in the heart of the great Grimpen Mire, down in the foul slime of the huge morass which had sucked him in, this cold and cruel-hearted man is for ever buried.

Here is no peaceful Nature! In good educated English of the day, before ecologists appeared on the scene, mire, bog and – where there is a floating raft of peat – quagmire are interchangeable terms. The fog heightens the tension, of course, but shows Dartmoor had fogs. The slime and mud presumably describe a soft peat. 'Foul' no doubt means marsh gases, including hydrogen sulphide. The plants named are:

rushes	*Juncus* spp?
green-scummed pits	algae or *Lemna* spp. Neither seem too likely. Perhaps algae growing in areas made very nutrient-rich by flocks of birds?
rank reeds	used in a general sense. *Carex rostrata*? Short *Carex* or *Juncus*?
cottongrass in a tuft	*Eriophorum* sp. *E. vaginatum* is more tufted, *E. angustifolium* grows in wetter places.
slimy water plants	slime – algae on the shoots (periphytic) of water plants? – *Utricularia* sp. would be likely.

This assemblage is not dissimilar to one of the bog pool communities in the National Vegetation Classification (*Eriophorum angustifolium* bog-pool community, M3).

These extracts go some way to explain the general dislike of wetlands, and therefore approval of their removal. When removal brings tangible benefits as well, then praise for the destroyers can become unanimous. The making of roads usable all year, though, is hardly controversial.

The magic of the wetlands begins to take hold in the nineteenth century (see Charles Kingsley, above), and becomes much stronger during the twentieth century. Firth's book on *Romney Marsh* (1984) and Linsell's on *Hickling Reserve* (1990) show an almost mystical relation between authors and the wetlands they guarded so long. 'As the sun finally sinks below . . . Hickling Broad, turning its waters ruby red . . . [the watcher] might be forgiven for thinking he is in paradise.' The recent Scots myth of tartan-clad Highlanders appearing down the mountain out of the mist has the peatland only as background: the interest lies in the people. Scots, Welsh and indeed English hill bogs have not yet attracted such literature, though they have the sentiment. These places are now as safe as other habitats. Not many wax lyrical when liable to death, injury, disease and other inconvenience – and in a dull, rather than dramatic way.

To relieve distress and danger is a proper aim. To increase the supply of food and other commodities is so also (subject to some reservations). To enrich oneself by doing these if people behave properly is not an unworthy objective. And in consequence: away with the wetlands!

Much of the rest of this book is devoted to discussing why wetlands should be maintained!

The creations of the waters

Wetlands occur where water is, or can potentially be, stored; where outgoing water does not, for most of the time, dry the habitat. This water comes from four possible sources: rain, run-off, groundwater and river water.

Rain (and other precipitation)

This is received by all habitats. The amount varies with region, but (in the smaller wetlands) is the same as on the dryland habitats alongside. To keep wet, therefore, a wetland supported solely or mainly by rain must be more efficient at retaining the rainwater than is the dryland. Water is lost by running out along or into the ground, and by rising into the air as water vapour.

Figure 1.2 Water held by *Sphagnum*.

Figure 1.3 Raised bog on flood plain. Dome shape and 'lagg' channel round edge.

Water evaporates from wet and water surfaces. The amount depends on the dryness of the air, a function of the absolute amount of moisture held in the air, and its temperature. Water also transpires from plants. Plants have open surfaces for gas exchange: and water is lost (loss of water from leaves means drawing water, and nutrients and other solutes up to the leaves). These two together, the total loss from both the ground and the vegetation it bears, are known as evapotranspiration. Transpiration is mostly in summer (see Chapter 6), when the plants are up (and, if evergreen, transpiring more). Evaporation has less seasonal variation. Decomposition rates increase in higher temperatures, and for bog to develop, peat production must exceed decomposition. Verhoeven (1992) cites July temperatures for temperate peatlands as being 8–20°C. Rainfall estimates allowing bog formation vary, but Verhoeven (1992) gives about 500 mm as the lower limit, rising to over 1000 mm in hills, more in lowlands, for blanket bog.

For rain to sustain a wetland – a bog – rain must stay there. And rain does (see Chapter 2). Once *Sphagnum* (or equivalent) is established, the water is held, just below the moss tips covering the surface (Figure 1.2). Rainwater lenses occur. Evapotranspiration is reduced because the main water is subsurface. Downwards loss is reduced because the substrate below (peat, rock, etc.) is hardly pervious. Sideways loss is reduced because of the low slope and the texture of the vegetation and peat.

Where these conditions are met, bog grows. Where people have drained the bog, it does not grow. Blanket bog is in west Norway and the west and north of the British Isles. This is a blanket, a sheet of varying thickness. In wet climates, it covers all except steep slopes. While its present climate can be defined, it has sustained itself through past climatic change, over post-glacial time, so the initiating and sustaining climate is less easy to define.

Although the distinction between blanket and raised bog is convenient, the two intergrade. Raised bogs (see Basic Figures 1–3 and Figure 1.3) are dome-shaped, and can form domes over, for example, blanket bog, lowland basins, flood plains and estuaries. They can be only a few hundred metres across, or plateaux up to hundreds of kilometres, so before man's destruction surely merged with blanket bog (e.g. Money, in Hughes and Heathwaite 1995). They are common in northern Europe, especially lowland temperate and boreal regions, and occur south to, for example, southern France (Moore and Bellamy 1974). Again, the climate of their present distribution can be specified, but that of past initiation, and sustainability, with or without impact cannot. In Britain they certainly occurred (growing until the Drainage) in the dry climate of the western Fens, and even now appear to be starting in the dry zone of Broadland.

Raised bogs are an ancient British habitat, some going back 10 000 years, to the end of the Ice Age. They probably are an 'end-vegetation' (climax) type, stable for many thousands of years. Blanket bog is a globally rare type, represented best in the British Isles. It also goes back to the end of the Ice Age. There is, however, some evidence of human activity in its formation: does it form without human impact (Lindsay 1995)? Certainly, once initiated in an area, it can continue to grow and to spread (blanket) ground without further impact.

Raised bogs often occur over infilled lake basins. Open water mud is deposited (Basic Charts E, H and L). As the lake shallows, reedswamp invades and forms peat, then fen. When the peat has built up to the level of the water table, no more can be deposited under water. There will be dry land, if rainwater cannot be retained in quantity, or raised bog, if it can. This raises the land level above the water level, and above the mineral soil land of the edge of the former lake.

Bog was developed over the fen peat of much of the western Fenland, of which Holme Fen is now a remnant. Interestingly, when the fens were wetter, it was noticed that bog areas were much healthier than fen and marsh ones (Miller and Skertchley 1878). Did the toxins of the bog make a less satisfactory habitat for mosquito larvae and pathogens?

Run-off

Although all land receives rain, only part receives much run-off. Rain (and melted snow, etc.) falling on land may stay at its surface, as in bog. It may sink into the ground, as in the porous soils of, for example, soft limestone and sandstone, less in harder rocks, and not at all in uncracked, impervious resistant rocks. Run-off water flows downhill, either above or in the ground, until it is slowed by flatter ground, or gathers with other run-off to form rills, brooks, rivers, flowing to reach the sea.

Run-off dominated wetlands are on flatter ground where run-off slows. They may be little areas in the downward path, fringes of larger plains that received but little river water: and, now, much of larger plains, where river water no longer floods (due to draining and embanking) (Basic Figures 7 and 16).

Run-off contributes to the sustaining water of all wetlands that are not at the top of the rain pyramid (Basic Charts B and C).

The water, of course, starts with the composition of rain (Chapter 7), so containing very little dissolved or solid material. On moving downhill, however, it picks up the dissolved substances of the surface, upper soil or subsoil in which it moves, and if on the surface, solids. It therefore takes on the characteristics of the substrate through which it flows, and transfers these to the wetland it sustains.

Therefore, run-off can lead to valley bog, poor fen, marsh, calcium-dominated marsh or fen, or rich fen – all depending on the nature of the ground. The amount and pattern of run-off depends, first, on the rain that sustains it, second, on the configuration of the ground, which determines where it runs, and on the soil and rock type, which determine how much of it sinks down. Run-off wetlands may build peat if (a) the water table is above (or at) soil level, and (b) incoming water is mostly subsurface, so with negligible incoming

(a)

(b)

Figure 1.4 Small and large flood plains: (a) narrow flood plain, stream at far side; (b) part of wide
flood plain, Fenland, with drainage channel (river out of sight).

sediment. Wetlands with surface water may or may not accumulate much sediment. This
is largely deposited when flow slows in the wetland, so soils may be part or mainly mineral.
If it now picks up pollutants, the wetland may be pollution damaged (Figure 1.4, Basic
Figure 14).

A wetland largely sustained by run-off must hold that run-off: it must not be on a steep
slope where water runs out fast; it must not be on a pervious substrate where water
quickly soaks underground, and it must not have a climate where evapotranspiration is too
great to keep the habitat wet (e.g. Basic Figure 8).

Groundwater

Groundwater is the most restricted of the four wetland water sources, being found only
where it upwells to the surface (Basic Figures 2 and 5; and see Figures 6.1–6.6, 7.3 and

7.4). Groundwater may come from deep aquifers and may be tens of thousands of years old, or it may be slow, deep run-off with a few days residence time in the soil. Subsurface run-off and shallow groundwater intergrade; separation is arbitrary. It may come to the surface in discrete and distinct large springs, or in scattered small springs or seepage areas, or may come up over a diffuse band or area. The sites are determined by the geology and topography, the placing and type of water-bearing rock. In practice, it means that most are at the edges of plains or near the sources of streams.

It is possible to have wetlands whose only effective water is from springs. Here run-off is little and nutrient-poor so that, like the rain, it is chemically submerged in the springwater, and is too little to significantly alter the spring-controlled water regime. Deep-water springs run all the time (unless stopped by abstraction) and maintain a very stable water level in the wetland. More superficial springwater may be more ephemeral, but is still likely to give a more stable water regime than run-off, and certainly much more so than a river flood pattern.

The nature of the groundwater of course depends on the rock or subsoil in which it was stored: whether limestone, sandstone, deep alluvium, etc. The largest groundwater wetlands in England and the Netherlands are in catchments of limestone or other calcareous sediments. There are, though, many poor fens. The calcium-dominated wetlands are usually fed by calcium-rich, low other-nutrient springs, e.g. low nitrate, phosphate. They may also arise from limestone run-off (where this is deficient in other nutrients and does not contain added agrochemicals, etc.). Pollution is not yet a serious problem in British groundwater wetlands.

Peat will build where the water table is above ground or nearly so. With negligible sediment from the incoming water, it will be mainly organic. Other springwater wetlands have also substantial run-off or river water or both, and the habitat, soil and water reflect these different influences (see Chapters 6 and 7).

Wetlands with rising groundwater (rather than constantly flowing of springs) can be more difficult to classify. These include the Breckland meres of East Anglia and the Irish turloughs. The turloughs are in limestone, with dramatic water fluctuations. In summers and dry times there is good grazing. In winter, habitat is good for waterfowl. The Breckland meres have longer and more mysterious water cycles (excluding damage by abstraction from the chalk below), and being smaller and less suitable for grazing, waterfowl habitat is good. Tall herb/short shrub vegetation is usual.

River water (including lake water)

The most widespread wetlands are those by flowing waters, the flood plains, which vary from a few metres or to many miles wide (Basic Figures 1–5, Figure 1.4). Before draining and embanking, streams (except in gorges and steep narrow valleys) overflowed their banks regularly in the high flows of winter, and storms in summer. These flood plains were therefore regularly flooded, and thus wetland. So much now is not just cut off from the river, but drained. It is difficult to assess past areas: surely there was more, over west Europe, than boglands? The most drained land is now arable, the intermediate, good quality grassland, and the damper (a smaller area) wet grassland. A little bears remnants of marsh, fen or raised bog communities.

River water quality reflects (1) the nature of the incoming run-off (and, if relevant, springwater) from the whole catchment above, and (2) the processes taking place within the river, which can alter chemistry considerably (see, for example, Chapter 8 and Haslam

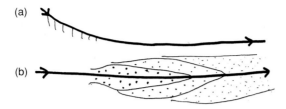

Figure 1.5 Sediment deposition downstream. Larger particles settle first. (Of course particle size also depends on water force and rock type: more large particles are derived from resistant rock than from clay, e.g. resistant rock is more likely to lead to a bouldery rather than a clay catchment.) (a) vertical; (b) horizontal.

Figure 1.6 Typical 1990s English flood: water level regulated to remain below the level of the houses (Haslam 1991).

1994, 1997). In recent centuries, effluents and other pollution have greatly increased. This means that the river water flooding a wetland from its centre is usually chemically different to the run-off water entering at its edge. The habitat will reflect this.

River water carries sediment, brought to it by run-off. The stronger the flow, the larger the particles. Deposition occurs as water force lessens, gravel (if present) being deposited first, silt last. Figure 1.5 shows typical patterns. The flood water may drain quickly, as is common on coarse sediments or over pervious subsoil or rock; or slowly, as on fine sediment and peat, or over an impervious base. The flood water may cover the flood plain for months, weeks, days or hours and be shallow or deep. Consequently, habitats vary. At one extreme are coarse sediments, mostly dry, deep-flooded with more deposition occasionally. At the other are fen peats, flooded by water that has already deposited most of its sediment, in habitats mostly under water and never dried. Both of these are building soil, be it noted: one by river deposition, one by peat growth.

With drainage and embanking, sediment deposition is now rare. It is found in some hill valleys, where rivers overflow their banks, and on washlands (areas designated to store flood water), etc. (Figure 1.6). Mostly, flood plain wetlands are cut off from the river that made them. This means stopping the fresh nutrient supplies brought with the silt, and drying. The wetlands receive run-off (and perhaps spring) water, perhaps river water seepage from under the river banks, and, locally, the occasional river flood. Water quality and quantity have been drastically changed. Any accumulating soil will differ from that building in the past. Habitat, and so vegetation and animals, change. So much river water is polluted that, in the few fen or marsh Reserves, river flooding must often be stopped. Flooding is polluting: such is the extent of the change from 'natural' times.

Flood plain water tables vary. Extreme fluctuations are found with unstable, spatey rivers and rapidly draining plains, giving deep floods and long dry periods. Extreme stability is where fen peat can form with water retention. Near-constant shallow-water lakes have only small fluctuations. Unstable habitats are characteristic of the lower parts of large mountain-rising rivers such as the Danube and the Rhône, stable, of Fenland and Broadland while their fen peat was developing. Most river-fed wetlands fall in between.

2   Wetlands matter

Wetlands are not wastelands.

(Young, in Mitsch 1994)

From painted Britain I a basket come
Imported and adopted here at Rome.

(Martial, in Miller and Skertchley 1878)

This tract abounds likewise with turf and sedge fen firing, reeds for thatching, alders and
other aquatic trees, especially willows.

(*The Fenland*, cited in Miller and Skertchley 1878)

They hang the man and flog the woman
Who steals the goose from off the Common
But let the greater villain loose
Who steals the Common from the goose

(Anon)

And he gave it for his opinion that whoever could make two ears of corn, or two blades of
grass, to grow upon a spot of ground where only one grew before, would deserve better of
mankind, and do more essential service to his country, than the whole race of politicians
put together.

(Dean Jonathan Swift, *Gulliver's Travels*)

The kidneys of the landscape

(Mitsch and Gosselink 1993)

Turf and peat and cowsheards are cheap fuel and last long.

(Francis Bacon, *Natural History*)

In Essex, moory ground is thought the most proper.

(Mortimer)

Introduction, and the World Charter for Nature

The World Charter for Nature, adopted by the General Assembly of the United Nations,
includes:

Nature shall be respected and its essential processes shall not be impaired.
   Civilisation is rooted in nature, which has shaped human culture and influenced all
artistic and scientific achievement, and living in harmony with nature gives man the
best opportunities for the development of his creativity, and for rest and recreation.

Every form of life is unique, warranting respect regardless of its worth to man, and to accord other organisms such recognition, man must be guided by a moral code of action.

Lasting benefits from nature depend on the maintenance of essential ecological processes and life support systems, and upon the diversity of life forms, which are jeopardised through excessive exploitation and habitat destruction by man.

The degradation of natural systems owing to excessive consumption and misuse of natural resources . . . leads to the break down of the economic, social and political framework of civilisation.

The genetic viability of the earth shall not be compromised; the populations of all life forms, wild and domesticated, must be at least sufficient for their survival, and to this end necessary habitats shall be safeguarded.

Activities which might have an impact on nature shall be controlled, and the best available technologies that minimize significant risk to nature or other adverse effects shall be used. Activities which are likely to pose a significant risk to nature shall be preceded by an exhaustive examination; their proponents shall demonstrate that expected benefits outweigh potential damage to nature, and where potential adverse effects are not fully understood, the activities shall not proceed.

Each person has a duty to act in accordance with the provisions of the present Charter.

At an international level, the ethical basis of conservation is agreed.

The sustainable development of a country needs its wetlands in satisfactory condition, because of their critical role in water quantity and water quality, apart from other values such as biological diversity. Although this is increasingly recognised by some sections of the population, it is not enough to halt destruction (Mitsch, in Mitsch 1994). Public and scientific opinion can be mobilised to stop wholesale destruction of striking areas (e.g. the Broadland, Halvergate Marshes) but who stops, or even knows about, the 'tidying up' of the nettle and scrub corner by the stream, the fertilising of that acre of moor to improve grazing, the further draining of grassland already partly drained and now only waterlogged, the continued or increased abstraction of groundwater for mains supplies from aquifers below wetlands? These, and many more, continue apace, and collectively result in rapid and shocking loss of wetland habitat.

Why, apart from ethical reasons, should wetlands be kept? What do they do, or have ever done, for people? Basic Figures 8 and 15–19 summarise these uses: water resources (supply, storage, regulation); products, past and present (both crops and other useful commodities); and recreation and study (leisure, explorations of the past and present).

So these can be listed (partly from Heathwaite and Hughes, in Heathwaite and Hughes 1995; Löffler, in Patten 1990; Mitsch and Gosselink 1993; Novitsky, in Mitsch 1993; Novotny and Olem 1994; Snardon, in Hook *et al.* 1988b; Whigham and Brinson, in Patten 1990). The human values and uses differ both regionally and locally. In North America, for instance, the monetary value for duck shooting is immense. In Europe, with most waterfowl grossly exploited earlier, they now need conserving, which uses money rather than, as in the United States, bringing it in for the conservation of the wetland and the increase of the waterfowl. A wet grassland, a blanket bog and Wicken Fen have values contrasting in many respects (including, respectively, pasture and hay, biodiversity, waterfowl; deer, grouse, global rarity, water supply; and an exhibit of semi-traditional fen habitat, to start with). A wetland is a **habitat between land and water** with some characteristics of each, and associated specialised organisms.

Values

Hydrological and physical

1   *Water supply (for mains).* Aquifers, bogs (much of Scotland's water supply, for instance, is trapped in the bog, and used as run-off or stream), flood plains. Use is too often overuse and leads to drying.

2   *Water supply for irrigation.* Wetter habitats can be used, but again this leads to drying unless only the collected natural run-out is used.

3   *Storage, dispersal and regulation of flood flows.* Flood water, dispersed on to a flood plain, and released slowly as the waters go down, reduces storm damage downstream (compared to a constricted river channel carrying the whole flow). Bog, fen, marsh and reedswamp outside larger flood plains may trap storm rainwater, and release it much more slowly. (As with anything ecological, there are exceptions: wetlands that increase flash flooding.)

4   *Long-term water storage* on flood plains at various levels, which contributes to stream flow in drier periods (regulation of flow, reservoirs, etc.).

5   *Lessening erosion and stabilising river banks:* by lessening the force of storm flows and the amount of sediment and detritus carried, as in point 3 above.

6   *Aquifer recharge or discharge.* Aquifers require to be refilled with water (particularly when used or overused for supply), and maximum water soaks in from surfaces above the aquiferous rock that are both porous and continuously wet. Alluvial deposit wetlands are the most porous, but fens and other marshes may also be good sinks.

Groundwater-fed wetlands are of course where water discharges up from the ground. The same wetland, however, can have water entering by springs and also leaving by soaking down.

Although it is obvious that aquifers need to be refilled, the fact is often forgotten: and folk wonder why the water supply is running short when its catchment has been built on or so converted to intensive agriculture that most rainwater runs off to the stream rather than soaking into the ground.

7   *Trapping and deposition of sediments.* When flood waters spill on to a flood plain, the sediment they carry is largely deposited. In modern terms this part-cleans the water though part-pollutes the plain. In traditional terms this made flood plain grassland the most fertile and valuable farm crop.

Landscape

1   *Water economy.* Because of the functions above, landscapes with wetlands have evolved with those wetlands playing a crucial, perhaps the most crucial part in the water economy of the region. Upsetting the balance may cause major difficulties and shortages, floods, or both.

2   *Vulnerability.* Wetlands are easily damaged or destroyed by simple land management techniques such as draining, which may have hydrological consequences distant from the treatment in both time and space.

3   *Landscape diversity.* Wetlands contribute substantially to landscape diversity, so to biotic, geomorphic and other habitat diversity.

4   *Areas of building soils,* whether bog peat, fen peat or by sedimentation.

5   *Maintaining topographical variation* due to stream meandering, building soils, etc.

6   *Trees and shrubs can abate noise* in urban areas.

Chemical and biochemical

1   *Clean water passing through.*
2   *Nutrient source and sink.*
3   *Immobilising contaminants*, e.g. *heavy metals, pesticides.*
4   *Provide buffer zones* to maintain water quality.
5   *Constructed wetlands for effluent, etc., purification*
     The chemical transformations (points 2–5) are discussed further Chapters 7 and 8. These functions are vitally important.
6   *Air quality improvement*, particularly by trees (which have the largest leaf area) but also by shorter vegetation, filtering particles, and degrading contaminants (as above) lifted from agricultural and urban areas.
7   *Accumulating organic matter*, and so provide a sink for atmospheric carbon dioxide. The total carbon store in peat is enormous. Drainage and those other impacts affecting organic-rich soils may alter the balance of carbon cycling. Release of sufficient quantities of carbon as carbon dioxide aids global warming (carbon dioxide is a 'greenhouse gas').
8   *Storing history.* Accumulating soil also stores and preserves the history of the time, such as artefacts showing human presence and activities, pollen showing vegetation of the time, plant remains showing the vegetation of the site, animal remains, etc.

Plant and animal

1   *Accumulating organic matter* (see above).
2   *Habitat for* endangered as well as common *plant and animal species*, as refuges for species from elsewhere, and, where relevant, as corridors for animal (and plant) movement.
3   *Gene pools* for wetland species.
4   *Higher primary productivity* than surrounding drier lands (see below).
5   Until overexploited, have *high secondary productivity* supporting fowl, fur, leather, feather, fish and other meat.
6   *Produce organic matter* for aquatic food chains.
7   *Export organic matter* to downstream ecosystems.
8   *Food chain support.*
9   *Maintenance of biological integrity.* Integrity is defined as having no part or element wanting, a material wholeness, an unimpaired or uncorrupted state. Such a biological integrity of wetland is of ethical and biological – and water resources, etc. – value. It is shown by the diversity, abundance representativeness, naturalness and rarity of the plants and animals present.
10 *Medicinal plants and leeches.*
11 *Crafts*, including thatching (needs sedges, heather, other small shrubs, tall grasses, rushes), baskets, matting, woven goods with innumerable uses especially from *Salix alba, S. purpurea, S. triandra, S. viminalis.* There are numerous local varieties for different purposes, e.g. *S. alba caerulaea* for cricket bats. Clogs (e.g. alder). Construction (large plants, soils). Stuffing (feathers, fruits, e.g. *Typha*, reed. Candle wicks (rush dips). Bedding, insulation, stewing floors. Timber, etc.
12 *Food and drink*, e.g. watercress, berries, reeds, *Typha.*
13 *Fuel.* Peat, charcoal, wood.

14 *Medicines*, for example, to disinfect, prevent infection and inflammation, and reduce pain and tumours (Neori *et al.* 2000).

Economic crops

1 *Plant crops from traditional wetlands*, e.g. reed, cranberries, rushes, withies.
2 *Animal crops from traditional wetlands*, e.g. deer, otter, fish, beaver, waterfowl.
3 *Peat, for energy production*, traditionally from both bog and fen. For the quantities needed for power stations, mostly bog, both raised and blanket bogs. Gas production is also from both fen and bog. Quick-growing willows (incorrectly termed biomass) are now also used.
4 *Peat, for horticulture*. For the large quantities needed, mostly from bog, especially (by geographic chance).
5 *Sand and gravel*, for construction, from alluvial deposits; and for their other constituents, where relevant, e.g. alluvial tin, Dartmoor.
6 *Cleaning water*, see above.
7 *Grazing and hay* from converted grassland, marsh hay (tall species) being perhaps the wettest type. Production increases with increasing drainage and intensity of management. As diversity decreases with these, diet may become less balanced, so less valuable.
8 *Arable*, from conversion to drained, tilled ground. Fen peat and silt wetlands are usually particularly fertile, with high yields.
9 *Forestry*, usually conifers on bog and hill peatlands, and poplar and willow on more fertile lowland areas. These are now grown primarily for timber and paper, but – especially willow – also for electricity generation, craft and general farm and country uses. Wood for electricity generation is currently and incorrectly known as biomass production. (Biomass is the total mass of the plant, under as well as over ground. 'Biomass' shoots are cut near ground level, so the stump and root remain to reshoot, while the wood is burnt.)

    Forestry of native species on naturally wooded wetlands is a valuable sustainable use. These include drier bogs, fens and marshes, and allow mature stands supporting a high and appropriate diversity of woodland plants and animals. Exotic tree species that usually support a poor fauna and flora, and forests frequently harvested the same year over large areas, are not satisfactory, and neither is draining or fertilising wetland so that it will bear trees.
10 *Turf*. The old and traditional meaning of 'turf' was peat, hence the many Turf Fens. However, turf in the modern sense (grass and the top soil it is growing in) can be grown in a grassland habitat, and sold for gardens, parks, etc.

Societal

1 *Important natural heritage*, particularly when scarce.
2 *Representative of personal intangible values*. Wetlands, for an increasing number of people, have a charm, an attraction of their own, one that differs with the wetland type; the vast expanse of brownish bog; the reedbed where visibility is 2–3 m and the sense of isolation is profound; the grassland by the river bank, with the green and diverse land, the flowing water and the panorama of the heavens above. These much benefit those who experience them.

3   *Aesthetic values* in a more abstract way, sensory experience.
4   *Education, research and teaching.*
5   *Art and literature.*
6   *Heritage*, natural and cultural (see above).
7   *Recreation and relaxation*, other than the above: active forms of leisure such as walking, birdwatching, sports; quiet ones such as picnicking, visitor centres, car parks.
8   *Sites for impoundments* for water supply, flood relief, recreation, etc.

Considerations of values

If a wetland is destroyed, its value to mankind is lost. Mitsch and Gosselink (1993) consider the cost of some of the same functions if done outside wetlands:

Hydrological functions	*Money-requiring replacement techniques*
1 Maintaining water quantity	Pipe from far away
2 Maintaining water table	Wells
3 Maintaining surface water level	Dams, pumping water, irrigation pipes, etc.
4 Regulating floods	Sluices, flood defence works, drainage

Chemical functions	
1 Maintaining drinking water quality	Purification plants, inspectorates, and laboratories, collection of run-off, small effluents, etc.
2 Cleaning effluents, urban run-off, etc.	Sewage treatment works

Biotic functions	
1 Food for people and domestic animals	Agriculture. Food imports
2 Thatching, livestock litter, crafts, etc.	Other, partly artificial roofing, bedding, craft materials
3 Maintaining species and genetic material	No replacement possible

Societal functions	
1 Heritage value	No replacement possible
2 Aesthetic and spiritual values	No replacement possible
3 Recreation specific to wetlands	No replacement possible

(See also Haslam *et al.*, in Westlake *et al.* 1998.)

The replacement techniques, where available, are expensive.

    Non-consumptive values of wetlands tend to involve higher aspirations, philosophy, beauty, learning, spiritual and humanitarian concerns. The importance of these is related to the type and amount of cultural interaction with wetlands (Sother *et al.*, in Patten 1990).

    Wetlands are capable of supporting a population. People can be fed, clothed, housed, heated and doctored by the wetlands' bounty. (Not every wetland provides for each purpose, of course.) The populations supported are small, however. They have, often over long periods, adapted to the isolated and very specialised life of the wetland, in the more obvious skills of catching fish and fowl, and in, say, building homes from reed (as recently in the Danube Delta), houses on stilts (e.g. Somerset levels), moving around on stilts

(Fenland until recent drainage), or on skates (e.g. the Fen Runner skates). Using water rather than land transport was normal in all larger marshy areas, of course.

When drainage comes, and the way of life is threatened, and disappears, this is tragedy for the population as well as for conservation. First, there is loss of livelihood. Second, there is loss of independence. The wetland dweller might have paid his rent, but he worked for himself, was responsible to himself, and did not want to become an employee (and might well not be wanted as an employee, either!). This change was noted in both the Somerset levels (Williams 1970) and the Fenland (Charles Kingsley). The independent peasants in both unfortunately got drunk, presumably because there was surplus disposable income and – at those times and places – little else to spend money on. Too much squalor and ill-health resulted. In fact, therefore, the change to dependence and earning a regular wage led to improved health and living conditions. Moral issues are seldom clear-cut.

A wetland now totally converted to arable or forestry may yet bear remnants of wetland vegetation, in flood plain dykes, in odd corners of the woodland. These are more suited to the care of the conserver, however, than for the livelihood of the cottager.

Wetland to dryland: the changing from fish and fowl to grain and vegetable

The history of fen and marsh is one of discontinuous progression to arable: discontinuous with many stops, starts and steps backwards. The history of raised bog has included its wide-scale removal, loss of wetland by loss of peat, rather than by loss of water. For blanket bog, large-scale removal (mostly Ireland) has so far been less than on raised bog. As fens and marshes are usually so fertile, there has been a tendency to drain these early for agriculture, then take peat for fuel and horticulture later.

Good drainage was developed in the Mediterranean in the fourth millennium BC. By the sixth century BC, Greek and Italian drainage of waterlogged land was well under way. The Romans developed good engineering in Republican and early Imperial days, for wetland as well as dry land. Their major projects included in Britain, Romney Marsh and the Fenland (the Car Dyke was cut round, to link Cambridgeshire, the Trent and the North Sea; it was a catchwater, and a canal transporting grain and other goods). After the second century AD, references to schemes lessen. Perhaps climatic decline made them more difficult (Potter 1981).

The story of drainage is the story of climatic change (drier, wetter, sea level higher, lower), interlinked with that of human impact (greater, lesser, differing types). Technology helps, but the human will is most. Abandon a cultivated wetland: it floods. Until, however, modern times when the maintenance of drainage is so easy that no massive effort or community involvement is required.

The Fenland was habitable for prehistoric man in the Bronze Age, although the silt fens were drained in the Iron Age. In Early Roman times, sea level fell, and the silt islands, at least, were habitable. There was much drainage and use. Saxon times were wetter and peat building through drainage also occurred. Meres were formed by shrinking peat (Hall 1981). In early Mediaeval days, the climate and land level again improved, so in 1250, Matthew Paris could write: 'The Fens were recently transformed from a haunt of devils to fertile meadows and fields.' The same sentiments could have been expressed by the Romans, the seventeenth-century drainers, or the late-nineteenth-century farmers! Under Charles II, in 1631 a consortium led by the Duke of Bedford (hence the 'Bedford levels' in the central Fenland) and employing the Dutch engineer, Vermuyden, made the Fens into the pattern we still see today. Vermuyden was unprepared for the peat wastage. The ground gradually sank below sea level (the fall at the marker post in the wet Holme Fen is 3.9 m from 1848 alone, and up to nearly 5 m in the drier land). The rivers became further embanked and above land level (Figure 3.3), and water from the wetland had to be pumped up into them: first by windmill, then steam, oil and electricity. The transformation was uneven. Areas reverted to wet land. The degree of seventeenth-century drainage left the land wet, foggy and unhealthy. Drying increased in the nineteenth century, and at an increasing rate in the twentieth, so that by 2000 it is really only the sparsity of settlements and the drainage works that show that this land, if abandoned, could still be under water.

Romney Marsh, Kent, had peat formed in the second millennium BC, and sediment was also deposited. The calcium in this has been much leached in the Old Marsh, while the New, receiving deposition much later, is still more calcareous. For a century each side of AD 1, the Marsh was dry and colonised, and was then drowned again. Most was reclaimed after the tenth century. In the *Domesday Book*, only one reclaimed Hundred was in the New Marsh, four in the Old, which was wealthy in late Saxon times. There was much drainage in the Mediaeval climatic optimum, the twelfth and thirteenth centuries (Brooks 1981). Since then there have, again, been ups and downs: but drainage enough for arable has now come.

Oxmoor, Oxon, has Neolithic, Iron Age and Middle Ages villages round the edge, and a Roman road going straight through: which tell their own tale of health, water and technology. The 'moor' was intensively used, under, as usual, strict and complex community control. The main drainage was nineteenth century (Bond 1981a).

The main Somerset wetland is composed of the inland moors, with bog peat and the seawards levels, with silt. Both are on the same large flood plain. And both, with the isles and edges, were unhealthy. There were summer grounds from the seventh to the ninth centuries, and a marked rise in wealth up to the fourteenth, reflecting drainage and the spread of most valuable grassland. The most important crops ceased to be fish, fowl, rush and reed, becoming pasture and meadow. (Meadow supplied winter feed, so was crucial to maintain the herds.) Turbary, turf (peat) cutting was minor, presumably only for local use. Reclamation was piecemeal, and generally avoided the acid infertile peat. After the Black Death, in the 1340s, there was a temporary setback, followed again by piecemeal but further drainage. Commoners were satisfied with pasture. By the seventeenth century most clay was cultivated, most peat was not. Efficiency varied with both time and space. The year 1770 was a turning point. Penned water meadows (like water meadows, see Chapter 1), ditches around fields, more drainage and agricultural improvement began to appear, followed by roads and settlements (many by 1822). The year 1840, however, marked a decline, not really reversed until 1939, when effective pumping and general intensification of agriculture started (Williams 1970).

The Broadland of East Anglia, like the Fenland, has a complex history of seawater coming inland and depositing clay (marine transgressions, sea level rising relative to land level), followed by the sea retreating, leaving freshwater wetland, which developed fen peat, with the nutrient- and lime-rich run-off from the catchment above (George 1992, modified). If the sea level remained unchanged, or went down, carr developed on top of the reedswamp, changing the (fen) peat type (Table 2.1). Going inland and upstream, there is more freshwater, and reedswamp peat with carr and wood (e.g. alder, oak), instead of the seawards saltwater and clay. Carr is on drier ground, so this is more likely to increase inland, although it does not necessarily do so.

Table 2.1 Peat histories (representative of numerous sites investigated)

1 The Fenland (Godwin 1978)

Most recent	Raised bog	}Acid bog peats
	Wood, pine, birch, *Sphagnum*	}
	Wood, oak, ash	}
	Wood, alder, birch	}Fen peats
	Sedge (*Cladium*) fen	}
	Reedswamp (*Phragmites, Typha*)	}
	Open reedswamp (*Scirpus,*	}
	Nuphar, Sagittaria)	} Organic
	Submerged aquatics	} lake muds
Oldest	Open (calcareous) water	}

2 Lonsdale Moors (Tansley 1911)

Most recent	Sphagnum	}		}
	Sphagnum and *Eriophorum*	} Moor	} Acid peats
	Wood, birch	}		}
	Sedge and reed	}		
	Brown-moss (*Hypnum*) peat	} Fen peats		
	Amorphous peat	} Organic		
	Shell marl	} lake		
Oldest	Lake muds	} muds		

3 Ranworth, Broadland (George 1992)

Most recent	Reed
	Clay
	Reed
	Carr, brushwood
	Reed
	Clay
Oldest	Reed

Sheep (at least) on Summer Land were abundant according to the *Domesday Book*, and finds suggest a water level at least 1 m below the present. Embanking and ditching were frequent by at least the thirteenth century, becoming more necessary as the water level rose. By the late sixteenth century there were more cattle: because the fields were wetter or just a change in fashion?

From the sixteenth to the eighteenth century, drainage was patchy, and got really under way only in the nineteenth century, when both law and technology favoured this. As usual, much has been done in the twentieth century. However, the natural importance of Broadland was recognised earlier than for most wetlands. The forerunner of the present Broads Authority was set up in 1974. Conservation and recreation together are beginning to equal the power of the drainers, and – as the battle of the Halvergate Marshes showed – it is possible for the former to win. Conservation *and* recreation: recreation can also be very damaging (mostly from George 1992).

Small as well as large wetlands may be of much importance. A small wetland may be very important during critical times of day (for various animals) or in the migration season for birds. A narrow strip along a stream may efficiently filter agrochemicals from farmland run-off.

Commercial values are finite in time, while wetlands provide values in perpetuity. Wetland development is often irreversible. People usually think in terms of 10 to 30 years, government and farmers, seldom over 50 to 100 years. If natural areas are destroyed for a 15-year aim, their public service value is removed for much longer, perhaps even for ever. Here wetlands are more vulnerable than many dry lands. An abandoned dry land typically reverts, over a century or two, to forest, with the native dominants, and perhaps eventually the total native plants and animals. A drained and developed wetland, on the other hand, has had its water regime altered, and while a shallow lake can be re-created (which, in centuries, may fill in as marsh or fen), re-creating blanket bog on denuded rock, or a *Schoenus nigricans* calcium-dominated fen, is more problematic. Economic analyses usually discount the value of future amenities. Wetland conservation stretches over generations and future generations are not here to present their case (based on Mitsch and Gosselink 1993).

Traditional management of the marshes was tuned to the finest nuances of the local water table. Each wetland evolved a landscape character as individual as the spirit of the people was independent. The real beauty of such places is not the actual components, but the system that underlies them: the harmony between man and nature that they represent (Purseglove 1989).

Wetland products

A few of the many products are described further in the following.

Peat (Basic Charts H and I)

Some fen peat, as well as bog peat, dates back to late Glacial, 10 000 BC. A climatic scheme is shown in Table 2.2. (Peat contains the pollen of the day, and the vegetation of the site.) Wet climates lead to increased bog peat, but, since the last Ice Age, British climates have always been wet enough for some bog.

In well-populated peat areas such as the Fenland and Broadland, each parish had land extending into the fen, as well as dry land. Therefore, the fen produce, including peat, was available to each village. The terms on which it was available varied, part or all for the commoners or the poor; with often very strict conditions laid down for the dates different crops might be cut, and how much (e.g. no employee to harvest). This ensured sustainable crops, the conservation of the produce.

Peat was cut in enormous quantities for local use as fuel. Peat was easy to obtain when reasonably near the water level (Figure 2.1). Broadland is the most dramatic, and first-studied, example of the effect of major peat-cutting. Dr Lambert, an ecologist, and Dr Jennings, a geographer, reported in 1951 that the Norfolk and Suffolk Broads, shallow lakes totalling, originally, 1057 ha (2611 ac), with a mean depth of 2.4 m (8 feet), were artificial, dug for peat extraction. In about AD 1100, land was at least 0.8–1.5 m higher than now, so cutting was easy (Linsell 1990). It is also easy with simple, older technology (e.g. Verhoeven 1992; Linsell 1990) and correct placing of baulks, pumps, etc., to extract to perhaps 3 m below the surface. Lakes are (or were) normally on the main course of their stream, be it large or small. The Broads, however, are off to the side, the river water not being wanted in a working peat quarry (Figures 2.2 and 2.3). (In Barton Broad, though, the river was later diverted through the Broad.) As this was a commercial enterprise, a parish has a Broad – Broads were not shared between parishes. (Parish boundaries

Table 2.2 Climate and geomorphic changes of the post-glacial era (from Gilman 1994; Godwin 1978; Higham 1986; Rieley and Page 1990)

(a) General

AD	1900	Warmer	
	1600	Maximum cold, climatic upturn	
	1400	Broads peat pits flooding	
	1200	Cooler, moister	
	1000	Warm	
	500	Wetter	Anglo-Saxon
BC	55	Drier, warmer, hospitable climate	Roman
	500	SUB-ATLANTIC, wetter, warmer. Much peat development	Iron Age
	2500	Climate much as now. *Sphagnum* covering sub-boreal bogs	Bronze Age
	3000	SUB-BOREAL, cooler, drier, *circa* 2°C. Warmer than now, more continental. Settled agriculture. Peat decline, pine spreading over bog	Neolithic
	5500	ATLANTIC, warm (2–4°C above present), wet blanket bog spreading over hills, replacing pine and birch	
		Raised bog growing in lowlands	
		Sea level rise, Britain an island	Mesolithic
	7000	BOREAL, cool, dry. Pine forest spreading over peat bog	
	9000	PRE-BOREAL, bogs spread (e.g. Teesdale)	
	10 000	Post-glacial. Cold	Upper Palaeolithic

(b) Fenland

Anglo-Saxon	Huge fens
Roman	Man-made watercourses, drainage
Iron Age	Build-up of silt, giving the present division of Silt and Peat Fens
1000 BC to AD 0	Sea incursion, extensive waterlogging, open sedge fen
To *ca.*2000 BC	Sea level drop, freshwater fen spread, and fen passing via carr to wood. Peat increase, trees invaded from the margin. Raised bog developing where alkaline flooding least (Middle Fens, and fen edge)
*ca.*2500–2000 BC	A vast brackish lagoon, 1–2 m deep. Coastal silt deposited, followed by *Phragmites* then inland sedge and woods far inland
*ca.*2500 BC	Rapid peat development, sedge fen, recent black peat. Sedge fen leading to carr then woods
*ca.*3000 BC	Sea incursion, waterlogged, freshwater ponded inland. Black peat developed in this
4000–3000 BC	Dry acid peat, south fens remained alkaline, middle, to raised bog (only marginally affected by fen clay)

were first set in Anglo-Saxon times.) Assuming each household used about 3000 'turves' a year, the Broads were dug over perhaps 350 years (George 1992).

Small broads would have been quickly colonised by reedswamp (e.g. *Typha angustifolia*), and new fen peat built up, and most pits refilled. These ones can be found because of the few centimetres of shelly lake mud above the cut peat surface, or by lines of bushes or trees around slightly lower areas.

In Broadland the peat was mainly for fuel for East Anglian use, Norwich, Yarmouth and smaller towns and villages, but some was exported further, and some was used in salt extraction, for burning. Peat was certainly being cut in the ninth century. The Danes, who were peat-cutting from 500 BC, were there in force from 865, and population and wealth were high in the *Domesday Book*, 1086. With the drying of the land cutting became even easier, and amounts extracted increased to the fourteenth century (e.g. cutting in 15

Figure 2.1 Peat-cutting.

parishes in the thirteenth century, in 25 in the fourteenth). Sea level rose in the thirteenth century, freshwater level rose and bad sea floods increased. In the fifteenth century most of the large peat-cuttings were flooded lakes, the Broads, and fishing rights had replaced turbary rights.

Peat was still cut into the twentieth century, but not on the large, crater-developing scale; though in the nineteenth century, individuals were still being authorised to cut, e.g. 34 000 peat turves a year.

A pause here for a conservation issue. Are the Broads worth having? Think of all the damage done during excavation! The twentieth century, which sees none of the damage, was heartily glad of the Broads. Will future generations be equally glad of gravel pits (Figures 2.4 and 2.5)? (This section mostly from George 1992.)

In the Fenland, vast quantities of peat were also cut for fuel. Here it was superficial, i.e. shallow over large areas, rather than local and deep, as in the Broads. In the nineteenth century, peat could be dug only about a hundred days a year. The moss (*Hypnum fluitans*) peat, as around the Isle of Ely, was firmer than the rest: rush, sedge, *Hydrodictyon* (alga) etc. (Miller and Skertchley 1878).

Pasture and peat could alternate as crops from the same land.

Some bog peat still remains in the western fens (e.g. Holme Fen). As there is fen peat below, the bog was raised bog. Old records from this area indicate bog species,

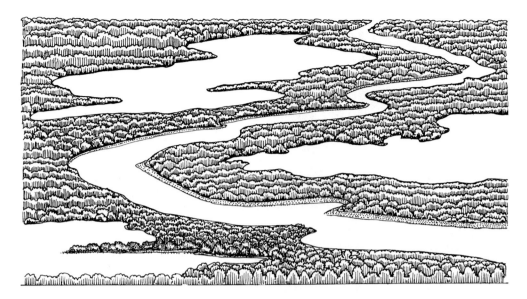

Figure 2.2 Lakes formed by peat-cutting, Broadland (modified). Broads (lakes) separated from river. This is not an island pattern (contrast Figure 4.28). Drawn with land entirely (rather than mainly) covered by carr, to emphasise pattern. Until recently the land would have been much used, with little woody vegetation.

abundant cotton grass (*Eriophorum* spp.), etc. – and the puzzlement and disbelief of late-nineteenth- and early twentieth-century botanists that these could have existed (partly from George 1992; Godwin 1978).

Both lake-forming and superficial peat-cutting were done in the Netherlands also (e.g. Verhoeven 1992).

Present technology allows peat to be removed very much faster, and this is being done for fuel for power stations (much in Ireland) and for substrates for horticulture. Vast tracts are being devastated in this way. In Britain, these are mainly of raised bog as this is easier for access and is thicker, e.g. Isle of Axholme, Thorne Waste, Hatfield Moor and Chase, all in the Trent-Ouse basin. The destruction is frightful, as there is no flooding to give the start to a new fen. With centuries, of course, anything can become of conservation quality, however unlike the original.

The other large-scale present loss of peat is in the cultivated peatlands, where, through being dried by drainage, and further exposed to the air much of the time, peat shrinks and wastes. The organic material of which it is composed mineralises, oxidises and vanishes (little ash, much carbon gases). Entering the Black Fens from the south now, the peat starts far from the old boundary, several metres below its seventeenth-century position. In the parts with thinner peat, this may now have gone and the underlying clay be at the surface. This is less disastrous for conservation: but what a pity if the black peat is present only in conservation areas! Wicken Fen is now a damp fen, not a wet one. It has, of course, had much less peat loss than the arable fields around, which now stand lower. Hence Wicken is becoming increasingly more difficult to keep wet, the surrounding land having deep agricultural drainage. (Landwards to the Fen, a large conservation area has

(a)

(b)

Figure 2.3 Wet grassland. (a) Near-traditional, species-rich, flower-rich, varied structure, ditches, trees; (b) intensively-managed, reseeded, species-poor, flower-poor, uniform short structure, ditches drained, tree line replaced by fence.

recently been proposed.) Peatlands bearing wet grassland are in a much better position. So is alluvial soil, which being mostly mineral, shrinks less, even when deeply drained.

The global warming changes of our day are within the climatic range of recent millennia. Plants and animals vary with climatic change, e.g. the bog to pine to bog noted in

Figure 2.4 Gravel extraction in flood plain forming lake, Northants.

Table 2.1. The effect of climate change now is much more acute, as so few areas of 'wild' vegetation remain to form a reservoir.

Thatching reed, Phragmites australis (*Figure 4.19*)   Reed is the most widespread European dominant of reedswamp, the most widespread constituent of fen and marsh. It is the principal constituent of fen peat. Its past performance was therefore as a dominant, and a dominant in good, peat-forming habitats. 'Natural' reedswamp is flooded all year, and in consequence may be difficult to harvest, and the cut reed difficult to move (even by boat) to dry land. Land dries either by build-up of soil or by lowering of water level (by drainage or by sea level dropping in relation to land level). In a natural situation carr invades and replaces the reedswamp (Basic Chart N). Where reeds are wanted for thatching (or fencing, insulation or other farm uses), the drying reedbed will be managed, and the reed crop retained, as long as there is a little flooding, or even just some waterlogging each year.

Reeds are more profitable than rough grazing, but need a different and more specialised skill. Reeds and unimproved grass occur in the same habitat, although reeds extend further into wetter places, and grass into drier ground. Good reed crops can be obtained from treated marsh and rich fen (poor and calcium-dominated fen probably need fertiliser or drying).

Management for yield is threefold:

1   *Winter cutting.* The bare soil in spring favours reed. Lack of insulation, and some frosting, encourage denser, more uniform and straight reed. They give reed, an early emerger, a competitive advantage, so allowing it to form an early canopy, shading many other species. Cutting may be annual or biennial.

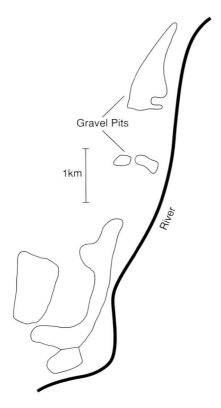

Figure 2.5 Distribution of gravel pits in River Great Ouse valley, near St Neots.

'Wild reed', that which is uncultivated, i.e. non-uniform, often sparse and bent, can be converted to cultivated (in 1–3 years) by winter cutting or, in the first year, by burning too.

2　*Regular water regime.* Regularity is important. It makes little difference whether the water level is high or low, stable or unstable, provided it is regular: the level of each spring should be the same, be it +75 cm, 0 cm, or −75 cm relative to ground level.

3　*Properly timed harvesting*, usually late December to March, after the shoots harden and before young shoots are over about 30 cm. Cutting should be at about 10–30 cm above ground. Shrubs and trees should be removed every few years. Cut reed must be stored dry and away from sunshine.

Good reed, well treated, used by a competent thatcher in an airy place, will last up to 80 years, equivalent to tile, and longer than straw thatch. A bad thatch (for whatever reason) may only last five years, but this is a legitimate ground for complaint. Maximum durability is obtained by using hardened, dead reed not yet breaking down in the reed bed.

Fen sedge, sedge, saw sedge (Cladium mariscus) (*Figure 4.18*)　*Cladium* is local compared to reed, usually occurring in calcium-dominated or calcium-rich fen or marsh. As a commercial crop, water depth varies from intermittently flooded (to about 0.5 m) to merely waterlogged, partly dry. (The driest *Cladium* beds are wetter than the driest *Phragmites* ones.) As with reedbeds, management means *Cladium* can occur in much drier conditions than it occurs 'wild': management prevents the invasion of carr. *Cladium* was formerly used for thatch, bedding, fuel, and general farm purposes, its commonest present use is in roof ridges and window ornamentation in thatch.

Management, within the correct fertility (plenty of calcium, medium other nutrients) and water regime is:

1   Cutting (usually about July) every 3–4 years. This depresses competition rather than favouring *Cladium*. *Cladium* stores much food in its evergreen leaves, so cutting at any season depletes the plant: hence the infrequent harvesting, allowing time for recovery. Most associate species are even more depressed than *Cladium* by summer cutting. They produce lower crops the following year. This then helps *Cladium* to regrow and regain dominance. There are habitats that can equally well bear commercial crops of reed, *Cladium*, marsh hay, litter (sheaf), grass, or indeed carr. The crop species depends on management, especially on the season and frequency of harvesting.

2   Control of water regime to avoid prolonged drying (regularity is less important than for reed).

3   Harvesting well above shoot bases and removing woody species regularly. The main commercial beds have dried over the past centuries, and are now – if not in the past – fully dry enough for carr.

The crop is now of little importance compared to that of reed.

Fowl

Duck, waders, other waterfowl and many land birds live in wetlands (and see Chapter 5).

The ordinary way of killing them now is shooting with guns. In the past, methods included nets, traps and decoys, perhaps with decoy ducks, shooting with arrows as well as bullets, spearing, liming, etc. Public opinion in Britain and Europe is increasingly against killing, as the waterfowl are now so scarce. In North America, with a combination of much less impact and the awareness of population decline before it became drastic, shooting is still plentiful, and the money paid for this contributes to the upkeep of both the reserves and the duck. Organisations such as Ducks Unlimited bought up and managed breeding grounds in the north, so the numbers of migrant waterfowl were increased – correct management of wetlands.

Fowl were a staple food from undrained fens, marshes and streams and, to a lesser extent, bogs. Fowl were regarded as a normal part of the assets of a Manor (Norden 1610). Gradually, wetlands were drained and the birds decreased. And communications improved, and more were killed, as they could be exported to London and other large cities (e.g. '12 12-horse wagons went from Peterborough to London twice a week, stuffed with fowl.', Defoe 1724–7). The numbers now seem incredible, and this overexploitation, naturally enough, much lessened the population. In the nineteenth century there was another great diminution in habitat, plus a great increase in sport killing. These birds were eaten (if edible), but the motive was not sale to London, but the sport. Some of the slaughter, for example that of the Norfolk bitterns and coots recorded by George (1992), again seems incredible. Added to habitat loss, it led to species decline and even loss.

The twentieth century had another vast habitat loss, and a change in public opinion: that wild birds should not be killed.

Unfortunately, the moral revulsion felt at thinking of a wild bird being shot does not extend to being revolted at loss of habitat or bird food. If effluents or pesticides prevent chick food developing, if wetlands are dried or dykes made bare (preventing waterfowl breeding, living or feeding), these are not revolting, these are modern progress and proper development. There is, fortunately, a move to conservation, but even nineteenth-century populations are but a pipedream. The twentieth century is the century of slaughter by habitat loss.

Sport shooting does occur on wetlands, and is a useful conservation tool. The shooting is mostly of birds that – like chickens – are specially bred for this, primarily grouse on moors, pheasants on lowlands, including various carrs and small wetlands, and partridge. These birds need good habitat, people pay for the sport – and the habitat with its conservation value is preserved. Wild-fowling is very strictly controlled to maintain populations.

Here the late-nineteenth- and earlier twentieth-century sportsmen-naturalists should be mentioned with praise. They did much to research and understand wildfowl population and behaviour. Through this understanding came a wider love of birds, and the movement for sustainable cropping and (with greater habitat loss) no cropping.

Marsh hay

With the development of silage and non-grass livestock feeds, and the virtual disappearance of the horses that were the main power for transport ('horsepower'), it is easy to forget the vast quantities of dead grass that were needed to feed horses, who never saw a field, as well as the general livestock that had inadequate outdoor winter feed. Some of this came from the marsh and fen. A good crop was *Glyceria maxima, Phalaris arundinacea* or a combination of both. Both have the English name of 'reed grass', showing this joint use. Both grasses are tall, usually 1–2 m. *Phalaris* needs to be dry at least in late summer, *Glyceria* can grow in permanent flood. Like reed and sedge, both can grow in habitats suitable for carr. Medium-sized grasses could also give a good crop (e.g. *Calamagrostis canescens*), and so could the rush *Juncus subnodulosus*. Regular mowing is needed to encourage the grass and discourage competitors.

Other mixes, and indeed shorter grasses, were also cropped. Anywhere not producing a more useful crop could be pressed into service for low-quality marsh hay.

With the sudden decline in demand in the early twentieth century, what was to be done with the land? Some was converted to grassland or arable, some was abandoned and, in the course of time, dried. Carr invaded part (particularly when it could do so early, see Chapter 10), tall dicotyledons such as *Filipendula ulmaria* and *Urtica dioica* invaded part, becoming what is now called the 'Tall Herb' community.

Grassland

Grassland develops on land flooded for half the year, and on land never flooded or even waterlogged. Typical forage grass has its growing point at soil level, its leaves growing above, so the leaves can be grazed or mowed without harming the growing points, which grow more leaves. (Wet grassland communities are described further in Chapter 4.)

Traditionally, wetlands with suitable water regimes and adequate nutrients, if grazed, will become grasslands. Liming, traditional fertilising, later modern agrochemicals and reseeding with forage grasses, have all increased the yield of grass, and lessened the biodiversity (Figure 2.3) and conservation value. There may of course be management specifically for wildfowl instead, as on the Ouse Washes Reserve in the Fens. This is similar to, but drier than, traditional management.

Cattle used to be tended by cattle boys, women and dairymaids, sheep by shepherds and shepherd boys, geese by goose girls on the open common or marsh. Enclosed wet fields were divided by dykes, and the herders were no longer needed there. With increased drying, dry ditches, fences or hedges are too often the dividers.

Cattle and sheep fall into dykes. Ponies, if the bottom of the dyke is firm, can climb out. Cattle and, particularly, sheep are liable to get liver fluke on wet grassland, so the balance between meadow and pasture must be set carefully in regard to both water and season.

Some wet grasslands are used as washlands: overspill during floods, to store the water for slow release when the danger has passed. Depending on the farmer's attitude, and the frequency of flooding, they may be anything from high-yielding grassland to open, ungrazed fen. Other grasslands are flooded despite some flood defence. As the first priority is always settlements, excess water will generally be diverted to, in order, washlands, wet grassland, other grassland, and arable. Frequently-flooded land is usually more efficient as grassland than as arable, although there are arable washlands (e.g. in Lincolnshire) let at lower rents because of the occasional crop failure from flood.

The criteria governing which livestock, which variety, and at what seasons vary with the period. National and EU policy are now almost as important, within a geographic region, as the details of the fields. Earlier, with no agrochemicals, each field had to be considered individually. London, the largest city of the Kingdoms, absorbed much meat. In the eighteenth century (Defoe 1924–7), the north Thames marshes fed sheep in winter. Romney Marsh sheep were particularly big, the soil being fertile, and bullocks were also fed there. The Somerset levels and moors were good for the breeding and feeding of cattle. Meadows fed by the flooding Severn were rich, as were the water meadows north of Salisbury. The grass there was 3–4 m high, presumably *Glyceria, Phalaris*, or both. The Silt Fens also produced large sheep and large oxen.

The main lowland livestock are now cattle and sheep, with a few horses, donkeys, etc., and the occasional wild deer (muntjack, Chinese water, sika, and the native fallow and roe deer). Formerly, as explained earlier, there were more horses (rural horses could feed at least partly in the fields). There were also rabbits (introduced by the Normans) being cultivated and fostered in warrens, grown for food and fur, as well as being caught countrywide for local use. Rabbit stew was a common and cheap (or free) country dish. Rabbits do not graze really wet land, but are sparse on poor grazing, and – if not kept down by either people or diseases such as the man-introduced myxamotosis – can substantially decrease the grazing for domestic animals.

These different animals, naturally enough, feed differently, and the resulting grassland varies in structure (Figure 2.4), always supposing the farmer has not allowed the field to be grazed bare.

Another formerly abundant grazer was the goose. Geese were grown not just for food but also for the quill pens that were, surprisingly, used into the nineteenth century. It is not easy to find a reliable picture of a goose-grazed pasture, but a description reads:

[I would like to be] a goose girl, and sit in the greenest of fields minding those delightfully plump, placid geese, whiter and more leisurely than the clouds on a calm summer morning . . . The fields geese feed in are so specially charming, so green and low-lying, with little clumps of trees and bushes, and a pond or boggy bit of ground somewhere near, and a profusion of those delicate field flowers that look so lovely . . .

(von Arnim 1899)

(Geese no longer roam with their guardian!)

This is grassland of much conservation value. Drainage is light, biodiversity is high, and structural diversity is high. It is a pity geese were no longer in demand when conservation arrived.

Bog and moor grazing is necessarily of low yield. Deer are mainly red deer. Public opinion – as with regard to birds – is turning against killing deer for sport and eating (except it is all right to keep deer in a field, and kill them for food). Excess numbers are therefore often 'culled'. It is far more profitable to conservation to maintain the traditional habitat and raise meat from it than to abandon both. Cattle, sheep and horses do well on hill and bog wetlands, at a low stocking rate. Culture rather than grazing quality determines which is the most abundant in each wetland and at each time.

Litter or sheaf (names used interchangeably, or 'sheaf' as vegetation taller than litter)

This is the other main crop, or, rather, former main crop mown from lowland wetlands. It is mown once a year, in summer, and is whatever useful comes up (on winter-flooded to winter-waterlogged land). It is tougher and less palatable than either marsh hay or grass. It is usually diverse, including thistles, rushes, sedges, Carices, medium-sized grasses (such as *Molinia carerulea* and *Deschampsia caespitosa*) and many more. The use was for animal bedding, strewing, fuel, and general use around the farm and house. At a pinch it could be used for temporary thatch, or roof ridges.

3 How wetlands work

The well-being of the water environment is fundamental to sustainable development.

(J. Gardiner 1996)

The sustainable development of a country requires wetlands in being.

(W. Mitsch, in Mitsch 1994)

The only biotic parameter that can be recorded consistently is the species composition of the vegetation.

(Barendregt *et al.*, in Vos and Opdam 1993)

There are exceptions to any ecological rule: principles alone have validity.

Introduction

Wetlands developed as wet land, and either they are now wet or they have been dried, usually by human impact removing water. These wet land conditions may be the flooded gravel pit abandoned less than a decade ago, or sites where drainage was impeded at the end of the Ice Age, 10 000 years ago.

Wetlands in being accumulate soil. This soil may be sediments deposited by rivers and run-off, it may be plant remains that, under the conditions of the wetland, do not decompose, oxidise and erode, but humify and develop into peat. Or it may be a mixture of the two. Either way, soil builds up. Some old wetlands have had soil rising intermittently or continuously for millennia, others have had erosion and rebuilding, and yet more have been lost – by earlier natural processes, or later human impact.

Peat is mostly dead vegetation. The two major types are *Sphagnum* spp. in bogs, and *Phragmites australis* in reedswamp, but all peatland species are incorporated into growing peat, including fen carr, large sedges, bog shrubs and cotton grass.

Wetlands are very sensitive environments. A slight change in environment leads to a corresponding change in the vegetation: a little drainage of a bog allows tree invasion, a little accumulation of rainwater on a fen surface allows *Sphagnum* to grow. The development of the different peat communities depends on the balance of plant growth and of change in water regime. Plant growth is an internal process: the plant has its own growth pattern. The water comes from outside the wetland, it is an outside force. When the balance of the two is changed, whether through natural climatic or geomorphic change or through human activities, the wetland communities respond; even small changes in water level lead to changes in the species composition of *Sphagnum* or in the species growing between black bog rush, *Schoenus nigricans* tussocks (partly from Gopal, in Patten 1990). The response may be rapid, or delayed by inertia of the community, but it is sure. Anyone altering the outside forces on a wetland, be they flooding, grazing or waste disposal, must expect the communities to change.

Change they will, whether it is with changes in climate (and, for low-lying land, sea level) over centuries or millennia (Table 2.2), or the variable impacts of people. Communities are in continual change, though the rate and type of change vary. Even in the bog that has been bog with similar species for millennia, last century's vegetation is this century's peat. Studying a wetland now is studying a photograph in time, not looking at a time-lapse video.

Integrating wetland processes

Water

Water in wetlands can move in any direction at any speed, and can come from rain, run-off, groundwater and surface river water (sea-formed wetlands are not considered here). Basic Chart A shows the main water regime resulting from each of the four sources. (Mixed-source wetlands have mixed water regimes.) Where the principal water source is rain, the resulting bog has a stable water level, with little fluctuation (if it dries substantially, the bog would be in decline). Only the surplus rainwater runs downslope off the bog. Where water soaks down in blanket bog peat, it may also flow through subsurface networks of permanent or short-lived 'pipes' on the rock or within the peat.

While rain falls on all the land (though varying geographically, with altitude and locality), run-off varies. The amount of rain running off depends on the amount of rain falling, the absorbency, at the time, of the land below, and the position of the site in the catchment (Basic Figure 2, Basic Charts E and F, Figure 6.1). Places receiving run-off receive far more water than those getting rain alone! As the rain varies, so does the run-off, filling up a wetland, raising its water level. Then, after the run-off, the water level falls, the water running further downslope or into the ground. Flushes are typical run-off wetlands. Marshes and fens often used to have river water or groundwater or both.

The water sources, so clearly separated in principle, are – as so often – difficult to separate in practice. Rain, falling vertically on the earth, turns into run-off, moving laterally down the slope. Run-off runs both overland and in the soil. Run-off in the ground soil may become groundwater (Basic Chart E, Figure 6.12). Water just into the soil, running freely, is run-off. That running deeper and slower, with a longer residence time, which enables the water to exchange chemicals with its surroundings and so take on their chemical nature, becomes groundwater. Water welling up from far below, with a residence time of centuries or more, is undoubtedly groundwater. But water welling from the hill above may have a residence time of days, even of hours.

Surface run-off becomes river water. If it runs all year, it is a stream. If for a few hours now and then, run-off or a run-off channel. In between is difficult to define. A stream bed in drained land now taking only a little storm run-off but marked on an Ordnance Survey map surely carries status as a stream!

Moving water, run-off or stream, delivers more chemicals than still water. If water moves, the soil, plants and animals in its path are bathed in its stream. This is why flushes (gathered run-off) are more nutrient-rich than the land around, why rivers are more nutrient-rich than their wetlands.

Spring-fed wetlands generally have a constant (or seasonal) inflow of water. As that cannot accumulate for ever, there is necessarily a means of water running out – a natural block or artificial sluice, so that water can rise to a given level, but (except in storms) no higher. Water can fall below this, but not rise above it.

Groundwater only rises locally to the surface. Most stays and moves below, and if below any influence on the wetland, is irrelevant to it. Groundwater, necessarily, is only in ground that can store water, that is, with pores or spaces, not in solid rock. Springs and seepages typically, but not always, are around the outer edges of wetlands, from limestone, sandstones, various types of glacial drift and alluvium deposits, and deep subsoils and soils. Groundwater, if moving, is necessarily under pressure – pressure from entering water, from compression, etc. – so it emerges under pressure also.

Finally, river water causes very large variations in water level in the alluvial plain through which it runs. It spreads over the whole in high floods, and is confined to part of the channel in drought flows. When the river, as is now common, is cut off from its flood plain, some water usually still seeps through the bed, but mostly the 'wetland' is dependent on run-off (and rain and, if present, springs). The water level fluctuations are generally greater than with other water sources.

Nutrients and sediments

Bog is rain-fed (Basic Chart B), although valley-bog is also fed from low-nutrient acid run-off. Poor fen has any low-nutrient water, and rainwater may be more important than in other fen types. Calcium-dominated fen must have calcium-rich other-nutrient-low inflow (and no other strong chemical influence). Occasionally this inflow comes from limestone run-off or from run-off moving through (and influenced by) calcium-rich peat. Usually it is from limestone springs.

Flooding with more river water leads to *marshes*, where silt or other sediment is deposited, and *fens*, if the water extends beyond where it has deposited its main inorganic load. Its nutrient status depends on that of the sediment and the water. Most wetlands are now removed from river flooding, and so from the incoming nutrients that sustained them. Their inflow is now too often polluted run-off, from fields, perhaps with road contamination. Vegetation changes with all such alterations.

These nutrient regimes are further clarified in Basic Charts C and D. In C, the lay-out is by water source, showing the very low nutrient status of rain, that springs depend on the aquifer, that run-off depends on the land over and through which it flows, and that rivers depend on the run-off and carried sediment of the whole upstream catchment (as modified by river processes). Depending on its type, run-off can give rise to valley bog, all fen types and marsh. There are some non-compatible combinations; for example, bog does not develop with sediment deposition (although bog can grow on currently non-flooded alluvial plains), and calcium-dominated fen does not develop with nutrient-rich river water.

Soil type is closely linked to the chemical status of the water and carried sediment: in chemical status, in composition and in its further development. Bog peat is necessarily very nutrient-poor. Made of *Sphagnum*, and to a lesser extent heathers, it also contains inhibitors. (The inhibition – as with calcium-dominated peat – is relaxed on disturbance and drying.) Bog soils were termed 'dystrophic' (deficient or ill-feeding), compared with the next up, 'oligotrophic' (low-feeding), before these growth-hamperers were known, so the term 'dystrophic' is now reserved for them. Water, with almost no nutrients, but with no hindrance either, is now at the bottom end of the oligotrophic group. Water affects soil composition, and so humification to peat in peatlands. The type and amount of solutes and sediment carried are crucial.

Basic Chart C also shows which soils are formed under, and which above, the general water table, and that the underwater ones are both peats and mineral soils, while the overwater ones are bog. In wetlands, calcium is in a special position, partly because of how it acts and partly because of its pervasive ability. Calcium carbonate affects pH, that crucial reflection of habitat status, alkalinity, acidity, base status and more. It has, in some ways, a greater influence than other nutrients. A middle-nutrient, mesotrophic fen with low calcium is transitional to bog, a similar one with high calcium is fen. High calcium plus high other nutrients gives a rich fen. With low other nutrients, calcium removes from availability much of the phosphate and other nutrients. This creates a separate habitat to poor fen or bog with equivalent levels of available nutrients (apart from calcium). In those, the calcium is low and suppressing availability. Nutrients are low. In soil made nutrient-poor by high calcium, these nutrients can be made available to plants by mineralisation (if dried and disturbed). Dry a piece of calcium-dominated peat: it is fertile. Calcium-influenced habitats are intermediate to rich fens where all nutrients are high. Calcium-influenced fens can, like calcium-dominated ones, bear fen sedge, *Cladium*. This does not grow in rich fen. *Phragmites* can occur but does not dominate in calcium-dominated fen. It can dominate in calcium-influenced and rich fens and marshes (and also in lower-nutrient habitats). *Glyceria maxima* grows in rich fen or, of course, in marsh.

Calcium-dominated fens are of much conservation interest, and have been studied particularly in Britain and the Netherlands. There is a sharp difference, though. In Britain the aquifer water is in limestone (mostly chalk), while in the Netherlands it is in lime-rich alluvial deposits, and the calcium influence, though enough to immobilise phosphate, etc., is much less.

Basic Chart D is divided by habitat. It stresses the water regime necessary for soil growth, varying from the sediment-laden flood once a decade to near-constant wet for peat growth (seasonally dry bogs and fens oxidise while dry, and so may make no net growth).

Sediment

The sediment content is highlighted in Basic Chart E, showing the soil formed, its sediment source, amount and nutrient status. With high sedimentation, the soil is, of course, largely mineral, deposited mostly during river flooding. In nutrient status, it reflects the catchment and also the size of particle deposited. Silt contains the highest nutrients of the catchment, gravel and sand washed free of silt contain the lowest. A silty gravel from a high-nutrient catchment may nevertheless have a higher nutrient status than silt from a low-nutrient one. Sedimentation is found in all types except blanket and raised bog. It dominates where river flow is rapid, or if the soil is too dry for peat formation. Flood plains used to be regularly flooded by the streams making them. The fertilising silt covered the plain. With time, sediment may build up to be out of reach of any but the highest floods. Naturally, with climate change or human impact the plain may become dried and above river level, or sunk below it and rebuilding.

The nutrient status of wetland soils varies greatly. Basic Chart G shows

1   the variation in inherent nutrient status;
2   the variation with drying and disturbance. Wet habitats have decreased available nutrients, and dried and disturbed ones have greatly increased nutrients (especially in calcium-dominated fens, see Chapters 6 and 7); and

3   additional factors. Bog peats are a particularly harsh environment. Some plants even get extra nutrients from insects, e.g. the sundews, *Drosera* spp., and butterwort, *Pinguicula vulgaris*. Such species extend into poor fens and calcium-dominated fens, which are also nutrient-poor.

Nutrients are added by human activities, by agrochemicals, by air pollution (especially nitrogen, see Chapters 6 and 7), by effluents and other incoming sources.

The nutrient regime shown in Basic Charts C–G integrate these factors.

Peat pattern

Peat fates, as well as types and ages, vary (Basic Charts H and J). Plant material humifies: its dry bulk density is increased, pore space decreases and the material gets darker, turning from the green of living plants and the brown of dead ones to dark brown or even black. This peat has a higher calorific value than the plants and humus that preceded it. The peat varies with its constituent plants (reed, sedge, heathers, Sphagna, different trees, etc.). It also varies with its mineral content (negligible, much clay, sand, etc.), the weight of 'soil' above it, transport, and microbial activities (Lüttig, in Fuchsman 1986).

The basic pattern for an infilled basin is shown in Table 2.1 and Basic Chart E. Open water silt is followed by fen peat grown under standing water, the plant constituents showing the shallowing of the water from reedswamp to wet fen to fen carr. Then, with the peat at water level, rainwater starts being held above the water table, and bog peat begins. Trees die, and only bog is left.

The open water–fen–bog cycle is repeated where water and ground level vary over time, or it may be cut short and fen, rather than bog, be left or over-flooded and have silt deposited (see Basic Charts K–O). Where the soil is lost it is, of course, no longer there to tell the tale. Dry land, without bog above, may occur either because the process has not reached the bog stage or because it cannot do so. Unless sufficient rainwater can be stored in the surface, there can be no bog. Inadequate stored water can occur with insufficient rain, insufficient rain for the incident temperature, with intermittent flooding (swamping the acid rainwater), or with a porous surface so water soaks down into the fen peat instead of starting a new bog. Drier peat becomes more textured and more porous.

Finally, there are the bogs that grow direct on the rock, subsoil or soil below, with no intervening reedswamp or fen. These are the covering bogs, the ones that can creep round, blanketing all except steep slopes as they pass. This is mostly blanket bog, but also raised bog when it spreads from the basin below or (as in Figure 1.2, Basic Figure 1) grows above a mineral soil of a flood plain or estuary: where it was possible to have impeded drainage and rainwater storage, but no lake beneath.

In wet places, peat often grows over open water. This is the 'quaking fen' of the Dutch, who still have plenty, the 'quagmire' of the English, who now have little, and seldom use the word in its literal sense. It is possible for plants to grow on peat so soft that no firm bottom is reached for several metres. In the Fenland, even in the nineteenth and early twentieth centuries, the 'bottomless' dykes were well known and best avoided, as drowning was near-certain unless help came. Peat may form a raft over water, or water level may rise, or light peat may occupy all the space available. A raft may be unable to bear even a sheep. Or it may be thick and firm enough to bear alder carr, as in Broadland, where the presence of water below can be detected only through heavy jumping and the resulting quiver. Rafts can form by new surface vegetation being flooded or by growth from firm edges. They are made of vegetation, becoming peaty with age. Therefore, most are on places sheltered from much disruption, including, of course, peat pits. Rafts are the commonest way that quaking bog or fen develops. The 'man-eating' mires mentioned in Chapter 1 may be either holes in this raft (either new holes, or where the new surface is not yet made) or separate pits, where, for some reason, water has collected. (In the Slough of Despond, the 'some reason' was no doubt the traffic, especially the many herds of bullocks that Defoe (1724–7) describes being driven to London through Bedfordshire.) Such water layers are not usually shown in peat cores. They are now scarce in Britain. It is easier to get cores from solid peat, and also water may be of no relevance to those collecting cores. Water layers were common before drainage, however, and can still be found, for example, in Wicken Fen (Friday 1997).

Soil-building differs between fen plus marsh, and bog (though the two integrade). Fen and marsh build their soils passively. Sediment is brought in and deposited. Plants die and their dead bits are deposited. The more the deposition, the quicker the standing water is filled up, and the quicker the end of the wetland – without additional habitat change.

In bogs, however, the build-up is by living *Sphagnum*, etc., the water level rises as the peat builds, keeping pace with the soil surface. The more the peat, the higher the water level. Bog growth may be stopped by climate change and human impact.

Not all wetlands grow soil. The habitat may be wet enough to be a wetland, but have no sediment deposited, no constant high water for peat to grow under and no long-term storage of rainwater. Under this heading come the nettle-patch in the depression at the bottom of the garden, the miry hollow where shoes stick in winter – and innumerable other examples, large as well as small, throughout the country.

Basic Chart I shows the date of peat formation. Most started in the late Ice Age, and – presumably – was then near-free of human impact. Bog development has varied with climate, and fen with the occurrence of shallow lakes that are almost unsilted. Drainage has certainly decreased fen peat growth, and total fen peat. Basic Charts H and I show the characteristic losses. Removal for parish use has usually been sustainable. Wetlands are seldom centres of dense population. Immense amounts have been removed both before and during the twentieth century. Bog peat – where there is a choice – is chosen for fuel, domestically, and for power stations, and for horticulture. The fibre is better. Fen peat, though, with its higher fertility, is chosen for arable cultivation. Peat, like other soils, is not destroyed by growing a cabbage in it. Fen peat, dried and exposed to the air, however, mineralises, oxidises and, having negligible ash, vanishes, so the peat is lost as surely as the raised bog is taken to the garden centre. 'As surely?' No, more surely, because there is a movement to limit peat-milling from bogs, little to wet and keep the peat in arable fens. Conservers think of the flora and fauna already gone from arable fens, rather than the peat itself.

Basic Chart J looks at soil composition, showing the material of which different peat and marsh soils are composed (*Sphagnum* and others for bog, *Phragmites* and others for fen and mineral particles for marsh), the nutrient regimes these give rise to, and their capacities for agriculture. Here marsh, once drained, behaves like any land soil (if with a low organic content), fen peat is good but wastes, and bog is too infertile for high yields anyway.

Vegetation

Basic Charts K and L bring in the existing (rather than past) vegetation. Basic Chart K shows water fluctuations, the great fluctuations of, say, the Rhône or Danube to the left of the chart (this is not now found in Britain) and a fen basin with but minor fluctuation to the right. (These have water above ground, so do not include bog.) Riparian woodland is the first stage on the bare gravel (or other sediment) with the high fluctuations. It is also the last stage: as the habitat dries, woodland stays, although its composition may change (more oak and elm in drier woods). In more stable water, there are more stages before the wood comes in. Riparian forest trees can stand submergence provided they get established (on intermittently dry land) and are dry for at least part of summer. Intermittently dry land means water fluctuations. On the right of the chart where drying is less, wood habitats start further up the stages of succession. Plants of more constant wet colonise first. *Phalaris arundinacea* reedswamp needs to be summer-dry. Those on its right are the always-flooded brigade of *Phragmites australis, Typha* spp., *Carex rostrata* and more. In a yet more stable water regime, a wet community may be interposed between the flooded one and the wood. However, such a community is usually due to impact: grazing, mowing or other treatment preventing tree growth. Finally, in really stable conditions, raised bog can develop above the woodland. Elsewhere, the others are flooded too much (and when too dry for flooding, have soils too porous to hold water).

Basic Chart L show how bog covers everything else, given suitable rainfall (and temperature). There is no consensus as to the minimum rainfall needed. Bog may be the last stage in a long and complex wetland succession, or the last in a short one, or the only phase. Once bog is growing well, the peat record shows it can go on growing, for millennium after millennium, with no significant change. Only two things halt it: climatic change (where a drier climate permits pine and birch to invade or, in a basin, a wetter one floods it), and human impact. The impact may lead to heather or grass moor instead of pine-birch forest, or may be drainage, erosion or extraction. Bog endures. British fen, in contrast, is transient and requires surface water.

Basic Charts M and N show plants and plant communities. Many wetland genera, indeed often wetland species, or look-alikes, occur throughout Europe, or further. *Phragmites*, though most important in Europe, is found in all five continents. So are tall, sward sedges, though the species may change (*Cladium mariscus* in Europe; *Cladium jamaicense* and *C. mariscoides* in the United States). The same applies to *Sparganium* (*Sparganium erectum* in Europe; *S. americanum* in the United States), *Sphagnum*, rushes, *Salix* and more. The wetland types are described for this book as bog, three types of fen (poor, calcium-dominated and rich) and marsh. The communities differ in the poor and calcium-dominated fen, but are similar in marsh, rich fen, and, roughly, medium fen. This similarity continues into the drier habitats, except for the more divergent habitats. Wood will develop unless prevented. In Britain, with drying (non-bog) wetlands that used to be managed to prevent tree invasion, there is in fact a lot of nutrient 'medium' open fen. Woods are there, or have been prevented. Birch and pine grow in bog and poor fen and (extreme) calcium-dominated fen. Alder and ash grow in all the rest. So do the *Salix* spp. but here the species may differ, e.g. the nutrient-poor *Salix pentandra* and the nutrient-rich *S. cinerea*.

When impact prevents trees and reedswamp, what remains? If managed for specific crops (Chapter 2), these or their derivatives occur: marsh hay, rushes, sedge-, rush- or grass-fen, wet grassland, or withy, reed or sedge beds. The variety is in fact wider as there are many different, say, rush-fens.

If wet heath and carr appear stationary, then helping rainwater to be retained may start the succession to bog. Such an 'accident' from rich to poor fen may also be rapid. Base-rich water can – quite quickly – no longer reach the building, or indeed the drying, peat. In the Netherlands, succession of *Caltha palustris*-type fen via *Molinia caerulea* fen then

Figure 3.1 Tall herb vegetation.

wet heath to bog has been documented in three centuries. There are also, and more recently, areas changing from heath to *Molinia* or *Deschampsia* grass-heath. This is attributed to greater nitrogen mineralisation (from, presumably, impact pollution) (Van Wirdum, in Wheeler *et al.* 1995).

Very abundant at present in marsh, rich fen, medium fen and (dry or disturbed) calcium-dominated fen are the tall herb communities. On poor fens they have more nutrient-poor species. They occur over acres in large wetlands, in hollows or ditch banks of wetland or arable, by garden ponds, and more. They have no doubt occurred for millennia. It is more doubtful how abundant they were before the twentieth century. Most have resulted from abandoning managed drier wetlands, or drying and abandoning wetter ones. At a distance (Figure 3.1) all look similar, one to two metres high, with wide-leaved dicotyledons, prominent. Grasses may be co-dominant or absent; the dominants may be ruderals, occurring in much-dried and abandoned marsh hay lands (e.g. *Epilobium hirsutum, Galium aparine, Urtica dioica*), or wetland species, occurring in drier reedswamps, and becoming prominent here on the open marsh or fen (e.g. *Peucedanum palustre, Lychnis flos-cuculi, Lysimachia vulgaris*). The tall herb community, structurally, is widespread as well as variable; variable with water regime, nutrient regime, past management, present management and perhaps the size of the site (more species-rich in the large Broadland areas than elsewhere).

Basic Charts M, N and O select some plant communities out of a vast continuum (see Chapter 4) and are incomplete. Human impact is shown as one of the three determining factors for wetlands: water regime, nutrient regime and impact. It shows short vegetation continuing with impact where trees would otherwise be.

Basic Chart O shows patterns in the alluvial marsh of the Rhône plain. In the plain, the waters of this large, mountain-rising river deposit sand (more water force) or silt (less water force). They may also sweep away (erode) the soil they had earlier deposited. On a smaller scale this occurs in Britain. Sand is quick-drying, easily flooded but not easily waterlogged for long periods. The first stage is as in Basic Chart K, *Phalaris* followed by *Salix*. In the Rhône plain the drier land has poplar and finally oak. On the silt and silt-clay, water can be retained, and the first stage is a sedge (*Carex*) marsh, leading to alder, then mixed-oak and finally oak-dominated forest. These contrast quick-drying with slow waterlogging. While both have reedswamp leading to wood, the species differ. When silt-clay is deposited above sand, it, when thick enough to hold water, forms the waterlogging habitat, and the vegetation swings over to that community. However, in the flooding area of a swift river, erosion can remove all.

There is now much short herbaceous vegetation in land dry enough for wet – or even dry – woodland. Basic Chart P shows some causes of the absence of trees – too wet for establishment, wrong nutrient status for the relevant species, habitat developed to raised bog, management, and draining. Overall, it is rare to find the particular combination of conditions necessary for a seedling of a particular tree species to become established, and to grow large enough to be no longer killed by ordinary wetland hazards such as 30 cm flooding, cattle grazing or spring frosts.

Distribution

Wetland occurrence (Basic Chart F) depends on lack of drainage. The natural drainage of the land must be impeded in some way, holding up water, under which fen peat and marsh soils are laid down. Alternatively, land and climate must be suitable for the holding of rainwater *in situ*, within which bog (blanket and raised) may grow. These bogs hold rainwater within, preventing its escape. The others have water outside. Figure 3.2 shows the distribution of bog now, after so much bog has been removed by man.

It would be difficult to construct it, but a map showing alluvial plains, particularly if it started as those not 10 m wide and ended with the Kent marshes and the Silt Fens, would show them far more widespread than peatlands. Fens are the least common type. Because of the intensive study, and extreme interest, it is easy to overestimate their abundance. Poor fens are scattered particularly in the North and the West. Calcium-dominated fens have been particularly studied in East Anglia, Anglesey and Oxon, where limestone springs emerge into basins that have been gradually infilled under lime-dominated water. Rich fens and mixed fens cover the main basins of Broadland, the Fenland and the Somerset levels, those of the East being more limey, more fertile and calcium-rich than, in general, those of the West.

Wetland landscapes

Basic Figures 1–20 illustrate a selection of wetlands in the landscape. Naturally there are many other patterns: flood plains can be of different water regime, different substrate texture and different width, from a few metres to the wide expanses of the Netherlands. The figures show water sources, water movements, and the resulting wetlands, from the small flush at the base of the hill, and the reedswamp fringing the lochan or tarn, to blanket bog and flood plain. The passage of run-off waters and the difference made underground by aquifers should be noted, as also should minor points such as the lagg channels round the raised bog and the acid springs in the valley bog.

The highland landscape of the north or west of Britain, to the left of the figures, is contrasted with the chalk or other lowland of the south and east of England to the right. On the highland landscape are bogs and moors, where bog peat stored rainwater, and

Figure 3.2 Areas of deep (over 1 m) peat in Britain (from Lindsay 1995).

grew. Acid New Forest sands have their valley bogs, not on the tops like blanket bog, but down the lower slopes and in the valleys. This is a water-vibrant landscape, integrated, with varied, renewable resources, valuable to people for the water, for the habitat and the flora and fauna they have.

A small lake has been put in the highland side. Tarns (lochs, etc.) are, as can be quickly seen from the map, much more frequent in the north and west with hard resistant rock and high rainfall. But there are ponds, village ponds, cattle ponds, manor ponds, dew ponds and so on in lowland country, and even in some parts there are natural lakes. The new lakes are the abandoned gravel and sand pits of mainly lowland river plains, and, to a lesser extent, reservoirs. Gravel pits have wetlands on shallow edges, where these exist or have been made. Their water table is relatively stable. It varies with run-off, less with river level: unless the pit is connected to the river, when it varies more. Reservoirs are scattered in both highlands and lowlands. Because of the great fluctuations in level, and long periodicity, the exposed shores are a harsh habitat for vegetation. These two, gravel pit and reservoir, are therefore very different in wetland habitat, resembling each other in being recent, and of mineral soil, differing in water regime.

Basic Figures 7, 14 and 16 show a dying system: dying only slowly, but dying, not sustainable. In place of natural variety there is man-made uniformity. The groundwater level is dropping with abstraction, the bog is no longer growing. Water is not being stored, the river level is lowered, ditches abound. (With further drying, ditches too are lost, as water level sinks below ditch level.)

In the flood plain there may be gravel, sand, silt or peat. Among the peats, the calcium-dominated fen occurs (mainly) by limestone springs, the rich fen receives fertile river flood water as well, and the poor fen, nutrient-poor river flood water. A raised bog has been placed growing on either poor fen or waterlogged marsh. It could have – and often has – grown over rich fen and marsh also. Raised bogs grow over substrates of any fertility.

In contrast to the mixed and variable pattern in the flood plain, the bog pattern above is simple: it is found on all ground of low slope. The steeper slope has peat too, though now nutrient-fed and dried through being on the slope.

The chemical Basic Figures 12 and 13 repeat the pattern of great variety in the traditional state. Water comes first. Variety in water gives variety in habitat chemistry. There is variability in the bog peat, with the active growing bog, the drier moors on hilltops, and the moors on slopes. Water differs, soil then differs. The flood plain has soils following the water patterns: spring-peats, calcium-dominated from limestone springs, valley bog from infertile oligotrophic waters from acid sands. Silting follows rivers, so marsh and fen soils do, too. In the same way impact has altered wetland water, so it may alter wetland chemistry, including by pollution (Basic Figure 14). It is at least pleasant to note that while water devastation is common, drastic chemical inputs are rare.

Short vegetation encourages high plant diversity. A community heavily shaded (Figure 3.3), whether by trees, tussock sedges, or reeds, is inhospitable for most shorter plants. A few, such as *Galium palustre*, and indeed as a sparse plant, *Phragmites australis*, can grow adequately in shade, most cannot. There are also far more wetland species under 50 cm high than over 150 cm high. Therefore, to get the diversity, fen and marsh vegetation must be short or be shortened. Therefore, management is necessary for high biodiversity. 'Natural' communities are unknown, so to conserve, maintain and sustain the traditional ones that have evolved over centuries in conformity with management – or the deteriorated habitats most close to these – is theoretically as well as practically sound.

Biodiversity

Biodiversity (mentioned earlier), or biological diversity, describes the amount of variation in ecosystems. It may be the diversity of habitats in a fenland, but is more usually used for

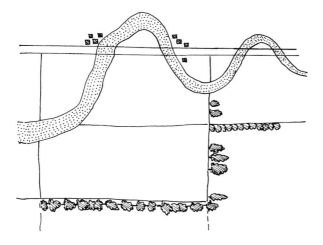

Figure 3.3 Roden, Fenland. Silted river bed wastes and erodes less than peat, so rises above the general fen level. Farms on the higher land.

the number of species in a habitat. This may be the number of all species, or the species of any one type: birds, fungi, dragonflies, bacteria, or whatever. (It may even be the number of DNA sequences in the genome of a species.)

As is only to be expected, there are no data available on the total number of all species in any British wetland (even, say, Wicken Fen has not had, for example, year-long bacterial surveys!).

Data are, therefore, patchy, local, out of date or with other disadvantages. Biodiversity is, however, of public concern. The Rio Convention on Biodiversity was drawn up and agreed in 1992. The reasons are (1) ethics – no life form should be allowed to become extinct (with the exception of such as the smallpox virus); (2) value to people, either now (e.g. rice is still the most important of the staple foods), or in the future.

In general, there is a perception that high biodiversity is Good: a woodland contains more species than a car park. Within wetlands, however, other criteria also apply. Reedbeds are species-poor in vegetation. One dominant species, others are few and sparse. However, they are a specific habitat, with much value in wetland development and their animal (particularly invertebrate and bird) species. Species-rich habitats often contain more rare species (on general proportionality), so have that value. Small variations within habitats create niche habitats in which other species can live, increasing totals.

Species diversity is, in general, largest in habitats with an intermediate to low nutrient status. Where nutrients are plentiful, a few highly productive and competitive species dominate, while in extremely nutrient-poor systems, lack of nutrients becomes limiting for many species and limits productivity (Junk, in Gopal *et al.* 2000). But none would compare 'value' of species-rich fen and species-poor bog or reedswamp. Species-rich for what groups, first. Then value to what group, for what purpose.

There used to be, for instance, a prevailing view that wetlands exist for waterfowl: that they should be managed for high biodiversity and high populations of these, and never mind the rest. Fortunately, wiser counsels prevailed, and it is now realised that, for maximum benefit to bird life, benefit to other biota must also be provided.

There are many difficulties in measuring biodiversity. Animals spending their lives in wetlands clearly count. Birds on migration are less obvious. The wetland is used, and may be most important to them, but for negligible time. Plants present only in seed banks, and invertebrates in resting stages are also unclear. They cannot be measured properly, using current techniques. Species could be counted per wetland, per habitat, or per sample size.

There is currently only one reasonably comparable dataset, that of the National Vegetation Classification (Table 3.1). Sample sizes do vary, but similar measuring techniques

Table 3.1 Species richness of plants in vegetation types of the (British) National Vegetation Classification (data from Rodwell 1991a,b, 1992, 1995)

Sample sizes vary: woodland (W), 50 × 50 m for tree layer, and 10 × 10 m or 4 × 4 m (for low vegetation) for the field layer; grassland (MG) 2 × 2 m; swamp (S), 10 × 10 m; mire (M) 2 × 2 m for most short herbaceous vegetation and dwarf shrub heaths, 4 × 4 m for taller and more open herb communities and low shrub heath, and 10 × 10 m for species-poor or very tall herbaceous vegetation and dense scrub.

The average number of species per sample cited are therefore not fully comparable, but are here treated as such.

Average species	Code	Name of community
40+	M13 ii, iii	*Schoenus nigricans–Juncus subnodulosus*
Total 2 types		
30–39	M8	*Carex rostrata–Sphagnum warnstorfii*
	M10 ii	*Carex dioica–Pinguicula vulgaris*
	M26 ii	*Molinia caerulea–Crepis paludosus*
	M28 i	*Iris pseudacorus–Filipendula ulmaria*
	W7 ii, iii	*Alnus glutinosa–Fraxinus excelsior–Lysimachia nemorum*
Total 6 types		
20–29	M7	*Carex curta–Sphagnum russowii*
	M9 i, ii	*Carex dioica–Calliergon* spp.
	M10 i, ii	*Carex dioica–Pinguicula vulgaris*
	M17 i, iii	*Scirpus caespitosus–Eriophorum vaginatum*
	M18 ii	*Erica tetralix–Sphagnum papillosum*
	M19	*Calluna vulgaris–Eriophorum vaginatum*
	M23	*Juncus effusus/acutiflorus–Galium palustre*
	M24 i, ii	*Molinia caerulea–Cirsium dissectum*
	M26 i	*Molinia caerulea–Crepis paludosus*
	M28 ii	*Iris pseudacorus–Filipendula ulmaria*
	S24 i, ii	*Phragmites australis–Peucedanum palustre*
	W2 ii	*Salix cinerea–Betula pubescens–Phragmites australis*
	W3	*Salix pentandra–Carex rostrata*
	W4 ii	*Betula pubescens–Molinia caerulea*
	W5 i, ii	*Alnus glutinosa–Carex paniculata*
	W6 ii, iii	*Alnus glutinosa–Carex paniculata*
	W7 ii	*Alnus glutinosa–Urtica dioica*
	MG8	*Cynosurus cristatus–Caltha palustris*
	MG12	*Festuca arundinacea*
	M4	*Carex rostrata–Sphagnum recurvum*
	M5	*Carex rostrata–Sphagnum squarrosum*
	M6	*Carex echinata–Sphagnum recurvum*
	M13 i	*Schoenus nigricans–Juncus subnodulosus*
	M14	*Schoenus nigricans–Narthecium ossifragum*

Table 3.1 (*Cont'd*)

M17 ii	*Scirpus caespitosus–Eriophorum vaginatum*	
M18 i	*Erica tetralix–Sphagnum papillosum*	
M19 i, ii	*Calluna vulgaris–Eriophorum vaginatum*	
M20 ii	*Eriophorum vaginatum*	
M21 i, ii	*Narthecium ossifragum–Sphagnum papillosum*	
M24 iii	*Molinia caerulea–Cirsium dissectum*	
M27 i, ii, iii	*Filipendula ulmaria–Angelica sylvestris*	
M28 iii	*Iris pseudacorus–Filipendula ulmaria*	
S7	*Carex acutiformis*	
S10 ii	*Equisetum fluviatile*	
S11 ii, iii	*Carex vesicaria*	
S14 ii	*Sparganium erectum*	
S18 i	*Carex otrubae*	
S21 iv	*Scirpus maritimus*	
S25 i, ii, iii	*Phragmites australis–Eupatorium cannabinum*	
S26 ii, iii	*Phragmites australis–Urtica dioica*	
W1	*Salix cinerea–Galium palustre*	
W2 i	*Salix cinerea–Betula pubescens–Phragmites australis*	
W4 i, iii	*Betula pubescens–Molinia caerulea*	
W5 iii	*Alnus glutinosa–Carex paniculata*	
W6 i	*Alnus glutinosa–Urtica dioica*	
MG9 i, ii	*Holcus lanatus–Deschampsia caespitosa*	
MG10 i, ii, iii	*Holcus lanatus–Juncus effusus*	
MG11 i, ii, iii	*Festuca rubra–Agrostis stolonifera–Potentilla anserina*	
MG12 i	*Festuca arundinacea*	

Total 44 types

1–9

M1	*Sphagnum auriculatum*	
M2	*Sphagnum cuspidatum*	
M3	*Eriophorum angustifolium*	
M20 i	*Eriophorum vaginatum*	
S2 i, ii	*Cladium mariscus*	
S3	*Carex paniculata*	
S4 i, ii, iii, iv	*Phragmites australis*	
S5 i, ii	*Glyceria maxima*	
S6	*Carex riparia*	
S8 i, ii, iii	*Scirpus lacustris* ssp. *lacustris*	
S9 i, ii	*Carex vesicaria – C. rostrata*	
S10 i, iii	*Equisetum fluviatile*	
S11 i	*Carex vesicaria*	
S12 i, ii, iii, iv	*Typha latifolia*	
S13	*Typha angustifolia*	
S14 i, iii	*Sparganium erectum*	
S15 i, ii	*Acorus calamus*	
S17	*Carex pseudocyperus*	
S18 iii	*Carex otrubae*	
S19	*Eleocharis palustris*	
S20 i, ii	*Scirpus lacustris* spp. *tabernaemontana*	
S21, i, ii, iii	*Scirpus maritimus*	
S22 i, ii	*Glyceria fluitans*	
S26 i, iv	*Phragmites australis–Urtica dioica*	
S28 i, ii, iii	*Phalaris arundinacea*	
MG13	*Agrostis stolonifera–Alopecurus geniculatus*	

Total 44 types

were used, and used throughout the country. The samples are grouped by community, not by region: though often the communities are in fact regional, e.g. blanket bog. Average species per community (per sample) range from one to 45.

Of the 44 least diverse, with average species under 10, four are very nutrient-poor with short species. Thirty-three are tall reedswamps, with nutrients ranging from poor to rich. The remainder are medium-sized grasses or Cyperaceae. This fits well with the general statement above.

Of the eight most diverse communities, with average species over 30, two are woodland, one is tall and luxuriant herbaceous, and the rest are more open, either shorter or with a semi-open canopy of tall species. Their nutrient status ranges from rather poor, through nearly calcium-dominant, to high. The herb layer communities mostly need management (e.g. grazing, peat-digging) to survive long term. The plant diversity drops substantially if large plants invade and form a dense canopy.

There are 26 communities with average species of 20–29, and 44 with 10–19. Eighty-eight out of 124 have therefore under 20 average species per sample. Communities of 20–29 are mostly of fairly low nutrient status or woodland. The low ones are usually those of open moor or moor-grassland, which are grazed. Or they are woodland. A very few are tall herbaceous (Norfolk Broads, with relatively low drainage, pollution and disturbance) or equally unusual grassland (one ex-water meadow, one estuarine). The large (44) group of 10–19 diversity are, as expected, mixed, nutrient-low mires, reedswamps nutrient-medium to high, herbaceous reed habitats, woods, scrub and grassland.

Generally, if short to medium communities only remain short because of management (not because of nutrient deficiency) they may be species-rich (with the exceptions always found in ecology). Wetlands, of course, have a history of such management.

To maintain biodiversity, as is internationally agreed and European Union (EU)-directed, active management must therefore be carried out. This should be the same as in recent centuries where practical, and something with as similar effects as possible, where not.

This is a major difference to countries such as the Americas, with a history of negligible management and disturbance. Enhancing biodiversity there may be just concentrating on avoiding damage: the 'put up a fence and leave it' plan.

Plants and animals come into existing habitats and, by their presence, can change diversity. A mammalian hoof print can bear aquatic invertebrates while flooded, and open-habitat plants later. (*Viola persicifolia* and *Carex viridula* ssp. *viridula* have both reappeared in opened habitats in Wicken Fen; Walters, in Friday 1997.) Taller plants may shade out shorter ones (see Chapter 10). A shrub provides habitat for different invertebrates, and different birds, etc.

There is, unfortunately, evidence that biodiversity recovery is not always compatible with the reason for renewing wetlands. Too often, 'restoration' means an abundance of a few disturbance-tolerant species (sometimes not even of local stock). 'Restoration' is often limited by lack of knowledge, lack of large sites, lack of co-ordination, and lack of appropriate strategy for introducing and sustaining biodiversity (Zedler, in Gopal *et al.* 2000).

4 In wetland wilds

I'm just in love with all these three,
The Weald and the Marsh and the Down countrie;
Nor I don't know which I love the most
The Weald or the Marsh or the white chalk Coast!

I've loosed my mind for to out and run,
On a Marsh that was old when Kings begun,
Oh Romney level and Brenzett reeds,
I reckon you know what my mind needs!

(R. Kipling)

A thing of beauty is a joy for ever

(J. Keats)

I am monarch of all I survey
My right there is none to dispute
From the centre all round to the sea
I am lord of the fowl and the brute

(W. Cowper)

There is no universal classification system for British wetlands, nor should there be: the scheme adopted will reflect the interests of the researcher and the aims of the study.
(Hughes and Heathwaite, in Hughes and Heathwaite 1995)

Continuities and discontinuities

Where does a wetland, let alone a type of wetland, such as a reedbed, begin and end? Sometimes there is a physical barrier, the hill rises steeply, the lake sinks deeply. Sometimes there is a man-made barrier, the arable, the road, the building, the fence and ditch, all separate off the wetland. Sometimes it is the plants themselves, along a lake shore the reedswamp grows thick, it is a wetland. Beyond, into deeper water it ends gradually. There is, in the same shallow water, a lake, not a wetland. Or many feet and boats have stopped the reedswamp being in the shallow water at all. People, by preventing plants, have prevented wetland. The land beyond is sedge swamp, wetland indeed. Far, the gentle hill bears dry grassland. But in between? At what point is the division drawn? Where the first dry-land species occurs in the wet? Where the last wetland one occurs in the dry? When the balance changes? But the balance of what; of the dominants, of the number of species, of the wetland values of the species as calculated by Ellenberg (1974) (Chapter 6)? There is no universal answer. Similarly within the wetland. Here is a bog pool, there a hummock.

Figure 4.1 What makes a unit of vegetation? Fen pattern to show inter-relationships. Which would you distinguish as communities? How many labels would you give?

Tomorrow, after the rain, the pool reaches over yesterday's hummock and only the tallest stand out of the water. Here in a dry fen is a patch of dominant *Phalaris arundinacea*. There, nearby, is a much older patch of *Phalaris* where the plant has spread much further and species from the surroundings grow in with it. Here is a *Carex paniculata* fen, flooded, sedge-dominant: but with a few young alders growing on its tussocks. There is the dark alderwood, nearly dry underfoot, but the trees still rooted to the now-dead *Carex* tussocks. Between, the alder gets denser, gets taller, the ground gets drier. Where does the *Carex paniculata* fen stop and the alder-carr begin?

All vegetation forms a continuum, in both space and time (Figure 4.1). Across the drying fen the *Carex paniculata* changes to alder wood. In the peat core, it did the same in centuries past. Along the continuum, whether of space – which we can study in some detail – or of time, which we can study much less, varies the driving forces of: water regime (patterns, levels and movement); nutrient and other chemical regime (patterns, concentration); and the vegetation is different in each combination (usually, reflecting the conditions in which each species became established).

The human mind cannot absorb so many variations in community, and so many habitats. Therefore, it imposes a grid on the continuum, a grid more like a colander than a net, with as much or more space blocked out between the holes – the seen communities – as within them. The transitions, and indeed some of the different communities, are in the blocked-off part. Naturally the grid can be placed differently by different researchers: they 'see', recognise, different habitats as the basic units of wetlands. This is irrelevant, as long as all understand the process.

There are far more classifications than can be given here. They fall into various groups. The botanical ecologists got in first and produced vegetation classifications, many of them, and vegetation is the feature most widely used. Then there are the hydrological and geomorphological ones, used by geographers. In the foreseeable future, there are unlikely to be animal classifications, the study is of the spiders in blanket bog; the bog, not the spiders, being the unit of classification.

Geomorphological, hydromorphic, and similar classifications

The geomorphological classification of the American Fish and Wildlife Service is given in Table 4.1. It is divided into river, lake and marsh, each subdivided by type of substrate or type of vegetation. This is widely used, but not very useful in Britain.

A British hydromorphic classification is shown in Table 4.2. This gives a clear division between rainwater and land water wetlands (bog and fen), and between flowing water fens and still water ones. Flowing water fens here include valley fens (including valley bogs), spring fens (including calcium-dominated ones) and flushes. Still water fens include basin ones, with river flood plains being intermediate. A more detailed hydrogeological

Table 4.1 The US Fish and Wildlife Service classification of wetlands (after Gopal, in Patten 1990)

Lake	Marsh
Rocky shore	Unconsolidated shore
Unconsolidated shore	Moss-lichen
Emergent	Emergent
	Scrub-shrub
	Forested

Table 4.2 Hydromorphological classification (after Gilvear *et al.* 1993)

Water input:–	Rainfall + surface water + groundwater	Rainfall only
Hydrological type:–	Minerotrophic (fens)	Ombrotrophic (bogs)
Nature of water:–	Standing water	Flowing water
Energy gradient class:–	Lotic (still waters)	Lentic (flowing waters)
Fen types:–	Open water transition fen	Valley fen
	Basin fen	Spring fen
		Flush fens
	River flood plain fen (intermediate)	

Table 4.3 Hydrological and hydrogeological classification for East Anglian wetlands (after Lloyd *et al.* 1994)

Class	Input	Topography	Geology in catchment
A1	Surface-water run-off only	Often in topographic hollow, also valley	Clay predominant
A2	Overbank flooding	Low relief adjacent to river	Clay predominant
B	Leaky aquifer and some surface-water	Shallow valley	Low permeability but mixed
C	Groundwater from superficial deposits	Shallow valley	Mixed typical clay–sand–gravel drift
D	Groundwater from superficial deposits and underlying main aquifer	Valley or closed depression	Sands and gravel over clays over major aquifer
E	Leaky aquifer	Closed depression, e.g. pingo (Breckland mere)	Clay overlying major aquifer, lateral isolated typical 'pingo'
F	Unconfined main aquifer	Wide range	No superficial deposits. Main aquifer rock outcropping
G	Unconfined superficial aquifer	Shallow valley	Superficial sands and gravels overlying clays

classification, of East Anglian fens only, is in Table 4.3. This gives a good grasp of water movement: but no inclusion of nutrient pattern, vegetation or human impacts.

Finally, Table 4.4 shows a geographer's classification that comes close to a botanical one (note it is not quite ecological: in ecology, carr is not just alder, but sallow, alderbuckthorn, buckthorn and others; reedswamp is not restricted to *Phragmites*, etc.).

Vegetation classification

Tansley

Botanists have adopted various approaches. The first major one is in Tansley's (1911) *Types of British Vegetation* (Table 4.5), expanded in his classic (1939) *The British Islands and their Vegetation* (Table 4.6) and still in use. It lists the principal species present, not just dominants but small and even sparse, though characteristic ones, using some measure of abundance. The commonest of these has come to be known as the DAFORL scale (an acronym, these days, lends credibility!): Dominant, Abundant, Frequent, Occasional, Rare and Local. The general meaning is obvious, the detailed is defined by those who need it. A dominant may, for instance, control the community, cover over, for example, 90 or 70 per cent of the ground, be the most abundant species there, etc.

Communities are fixed by expert field observation: this is a typical bog, what is in it? This is an unusual bog, what is in that and why is it unusual? Communities are linked, bog to moor, reedswamp to carr, for example along a shallow lake shore (as on Esthwaite Water). It is a pragmatic, thoroughly British type of classification, and has a universal validity, as far as it goes.

Table 4.4 Geographic classification (after Hughes and Heathwaite, in Heathwaite and Hughes 1995)

CLASSIFICATION A

1	BOG		Acid or almost entirely organic, peat developing rapidly, limited species type
2	FEN		Organic and/or inorganic, peat developing slowly
		(a) TREELESS FEN	Grass or herb-rich
		(b) CARR	Dominated by *Alnus glutinosa* (alder)

CLASSIFICATION B

1	MIRE		Peat soils, stagnant or slow-moving water
		(a) Soligenous mires	Small in extent with slow peat development
		(b) Basin mires	Formed in topographic hollows and isolated from groundwater
		(c) Valley mires	In river valleys with variable base status
		(d) Flood plain mires	In alluvium and with variable nutrient status
		(e) Raised mires	Developed from basin mires, isolated from groundwater, nutrient-poor
		(f) Blanket mires	Cover large area and nutrient-poor
2	WET HEATH		Mineral-based, acidophilous vegetation
3	MARSHES AND MEADOWS		Predominantly mineral substrate, not accumulating peat, regular inundation with surface water (riverine and/or lacustrine), vegetation grass or herb-rich
		(a) RE-CREATED MARSHES	Sedge and herb communities
		(b) REEDSWAMP	Reedswamp is a marsh-type dominated by *Phragmites australis*
		(c) WASHLANDS	In East Anglia, largely drained
		(d) WATER MEADOWS	Confined to chalk stream, artificially created
		(e) FLOOD MEADOWS	Periodic inundation

Table 4.5 Grass Moor Community, Pennines (Tansley 1911)

Molinietum caeruleae

Molinia caerulea	d	*Parnassia palustris*	f
Ranunculus flammula	f	*Eriophorum angustifolium*	la
Hydrocotyle vulgaris	f	*E. vaginatum*	la
Galium palustre	f	*Carex echinata*	a
Oxycoccus quadripetala	f	*C. goodenowii*	a
Erica tetralix	la	*C. dioica*	f
Pedicularis palustris	f	*C. panicea*	a
Myrica gale	la	*C. curta*	f
Narthecium ossifragum	f	*C. flava*	f
Juncus effusus	ld	*Agrostis alba*	a
J. articulatus	ld	*Deschampsia caespitosa*	ld
Triglochin palustre	f	*Sphagnum* spp.	a
Scirpus caespitosus	la	*Aulacomnium palustre*	a
Potentilla palustris	f		

---

d, dominant; a, abundant; f, frequent; o, occasional; r, rare; l, local.

Table 4.6 Bog Communities, Rossshire (Tansley 1939)

Regeneration complex

(a) In pools and hollows		Forming hummocks	
Sphagnum cuspidatum	a	*Sphagnum papillosum*	a
S. inundatum	r	*S. plumulosum*	f–a
Rhynchospora alba	a	*S. magellanicum* (medium)	o
Drosera anglica	a	*S. rubellum* (summits)	f
Eriophorum angustifolium	f–a		
Carex pulicaris	o		

Sphagnum tenellum (*molluscum*), *Rhacomitrium lanuginosum* and *Hypnum cupressiforme* were scattered on the hummocks

(b) More static area. This contained the following additional species

Eriophorum vaginatum	a–ld	*Erica tetralix*	f
Narthecium ossifragum	a	*Pedicularis sylvatica*	o–lf
Molinia caerulea	f–la	*Juncus squarrosus*	o
Potentilla erecta	f–la	*Scirpus caespitosus*	r–o
Drosera rotundifolia	f	*Carex panicea*	o
Calluna vulgaris	f	*C. stellata*	r

d, dominant; a, abundant; f, frequent; o, occasional; r, rare; l, local.

Plant sociology

The plant sociological approach, the next major one in time, was developed on the continent, and appropriately reflects the ideal in vegetation patterns, and the Central European desire to place everything in categories.

The classic work is Braun Blanquets' *Pflanzen Soziologische* (1928, translated as *Plant Sociology* in 1932). Much more detailed and more extensive work has been done since, though less in Britain where, naturally enough, the Tansley approach appeals more (Tables 4.7 and 4.8).

The immediate difference is that a hierarchy is present. There is not *Phragmites* reedswamp, *Typha* reedswamp, etc., there is a Class, Phragmitetea, in which there are several Orders, such as Phragmition. And within that are the basic units, the associations, e.g. the *Phragmites* or *Typha* reedswamps. The association is the basic unit, or plant community of definite composition, presenting a uniform physiognomy, and growing in uniform habitat conditions. It is thus an *abstract* not a *seen* concept, unlike the Tansley reedswamp. The table defining an association is built up of rélevees. Vegetation not fitting a described association, or if not homogeneous in appearance is not sampled, or is discarded (note Figure 4.1: those between the chosen named units are discarded). Once an association or alliance has been defined, it exists: even if none of the specified species is present. This of course has its illogical and non-ecological side. It also has the usefulness of, say, the Magnocaricetum, comprising all wetland communities of large *Carex* and allied spp., which are mostly nutrient-rich habitats, and the mainly nutrient-poor Parvocaricetum, which includes not just small sedge communities but also rush–grass or very mixed ones of similar habit though with some basic species. These are most valuable concepts – and easily recognisable and describable in the field. In Britain, describing a community without a *Carex* in sight as a Parvocaricetum is not done: and there is no comparable term for the group.

Table 4.7 Wetland Plant Sociology categories of Britain (from Riley and Page 1990)

II Emergent vegetation of freshwater swamps springs and flushes

Class 5		*Phragmitetea – Reed-grass and tall sedge vegetation of shallow water in lakes, rivers, canals and coastal fresh and brackish water marshes*
	O	Nasturtio-Glycerietalia – eutrophic drainage channels and shallow pools
	A	Glycerio-Sparganion
	A	Apion nodiflori
	O	Phragmitetalia – tall reed swamps
	A	Phragmition
	A	Oenanthion aquaticae
	O	Magnocaricetalia – tall grass and sedge beds
	A	Magnocaricion

Class 6   *Montio-Cardaminetea – Spring-heads and flushes of varying trophic status*
- O Montio-Cardaminetalia – oligotrophic springs
  - A Cardamino-Montion
  - A Mniobryo-Epilobion
- O Cardamino-Cratoneuretalia – calcareous and minerotrophic springs
  - A Cratoneurion commutati
  - A Cratoneureto-Saxifragion aizoidis

Class 7   *Parvocaricetea – Low-growing sedge communities of transition mires, calcareous fens and minerotrophic flushes*
- O Caricetalia nigrae – mires and mesotrophic flushes
  - A Caricion curto-nigrae
- O Tofieldietalia – calcareous fens and minerotrophic flushes
  - A Eriophorion latifolii

III River bank willow woodland, fen and bog carr

Class 8   *Alnetea glutinosae – Alder swamps on organic substrates*
- O Alnetalia glutinosae
  - A Alnion glutinosae

Class 9   *Franguletea – Alder buckthorn and willow carr*
- O Salicetalia auritae
  - A Salicion cinereae

Class 16   *Salicetea purpureae – River bank willow scrub*
- O Salicetalia purpureae
  - A Salicion albae

IV Acid bog and wet heath

Class 11   *Scheuchzerietea – Sphagnum-dominated communities of peat bog pools and hollows*
- O Scheuchzerietalia palustris
  - A Rhynchosporion albae

Class 12   *Oxycocco-Sphagnetea – Peat-forming vegetation of ombrotrophic mires and wet heaths*
- O Ericetalia tetralicis – wet heath
  - A Ericion tetralicis
- O Sphagnetalia magellanici – hummock-forming Sphagnum communities
  - A Erico-Sphagnion
  - A Sphagnion fusci

O, order; A, alliance.

Table 4.8 Example of a phytosociological community description from the Scottish Highlands (shortened) (from McVean and Ratcliffe 1962)

Schoenus nigricans provisional *nodum*
(with four site species lists)

The vegetation is dominated by dense tussocks of *Schoenus nigricans* and most of the associates have a low cover value. *Eriophorum angustifolium* and *Campylium stellatum* are the only other species present in all four analyses, and with an average of only 19 species per stand, the *nodum* is floristically poor compared with most eutrophic mires.

Schoenus nigricans mires have been found only in the Western Highlands and even here they are rare as a highly calcareous substratum is evidently essential. All occur at low levels (below 183 m) and close to the sea. The soils are calcareous peats or marls with a high pH (8.0 in one sample).

Despite its lower altitudinal range the *Schoenus* mire seems to be the western counterpart of the *Carex rostrata*-brown moss *nodum* of calcareous waterlogged hollows in the Central Highlands. Both types meet at Bettyhill in Sutherland.

The present *nodum* may pass into *Carex panicea–Campylium stellatum* mire or species-rich *Agrostis–Festuca* grassland as the drainage improves, or into the low-level facies of Cariceto-Saxifragetum aizoides where the flow of water becomes rapid. Sometimes it appears as part of a complex of mire and fen vegetation, where enriched drainage water emerges on moors otherwise covered with Trichophoreto-Eriophoretum typicum. The *nodum* is then associated with *Carex lasiocarpa* and *C. rostrata* communities of open water.

In Scotland, the presence of *Schoenus* in blanket bogs would seem usually, if not always, to indicate an irrigation effect in the peat. *Schoenus* grows most abundantly and luxuriantly in hollows, channels and flushes where enrichment by irrigation is undoubtedly strong, but it appears to be less strictly calcicole in the Western Highlands than in lowland England. The connection with irrigation is evidently a real one quite apart from the possibility that sea spray is another central factor determining the occurrence of *Schoenus* in Scottish blanket bogs (Tansley 1949).

International Biological Programme (IBP) Classification (Hejny and Segal, in Westlake et al. 1998)

This (see Table 4.9) system has the unusual merit of stating it 'is probably no more consistent or better than many others'. It tries to provide a key to wetlands that has some functional aspects, and no strong regional bias (although this bias has been reinserted in the table). Four groups are recognised, wetlands associated with, respectively, permanent standing waters, permanent running waters, periodic freshwaters, and periodic salt waters. The end classes are defined by processes, dominant vegetation and perhaps geography.

The IBP classification gives a different perspective, and shows how comprehensive it is. All wetlands just fit in. None is discarded, no computer-bred errors can creep in. On the other side, as all world-wide wetlands, freshwater and salty, fit into just 26 types, there is a wide range of variation within the broader categories.

National Vegetation Classification (NVC)

In Britain's NVC (Rodwell 1991a,b, 1992, 1995), the unit is the smallest easily recognisable plant community, e.g. those dominated by *Phragmites, Typha*, etc. (Tables 4.10 and 4.11). As with the Tansley method there are habitat data and no hierarchy, but as with

Table 4.9 International Biological Programme Classification (relevant parts extracted from Hejny and Segal, in Westlake *et al.* 1998)

Class 1. Wetlands related to permanent standing waters (types intergrade, and intergrade with types from other classes)

1 Littoral belts alongside lakes and ponds
Typical temperate species are from *Scirpus* section *Schoenoplectus, Eleocharis, Typha* and *Phragmites*
Soil covered with water or exposed for short periods, remaining waterlogged
2 Swamps (including some marshes and fens)
Waterlogged or shallowly flooded, may become dry, e.g. in summer in temperate plants.
Closed vegetation, mostly of tall graminoid halophytes
 (a) Reedswamps
Shallow, flooded for most of year
Typical dominants include *Phragmites australis, Cyperus papyrus, Typha* spp., *Glyceria maxima*, Tall *Scirpus* and *Carex* spp. Relatively poor in species, including bryophytes
 (b) Reedlands
Without standing water for most of the growing season. Typical dominants often as in (a), also species-poor, but bryophytes are usually present
3 Transitional (valley) fens and bogs
Waterlogged, usually without standing water for most of the growing season, characterised by medium-sized graminoids and the accumulation of more or less mineral-saturated peat.
Bryophytes, often including *Sphagnum*, well-developed. Species-rich
 (a) On peat ridges, transitional to bog
 (b) On peat moving up and down above water (Quaking fen)
 (c) On drying up reedlands or transitional bogs. With tall dicotylenous dominants, e.g. *Filipendula ulmaria, Lysimachia vulgaris, Epilobium hirsutum*
 (d) On heavily grazed or mown transitional bogs. Graminoid vegetation with many wetland species
4 Lagg zones
Between mineral-poor waters and higher flood-formed (diluvial) soils, e.g. alongside blanket bogs, as successional stages of heath pools. Very sensitive to fertilisation of higher-situated surrounding areas
Mainly small Cyperaceae and other monocotyledons, mostly with a well-developed bryophyte layer of *Sphagnum* and tiny leafy liverworts

Class 2. Wetlands related to running waters

1 Riparian belts
Vary with (a) level in relation to water; (b) amplitude of water level fluctuation; and (c) intensity of water current. Peat accumulation varies
2 Flushes (springs, rills)
With a rapid and localised flow of drainage water near the ground surface, which varies greatly in volume and flow rate with rainfall
Peat accumulation varies
Usually one layer of small herbs and one of bryophytes or two of both

Class 3. Wetlands related to periodic waters

1 Periodic dry lands (flood plains)
Several hectares or more, with marsh vegetation, as nearly flat areas by lakes, rivers, etc., mostly in alluvial plains

---

Table 4.10 National Vegetation Classification types grouped into wetland habitat types (from Rodwell 1991, 1992a,b, 1995)

Nutrient status increases in fens with drying and disturbance, so the same fen may bear low-nutrient types in wetter places, higher-nutrient in drier.
M, mire; S, swamp; W, wood; MG, mesotrophic grassland.

BOG
Sphagnum auriculatum bog pool community **M1**
Sphagnum cuspidatum/recurvum bog pool community **M2**
Eriophorum angustifolium bog pool community **M3**
Scirpus caespitosus–Eriophorum vaginatum blanket bog **M17**
Erica tetralix–Sphagnum papillosum raised and blanket mire **M18**
Calluna vulgaris–Eriophorum vaginatum blanket mire **M19**
Northecium ossifragum–Sphagnum papillosum valley mire **M2**
Betula pubescens–Molinia caerulea woodland **W4**

POOR FEN (for reedswamp in all fen types, see below)
1   Poorer fens, equivalent to continental *Carex nigra* Parvocaricetum types
Carex rostrata–Sphagnum recurvum mire **M4**
Carex rostrata–Sphagnum squarrosum mire **M5**
Carex echinata–Sphagnum recurvum/auriculatum mire **M6**
Carex curta–Sphagnum russowii mire **M7**
Carex rostrata–Sphagnum warnstorfii mire **M8**
Carex rostrata–Calliergon cuspidatum/giganteum mire **M9**
2   Richer fens, equivalent to continental *Carex davalliana* Parvocaricetum types
Carex dioica–Pinguicula vulgaris mire **M10**
Schoenus nigricans–Narthecium ossifragum mire **M14**
Molinia caerulea–Potentilla erecta mire **M25 ?**
Molinia caerulea–Crepis paludosus mire **M26 ?**
3   Other
Carex rostrata swamp **S9**
Carex rostrata–Potentilla palustris fen **S27**
Betula pubescens–Molinia caerulea woodland **W4**
Salix pentandra–Carex rostrata woodland **W3**
Alnus glutinosa (–Urtica dioica) woodland **W6**

CALCIUM-INFLUENCED FEN Wetter herb communities only; drier and disturbed ones are in tall-herb communities, below (includes Parvocaricetum (*Carex davialliana* type) and Magnocaricetum)
Schoenus nigricans–Juncus subnodulosus mire **M13**
Schoenus nigricans–Narthecium ossifragum mire **M14**
Juncus subnodulosus–Cirsium palustre fen meadow **M22**
Cladium mariscus sedge-swamp **S2**
Carex appropinquata in fens
Phragmites australis–Peucedanum palustre fen **S24**
(*Carex elata* sedge-swamp S1)
(*Salix cinerea*)–*Betula pubescens–Phragmites australis* woodland **W2**
Alnus glutinosa–(*Carex paniculata*) woodland **W5**
Alnus glutinosa–Urtica dioica woodland **W6**

MEDIUM FEN AND MARSH
Carex elata sedge-swamp **S1**
Cladium mariscus sedge-swamp **S2**
Carex paniculata sedge-swamp **S3**
Carex appropinquata in fens
Phragmites australis reedbed **S4**

Table 4.10 (cont'd)

Carex vesicaria swamp **S11**
Glyceria fluitans swamp **S32**
Juncus subnodulosus–Cirsium palustre fen meadow **M22**
Molinia caerulea–Cirsium dissectum fen meadow **M24**
Molinia caerulea–Potentilla erecta mire **M25**
Salix cinerea–Betula pubescens–Phragmites australis woodland **W2**
Salix pentandra–Carex rostrata woodland **W3**
Alnus glutinosa–Carex paniculata woodland **W5**
Alnus glutinosa–Urtica dioica woodland – as willow **W6**
Alnus glutinosa–Fraxinus excelsior–Lysimachia nemorum woodland **W7**

RICH FEN AND MARSH
(*Carex paniculata* sedge-swamp **S3**)
Phragmites australis reedbed **S4**
Glyceria maxima swamp **S5**
Carex riparia swamp **S6**
Carex acutiformis swamp **S7**
Typha latifolia swamp **S12**
Typha angustifolia swamp **S13**
Carex pseudocyperus swamp **S17**
Phragmites australis–Eupatorium cannabinum fen **S25**
Phragmites australis–Urtica dioica fen **S26**
Phalaris arundinacea fen **S28**
Salix cinerea–Galium palustre woodland **W1**
Salix cinerea–(Betula pubescens)–Phragmites australis woodland **W2**
Alnus glutinosa–Carex paniculata woodland **W5**
Alnus glutinosa–Urtica dioica woodland **W6**
Alnus glutinosa–Fraxinus excelsior–Lysimachia nemorum woodland **W7**

REEDSWAMP including sedge-swamp. All habitat types (with notes on relevant nutrient status and water regime)
Carex elata sedge-swamp **S1** (medium+ nutrients, water level varies)
Cladium mariscus sedge-swamp **S2** (medium nutrients)
Phragmites australis swamp **S4**
Glyceria maxima swamp **S5** (nutrient-rich)
Scirpus lacustris ssp. *lacustris* swamp **S8** (deep water)
Carex rostrata swamp **S9** (fairly low nutrients)
Equisetum fluviatile swamp **S10** (medium nutrients)
Carex vesicaria swamp **S11** (medium+ nutrients)
Typha latifolia swamp **S12** (not low nutrients)
Typha angustifolia swamp **S13** (not low nutrients)
Sparganium erectum swamp **S14**
Acorus calamus swamp **S15** (nutrient-rich, usually)
Carex pseudocyperus swamp **S17** (medium+ nutrients)
Eleocharis palustris swamp **S19** (low to medium nutrients)
Glyceria fluitans swamp **S22** (medium+ nutrients)

TALL HERB COMMUNITIES IN FEN AND MARSH (relatively dry habitats) (with notes on relevant nutrient status)
Filipendula ulmaria–Angelica sylvestris mire **M27** (medium nutrients)
Iris pseudacorus–Filipendula ulmaria mire **M28** (medium nutrients)
Carex riparia swamp **S6** (medium+ nutrients)
Carex acutiformis swamp **S7** (medium+ nutrients)
Phragmites australis–Peucedanum palustre fen **S24**) (medium nutrients)
Phragmites australis–Eupatorium cannabinum fen **S25** (nutrient-rich)

Table 4.10 (cont'd)

Phragmites australis–Urtica dioica fen **S26** (nutrient-rich)
Phalaris arundinacea fen **S28** (usually nutrient-rich)

CARR AND WOOD
Salix cinerea–Galium palustre woodland **W1**
Salix cinerea–Betula pubescens–Phragmites australis woodland **W2**
Salix pentandra–Carex rostrata woodland **W3**
Betula pubescens–Molinia caerulea woodland **W4**
Alnus glutinosa–Carex paniculata woodland **W5**
Alnus glutinosa–Urtica dioica **W6**
Alnus glutinosa–Fraxinus excelsior–Lysimachia nemoreum woodland **W7**

GRASSLANDS (with habitat notes)
Juncus effusus/acutiflorus Galium palustre rush pasture **M23** (light grazing, not high nutrients)
Molinia caerulea–Potentilla erecta mire **M25** (light grazing, not high nutrients)
Molinia caerulea–Crepis paludosus mire **M26** (light grazing, not high nutrients)
Glyceria fluitans swamp **S22** (small, wet areas)
Cynosurus cristatus–Caltha palustris grassland **MG8** (includes water meadow type)
Holcus lanatus–Deschampsia caespitosa grassland **MG9** (medium+ nutrients)
Holcus lanatus–Juncus effusus rush pasture **MG10** (not high nutrients)
Festuca rubra–Agrostis stolonifera–Potentilla anserina grassland **MG11** (medium+ nutrients)
Agrostis stolonifera–Alopecurus geniculatus grassland **MG13** (medium+ nutrients)

---

phytosociology, sites of fixed but variable size are used, and it is possible to have a community not really fitting with the name; for example, an *Eleocharis palustris* community variant where *Agrostis stolonifera* is the dominant as a mat.

The NVC has the great advantage of having a wide range of named communities and variants, so most British vegetation can be fitted in: even where there was no such intention (see *The Hound of the Baskervilles* extract in Chapter 1!). There are disadvantages. Communities may be defined on less than 10 (instead of over 200) sites, so further collections may well alter the defined communities. Site size varies (50 × 50 m in woods for the tree layer and 2 × 2 m for grasslands, bogs, dwarf shrub heaths, etc.). Therefore, the given site diversity is not comparable between communities of different heights.

Having new combinations means older communities do not always fit well (for Broadland fens, see Wheeler 1980a,b). The overall pattern, carr, tall herbs, etc., is necessarily the same, but computer-made communities do not necessarily reflect the expertise of the national experts on the subject (Figure 4.2), and the older described communities – whether phytosociological, Tansley or other – often do not fit properly into this scheme.

The NVC types necessarily depend on where the samples were taken. For instance, M13 is *Schoenus nigricans–Juncus subnodulosus*. *Juncus subnodulosus* increases to co-dominance only when the habitat moves away (spatially or by drying) from the most calcium-dominated habitats. On numbers, the M13 type may be the most frequent, but the interpretation gets truncated. Again, no difference is made between the status of communities. This is advantageous in that it records what is, but unhelpful in the immediate recognition of, say, M20 *Eriophorum vaginatum* blanket and raised mire as a badly degraded community, or realising that M8 *Carex rostrata–Sphagnum warnstorfii* mire is restricted to small stands in hollows, among other communities, rather than covering sizeable areas.

Given all this, the NVC is an excellent attempt at complete classification of a country's vegetation. It gives information on a remarkably wide range of features, physiognomy,

Table 4.11 Example of National Vegetation Classification description, shortened

M6 *Carex echinata–Sphagnum recurvum/auriculatum* mire

Physiognomy Distinct in character but variable in composition: a poor fen with small sedges or rushes dominating over oligotrophic Sphagna
Floristics
***Carex echinata* subcommunity**
Sedge-dominated with a luxuriant *Sphagnum* carpet
Average species per sample: 13
Frequent species:

Polytrichum commune	*Agrostis canina canina*
Carex echinacea	*Sphagnum auriculatum*
C. panicea	*S. palustre*
Eriophorum angustifolium	*S. recurvum*
Molinia caerula	*Viola palustris*
Potentilla erecta	

***Carex nigra–Nardus stricta* subcommunity**
Sedge-dominated, with a variable *Sphagnum* carpet
Average species per sample: 19
Frequent species:

Agrostis canina canina	*Juncus squarrosus*
Carex echinata	*Nardus stricta*
C. nigra	*Potentilla erecta*
C. panicea	*Sphagnum recurvum*
Eriophorum angustifolium	*Viola palustris*

***Juncus effusus* subcommunity**
Juncus effusus dominated with sparser sedges and a luxuriant *Sphagnum* carpet
Average species per sample: 16
Frequent species:

Polytrichum commune	*Agrostis canina canina*
Carex echinata	*Sphagnum recurvum*
Juncus effusus	

***Juncus acutiflorus* subcommunity**
Rush-dominance over an extensive *Sphagnum* carpet
Average species per sample: 17
Frequent species:

Potentilla erecta	*Agrostis canina canina*
Anthoxanthum odoratum	*Polytrichum commune*
Carex echinata	*Sphagnum palustre*
Juncus acutiflorus	*S. recurvum*
Molinia caerulea	*Viola palustris*

Habitat The major soligenous community of high water table peats and peaty gleys irrigated by rather base-poor but not excessively oligotrophic (or mountain) waters. It is common in unenclosed pasture on highland fringes. The throughput of waters provides nutrients. Changing from sedge to rush dominance may be due to burning and drainage. Subcommunities may reflect waterlogging. The *Juncus acutiflorus* one is more westerly and lower level
Zonation and succession Typically as small stands in mires, grasslands, heaths, swamps, etc. Over-grazing leads to grassland. In mires, occurs with channelling of nutrient-laden waters. It may be associated with *Erica–Sphagnum*, *Scirpus–Eriophorum* (oceanic), *Calluna–Eriophorum* (easterly), *Scirpus–Erica* (west, degraded or disturbed peat).
Distribution Ubiquitous in the highland fringes
Affinities This category comprises a variety of poor fens between, in base status and calcium content, *Rhynchospora* and *Carex davalliana* communities. It is also related to *Juncus*, *Juncus–Molinia* and *Nardus–Galium*

Figure 4.2 What happens to vegetation in a computer?

floristics, habitat, zonation and succession, distribution and affinities with other NVC communities (e.g. forms patches in such a one), and to communities classified in other ways, particularly with the phytosociological continental ones, e.g. falls within *Carex davalliana* fens. There is no equivalent in the British literature. It is most valuable for reference, but should not become the only classification used. Vegetation is a continuum that can be broken in many different ways, each equally valid. Looking at different viewpoints adds to the understanding of plant behaviour as a whole.

CORINE

The final classification described here is that of the European Union, the CORINE biotope project (Commission of the European Communities 1992) (Tables 4.12 and 4.13). This is currently used by the European Parliament, the Commission and suchlike, but the Commission would like this to be better known and its database to be more accessible. As the EU take a long view, this classification will no doubt be in common use within a couple of decades or so. The term 'biotope' has been dropped except in the formal title, as it is yet another name with many definitions, and here it is unnecessary. It is neither universal, like Tansley, nor redefinable in detail into systems for each region, like phytosociology, nor British (though capable of redefinition), like the NVC.

A site is 'an area of land or a body of water which forms an ecological unit of community significant for native conservation, regardless of whether this area is formally protected by legislation.'

The CORINE authors describe the scheme as unique in breadth of coverage of geography and subject, a valuable source of information for scientific research and environmental protection. The NVC recognises about 70 wetland vegetation types in Britain; CORINE, about 270 less precisely defined categories for the (at present) 11-nation European Community. Comparing the tables shows the difference in approach. Memorising or serious study is not needed at this stage, but it is necessary to understand the basic plan of both. NVC is far more detailed, even though it suffers from inadequate data. CORINE suffers even more from this: but for both schemes, changes can be expected later. While NVC has depth, CORINE has breadth. Table 4.12 gives an impression of the range of European wet habitats not obtainable elsewhere. The CORINE types can be easily fitted within the general categories already discussed. The major separation between blanket and raised bogs – given as a separation equivalent to that between bog and reedswamp – is a more

Table 4.12 Outline of CORINE Wetland Classification (from Commission of the European Communities 1992)

---

51   RAISED BOGS
51.1   Near-natural, often domed, very rare. Communities interrelated, functioning as a unit
    51.11   Hummocks, ridges and lawns, higher and drier, e.g. *Sphagnum magellanicum, Oxycoccus, Erica tetralix* (14 subdivisions follow)
    51.12   Hollows, rain-filled, temporary or permanent (2 subdivisions follow)
    51.13   Pools, large, permanent, near the centre of bogs or along tension lines. Characteristically with plankton, often e.g. *Sparganium minimum, Utricularia* spp., *Nymphaea* spp. (2 subdivisions follow)
    51.14   Seeps and soaks, run-off paths to laggs around the bogs (3 subdivisions follow)
    51.15   Lagg rings of water, intermediate mire, acid fen, sometimes e.g. *Eriophorum angustifolium, E. vaginatum, Carex rostrata, C. flava, Parnassia palustris*
    51.16   Pre-woods. Colonisation by e.g. *Pinus* spp., *Betula* spp.
51.2   Purple moor grass bogs, drying, mowed or burned

52   BLANKET BOGS
52.1   Lowland, west coastal, e.g. *Sphagnum auriculatum, S. magellanicum, S. compactum, S. papillosum, S. nemoreum, S. rubellum, S. tenellum, S. subnitens,* other mosses, *Molinia caerulea, Eriophorum angustifolium, E. vaginatum, Scirpus caespitosus, Rhynchospora alba, Narthecium ossifragum, Erica tetralix, Myrica gale, Drosera rotundifolia, Calluna vulgaris*
    52.11   Black bog-rush swards, with *Rhacomitrium lanuginosum*
    52.12   *Sphagnum*-algal carpets, waterlogged pool edges, bog surfaces
    52.13   Deer grass swards
    52.14   Oblong-leaved sundew (*Drosera intermedia*) communities, slopes with surface water movement
    52.15   Bulbous rush communities, shallow drainage channels and pools
    52.16   Flushes, deep hollows and pools, *Potamogeton polygonifolius,* etc.
52.2   Hill, *Sphagnum* abundant, and e.g. *Eriophorum vaginatum, Calluna vulgaris, Erica tetralix, Rubus chamaemorus, Narthecium ossifragum, Scirpus caespitosus, Drosera rotundifolia, Rhacomitrium lanuginosum*
    52.21   Cotton-grass–heather(ling)
    52.22   Cotton-grass (*E. vaginatum*), species-poor
    52.23   Hill *Sphagnum* spp. hummocks of the preceding
    52.24   Dwarf shrub: cotton-grass
    52.25   Woolly fringe-moss (*Rhacomitrium lanuginosum*)
    52.26   Wet heath
    52.27   Flushes, deep hollows and pools

53   WATER FRINGE VEGETATION, REED AND SEDGE SWAMPS
53.1   Reedbeds, usually species-poor, often monodominant, stagnant, slow-flowing or waterlogged
    53.11   Common reed beds (*Phragmites australis*) (3 subdivisions follow).
    53.12   Common clubrush beds (*Scirpus lacustris*), intolerant of drying, tolerant of water movement
    53.13   Reedmace beds (*Typha* spp.), tolerant of drying and pollution
    53.14   Medium to tall waterside communities (10 subdivisions follow)
    53.15   Reed sweetgrass (*Glyceria maxima*), ditches, brooks, grassland, flooded with eutrophic water, rich associated flora
    53.16   Reed canary grass (*Phalaris arundinacea*), tolerant of drying, pollution, disturbance, land edges, often degraded
53.2   Large sedges, Magnocaricetum, *Carex* or *Cyperus,* damp depressions from oligotrophic mire to rich fen, wet or partly dry, towards land
    53.21   Large *Carex* beds (18 subdivisions follow, with different dominants)
    53.22   Tall galingale beds (*Cyperus* spp. except *C. papyrus*) (2 subdivisions follow)
53.3   Fen sedge beds (*Cladium mariscus*), calcium-rich, etc.
    53.31   Fen *Cladium*, species-rich, fairly open
    53.32   Valencia
    53.33   Mostly Mediterranean waterside, species-poor
53.5   Tall rush swamps. *Juncus* invading grazed and trampled poor fen, marsh, etc.
53.6   Riparian core formations, along permanent or temporary Mediterranean watercourses (2 subdivisions follow)

Table 4.12 (cont'd)

54   FENS, TRANSITION MIRES AND SPRINGS
54.2   Rich fens
    54.21   Black bog–rush fens (*Schoenus nigricans*), low altitude, declining or extinct
    54.22   Brown bog–rush fens (*S. ferruginosus*) (3 subdivisions follow)
    54.23   Davalliana sedge fens, diverse, often extensive, mostly near or in Alps, refuge of many rare species (2 subdivisions follow)
    54.24   Pyrennean rich fens
    54.25   *Carex dioica, C. pulica* or *C. flava* agg. fens (Dioecious, flea, and yellow sedge fens) (6 subdivisions follow)
    54.26   Black sedge alkaline fens (*Carex nigra*), calciphile spp., brown moss
    54.27   Russet sedge fens (*Carex saxatilis*), high calcareous mountains
    54.28   Ice sedge fens (*Carex frigida*), seepages etc. in Alps, etc.
    54.29   The same in Britain
    54.2A   Spike rush fens, *Eleocharis quinqueflora*
    54.2B   Greek flat sedge (*Blysmus compressus*) fens
    54.2C   Bottle sedge (*Carex rostrata*) alkaline fens, brown mosses
    54.2D   Alpine deer grass (*Scirpus hudsonianus*) fens
    54.2E   Deer grass (*S. caespitosus*) fens
    54.2F   Middle European flat sedge fens
    54.2G   Small-herb alkaline fens
    54.2H   Calcareous dune slacks, rush-sedge fens
    54.2I   Tall-herb fens
54.3   Arcto-alpine riverine swamps, rare glacial relics, neutral to basic, *Carex* spp., *Juncus* spp. diverse
    54.31   Riverine swards, *Kobresia simpliciuscula*, etc.
    54.32   *Carex bicolor*, etc., swards
    54.33   *Typha* spp.
    54.34   British mica flushes, *Carex* spp., *Juncus* spp.
54.4   Acidic fens
    54.41   Alpine cotton-grass (*Eriophorum scheuchzerii*) lake girdles
    54.42   Star sedges fens *Carex nigra, C. canescens, C. echinata*, etc. (10 subdivisions follow)
    54.43   Apennine acidic fens
    54.44   *Carex intrisata* Mediterranean, etc., fens (3 subdivisions follow)
    54.45   Deer grass (*Scirpus caespitosus*) acidic fens (5 subdivisions follow)
    54.46   *Eriophorum angustifolium* mires with a *Sphagnum* carpet
    54.47   Dunal sedge acidic fens
54.5   Transition mires. Mostly small sedge, floating, *Sphagnum* or brown moss. Important refuge of threatened animals and plants
    54.51   Floating meadows of slender sedge (*Carex lasiocarpa*), *Sphagnum* or brown mosses, etc. (2 subdivisions follow)
    54.52   *Carex diandra* quaking mires, open diverse swards
    54.53   Bottle sedge (*Carex rostrata*) quaking mires (3 subdivisions follow)
    54.54   Mud sedge (*Carex limosa*) swards (2 subdivisions follow)
    54.55   String sedge (*Carex chordarrhiza*) swards
    54.56   Peat sedge (*Carex heleonaster*) sedge hollows
    54.57   Beak sedge (*Rhynchospora alba*) quaking bogs
    54.58   Sphagnum and cotton-grass rafts
    54.59   Bog bean (*Menyanthes trifoliata*) and marsh cinquefoil (*Potentilla palustris*) rafts
    54.5A   Bog arum (*Calla palustris*) mires
    54.5B   Brown moss carpets, diverse, not acid
    54.5C   Cotton-grass (*Eriophorum vaginatum*) quaking bogs
    54.5D   Purple moor grass (*Molinia caerulea*) quaking bogs
    54.5E   Narrow smallreed (*Calamagrostis stricta*) quaking bogs
    54.5F   Alpine deer grass (*Scirpus hudsoniana*) quaking bogs
    54.5G   Iberian deer grass (*Scirpus hudsoniana*) quaking bogs
54.6   White beak sedge (*Rhynchospora alba*) communities, dug peat, flushes, etc.

37   HUMID GRASSLAND AND TALL-HERB COMMUNITIES
37.1   Meadowsweet (*Filipendula ulmaria*) and related communities, fertile alluvium, bog pasture, often colonised after disuse, with e.g. *Angelica sylvestris, Cirsium palustre, Deschampsia caespitosa, Epilobium hirsutum, Eupatorium cannabinum, Lysimachia vulgaris, Phalaris arundinacea*

Table 4.12 (cont'd)

37.2 Eutrophic humid grassland, nutrient-medium to -rich, often winter-flooded, wet or damp, may be fertilised
   37.21 High-level humid meadows, very diverse, including *Caltha palustre, Lychnis flos-cuculi, Trollius europeus, Lotus uliginosus* (9 subdivisions follow)
   37.22 Sharp-flowered rush (*Juncus acutiflorus*) meadows
   37.23 Subcontinental *Cnidium dubis* meadows, most flooded
   37.24 Flood swards and related communities, may be intensively grazed (2 subdivisions follow)
   37.25 Transitional tall herb, abandoned hay meadows
37.3 Oligohumid grassland
   37.31 Purple moor grass (*Molinia caerulea*) meadows and related communities (2 subdivisions follow)
   37.32 Heath rush meadows, humid mat-grass swards, peaty, *Nardus stricta, Juncus squarrosus*, etc.
37.4 Mediterranean tall humid grasslands
37.5 Mediterranean short humid grasslands
37.6 Eastern supra-Mediterranean humid meadows
   37.61 Greek
   37.62 Apennine
37.7 Humid tall-herb fringes
   37.71 Watercourse veils, screens of e.g. *Calystegia sepia, Cuscuta europea* and ruderals lining watercourses (5 subdivisions follow)

44 ALLUVIAL AND VERY WET FORESTS AND BRUSH
44.1 Riparian willow formation
   44.11 Pre-alpine willow brush (2 subdivisions follow)
   44.12 Willow brush (other) (7 subdivisions, divided on species and location, follow)
   44.13 White willow (*Salix alba*) gallery forests
   44.14 Mediterranean tall willow galleries (5 subdivisions follow)
   44.15 Canarian willow (*Salix canariensis*) galleries
44.2 Grey alder (*Alnus incana*) galleries (2 subdivisions follow)
44.3 Medio-European stream ash–alder (*Fraxinus* spp., *Alnus* spp.) wood, intolerant of permanent wet, alternately flooded and drained
   44.31 of brooks and springs (5 subdivisions follow)
   44.32 of fast-flowing rivers
   44.33 of slow-flowing rivers (2 subdivisions follow)
   44.34 N. Iberian alder galleries (4 subdivisions follow)
44.4 Mixed oak–elm–ash (*Querus* spp., *Ulmus* spp., *Fraxinus* spp.) of great rivers. Flooded in high floods only. Very diverse – the most diverse European woods, closest in type to Pleistocene ones. Mainly Rhine, Danube and Elbe
   44.42 Residual mid-European, degraded and species-poor
   44.43 Balkan (2 subdivisions follow)
   44.44 Po, relicts but diverse
44.5 Southern alder–birch (*Alnus* spp., *Betula* spp.) galleries (10 subdivisions follow)
44.6 Mediterranean poplar–elm–ash (*Populus* spp., *Ulmus* spp., *Fraxinus* spp.) forest, multilayered, alluvial (18 subdivisions follow)
44.7 Oriental plane (*Platanus orientalis*) and sweet-gum woods (*Liquidamber orientalis*), mostly riparian (5 subdivisions follow)
44.8 Southern riparian galleries and thickets (6 subdivisions follow)
44.9 Alder (*Alnus* spp.), willow (*Salix* spp.) and bog myrtle (*Myrica gale*) swamp woods, fens and marshes
   44.91 Alder swamp woods, usually sallow (*Salix* spp.) below (5 subdivisions follow)
   44.92 Mire willow (*Salix* spp.) scrub of fens, flood plains, lake edges, etc. (4 subdivisions follow)
   44.93 Swamp bog myrtle (*Myrica gale*) scrub, drying fens, regenerating bogs
44.A Birch (*Betula* spp.) and conifer swamp woods
   44.A1 Sphagnum-birch woods, peaty, acid fen or bog (if bog, little peat building) (3 subdivisions follow)
   44.A2 Scots pine (*Pinus sylvestris*) bog woods, bogs and transitional mires
   44.A3 Mountain pine (*Pinus rotundata*) bog woods. Alps, etc.
   44.A4 Sphagnum-spruce (*Picea* spp.) boggy woods (2 subdivisions follow)

Table 4.13 A (shortened) example from the CORINE classification (from Commission of the European Communities 1992)

(In: 54 Fens, transition mires and springs)

54.2   RICH FENS. Mostly peat- or tufa-producing small sedge and brown moss communities, permanently waterlogged, soligenous or topogenous, with a base-rich nutrient-poor often calcareous water supply, with some peat formation. *Carex davalliana*-type vegetation (nutrient-rich Parvocaricetum). Various subunits. In serious decline, although a few large examples are left, found in pre-Alpine Bavaria, pre-Alps Italy, mountain east France, north east Germany, coastal marshes of north France, south east and north England, Wales, and Ireland

    54.21   Black bog-rush (*Schoenus nigricans*) fens, not at high altitude. *Juncus* may be abundant, other species include *Carex lepidophylla, C. hostiana, C. panicea, C. pulicaris, Eriophorum latifolium, Molinia caerulea, Dactylorchis incana, D. praetissina, D. purpurea, D. transiana, D. trans-vides, Epipactis palustris, Parnassia palustris, Pinguicula vulgaris,* brown mosses. Declining or extinct

    54.22   Brown bog-rush (*Schoenus ferrigeneus*) fens

        54.221   Peri-alpine brown bog-rush fens, with grasses, small *Carex* spp., *Eriophorum latifolium, Drosera* spp. *Pinguicula vulgaris, Parnassia palustris,* etc.

        54.222   Scottish brown bog-rush fens, base-rich Perthshire flushes

        54.223   Baltic brown bog-rush fens. North east Jutland, Isles, north east Germany

    54.23   Davalliana sedge fens, diverse, often extensive, including numerous *Carex* spp., *Juncus articulatus, Scirpus caespitosus, Molinia caerulea, Potentilla erecta, Parnassia palustris, Pinguicula vulgaris* and many more, and the moss layer with *Drepanocladus intermedia, Cratoneuron glauca, Campyllium stellatum.* Most Alpine or peri-Alpine, where they are the refuge of many rare species, elsewhere in decline

        54.231   Species-rich Davalliana sedge fens

        54.232   Deer grass (*Scirpus caespitosus*) sedge fens, *S. caespitosus* dominant, species-poor

    54.24   Pyrennean rich fens, uncommon calcareous fens in the Pyrennees, with small *Carex* spp., *Eriophorum latifolium, Juncus articulatus, Parnassia palustris,* etc.

continental one: Britain has both bog types, and the two are blurred now. In the past it is thought they intergraded both spatially and in time (raised bog over blanket bog). Reed and sedge swamps are placed with water fringe vegetation: communities of, say, *Phragmites communis* or *Glyceria maxima* may cover acres or may be less than 0.5 m strips along, for example, lakes and rivers. The acidic fens (54.4) are clearly equivalent to poor fens, although the definition here differs, the acidic fens containing both poor fen and bog communities.

A more major difference is putting wet grasslands and tall herb communities in the same category (37). Abandoned wet grassland will often become tall herb: but then so will wetter abandoned fen and marsh types. Looking at the alluvial and very wet forest and brush categories (44) shows how deficient Britain is in these. Compare the seven NVC units, with only six trees in their names, with the 25 in CORINE, naming over two dozen! And compare the restricted set of NVC habitats, on dry bogs, fens, marshes, with the diverse ones of CORINE! This is not a difference in definition and interpretation. It is an absence, in Britain, with its but small rivers and intense human impact, of types occurring in similar latitudes on the continent.

Another difference is in the quagmire, quaking bog or mire on the continent, as shown by the vegetation categories found. This demonstrates another facet of the observable fact that Britain has been drained much more than the comparable European countries. There, a water band in the middle of the peat is often still noticeable.

The sites have been chosen unusually, not to try to find what is there, but because they have EU threatened species, or particularly diverse habitat types. It (the site) is an area of land or a body of water that forms an ecological unit of community significance for nature conservation, regardless of whether this area is formally protected by legislation. The size and frequency vary. There are, in Scotland, about 120, in England, about 100, about 30 in Wales, and four in Northern Ireland. The sites may be large existing reserves or blocks of country. There is no uniformity. This means a site could be chosen for, say, a bird species, and then its vegetation be used as representative: which of course it may not be. And a large reserve may include motorways, council tips, industrial estates, etc.

Site selection was on the following criteria:

1 **Threatened species present**: one of 100 or fewer sites in the EU or five or fewer in a region supporting a threatened species.
2 **Sensitive habitats present**: one of the 100 most important sites in the EU, or one of the five most important sites in a region for threatened species.
3 **Richness for a taxonomic group such as birds or orchids**: one of 100 or fewer sites in the EU or five or fewer in a region for the group.
4 **Richness for a collection of habitat types**: one of the 100 most important or representative sites in the EU or one of the five most important or representative sites in a region for the habitat type.
5 The site supports at least 1 per cent of the EU population of a threatened species.

It did not prove easy, or indeed feasible, to get data comparable in quantity and quality from the different countries. In addition, criteria depending on the richness of a group or habitats were usually assessed by individual researchers using their own expert judgements. For identifying sites, it was important to find: numbers, extent, rarity of the ecological characters demonstrated, representativeness, freedom from damaging impact, and the existence of scientific observations in the site. A large site containing the specified features will necessarily also contain those specified against these factors.

Sites are first classed not as, for example, bog, but as country (e.g. Denmark or England), then by region (e.g. Amtskommune) or county. Then, sites are classed by category, which wisely includes 'status unknown' but starts with categories such as special protection area, special landscape area, green belt, national park, nature reserves (various, including EEC Birds Directive, UNESCO biosphere, Council of Europe biogenetic, World Heritage). Sites are also coded by the reason for their importance, plant and animal categories, which habitats, communities or ecosystems, whether there are species that are rare, threatened or in danger of extinction. Then, coding is by the type, whether importance for migrants, landscape geology or geomorphology, endemic species, etc.

CORINE uses a hierarchy of coded classifications. For wetlands these come in 51 Raised bogs, 52 Blanket bogs, 53 Water fringe vegetation, reed and sedge swamps, 54 Fens, transition mires and springs, 37 Humid grassland and tall herb communities, and 49 Alluvial and very wet forests and brush. The corresponding NVC codes are, roughly, M (Mire) for 51 and 52, S (Swamp) for 53, 54 and part of 37, MG (Mesotrophic grassland) for part of 37, and W (Woods) for 49. Basically the same, as both are in essence the same as Tansley and phytosociology. Then the numbered class, e.g. 52 Raised bog, is subdivided as 52.1, near-natural, 52.2 drying, mowed or burned, and 52.1 is subdivided into six categories (52.11, 52.12, etc.) of hummocks, hollows, pools, seeps and soaks, lagg rings around and pre-woods. Here the CORINE division is based more on what might be called 'geography' (the laggs of raised bog, etc.) and NVC on what might be called 'taxonomy' (*Sphagnum auriculatum* bog pool community). While NVC takes this approach throughout, CORINE – with its more difficult task – varies. Within 53.1, the reedswamps are divided on dominants (*Phragmites australis, Typha* spp., etc., exactly equivalent to NVC). Within 53.3, fen sedge, *Cladium mariscus*, subdivision is, for 53.3, by habitat (53.21, fen), by location (53.22 Valencia), or both (53.23, mostly Mediterranean waterside, species-poor). *Phragmites* is the reedswamp dominant in Europe, with, necessarily, many subtypes. NVC gives it the same standing as a *Cladium* community (S4 *Phragmites*, S2 *Cladium*), while CORINE gives *Cladium* a three-figure code (53.3), and *Phragmites* is downgraded in the hierarchy to be merely a four-figure one (53.11). Both have subdivisions within the main coding.

There are many small points of interest in CORINE, such as fen meadows, e.g. M22, M24 in NVC defined as mire, coming as humid grassland, etc., still as boggy, but here listed differently. Reading the two adds a further dimension to understanding.

The further disadvantages of CORINE are all too obvious; basically that ambition far outstripped the data available. There are no specified numbers, or sizes of sites, covers or frequency of species (whether dominants or associates). It seems unfortunate using American terms such as brush (rather than the English shrub, grove, etc.) in a European context: and giving vernacular plant names in English but not other languages is perhaps unfairly biased (particularly for the more major categories). 'Gallery' needs defining as a new term. Using different separation criteria – location here, plants there – is inconsistent. A major snag is that sites are quite specifically not selected as representative.

Conclusions

These four methods, Tansley, phytosociology, NVC and CORINE, are all valuable, all excellent and most useful within their own parameters. All can satisfactorily be used in the twenty-first century: it is merely necessary to fit the classification(s) used to the purpose of the use. The strengths and weaknesses differ (and mention here of the weaknesses does not detract from strengths: no human scheme is, or can be, perfect). When purposes change, so new classifications have been developed: and will be developed. It must never be supposed that one scheme is the final one. Each scheme is authoritative, but that is a different concept.

Wetlands in the landscape

Classifications are theory; vegetation, landscape and hydrology are fact. Basic Figures 1–4 demonstrate one set of these. There are, of course, various other possibilities for each habitat on the figures: these are typical ones that may exist in these circumstances. Figures 4.4–4.28 show actual vegetation. This book does not provide species lists of the vegetation types discussed (see reference list at end of book).

Vegetation is a very sensitive indicator of habitat. It is in equilibrium with the conditions under which it developed. For instance, three species of *Galium* occur in fens. *G. aparine* is in nutrient-rich dry places, where it can be co-dominant. The other two have similar water requirements, typically growing in constantly waterlogged, often intermittently flooded places. *G. palustre* is the more widespread, and grows in more nutrient-rich places. Wet rich fens have much lower nutrient standing than dry rich ones. *G. palustre* and *G. aparine* can grow in wet and dry parts of the same habitat, on *Phragmites* peat: but *G. aparine* extends into far more eutrophic places. *G. uliginosum* grows in yet lower nutrient status (but not in completely calcium-dominated peat). Seeing any one of these three species gives an idea of habitat. Seeing *Phalaris arundinacea* in some wet part of a river, lake or marsh in winter immediately informs the observer that that place is above water level at least in late summer. The presence of *Glyceria maxima* means that the site is at least silt-rich, often nutrient-rich. *Cirsium arvense* and *C. dissectum* form a useful pair: *C. arvense* in definitely nutrient-rich places, *C. dissectum* moving into lower nutrients, and usually compacted soil. So the list could go on. The presence of one species indicates a certain range of conditions, and that range may be very wide, as with *Phragmites australis*, or very narrow, as now in Britain with *Parnassia palustris*. When a group of species occurs together, however, the ranges of each being different, the habitat can be deduced with accuracy – the water regime requirements of a given 10 species overlap only in such-and-such a water pattern. The same applies to nutrient pattern and all other habitat factors. Even wide-ranging monodominants such as *Phragmites australis* differ in behaviour in different conditions of monodominance.

Bog (Basic Figures 1–3; Figures 4.3–4.12, 4.23, 4.26, 4.29; Tables 4.14–4.18)

Sphagnum mosses grow upright, growing from the top and dying a few centimetres down the stem. In their leaves, alternating with the photosynthesising cells, are large cells with pores. These die when mature, and can absorb and retain much water. In fact, over 90 per cent of the bog surface – sphagnum and plant roots – is often water. Sphagnum can retain up to maybe 40 times its own weight of water (Barkmann, in Verhoeven 1992). Then water can rise by capillary action from the peat. Bogs have a high cation-exchange capacity, which means they can take up cations, mineral ions from water (releasing hydrogen). The cations are balanced by humic anions, as well as strong-acid anions. The ions vary with damp (rain, evaporation) and sphagnum properties, ion exchange, etc., processes. This

Table 4.14 Typical *Sphagnum* habitats (mainly from Tansley 1949)

1   Peat formers, main
     S. cuspidatum, S. imbricatum, S. magellanicum, S. papillosum, S. plumulosum, S. rubellum
     (*Note*: do not always form peat)
2   Most acid, hummocks
     Drier: *S. acutifolium, S. fuscum, S. imbricatum, S. rubellum;* less dry: *S. compactum,*
     S. magellanicum, S. plumulosum; wetter: *S. cuspidatum*
3   Intermediate
     Less mineral: *S. apiculatum, S. balticum, S. tenellum;* more mineral: *S. angustifolium,*
     S. cymbellifolium (wide-ranging), *S. papillosum*
4   'Fen Sphagna'
     S. amblyphyllum, S. contortum, S. fimbriatum, S. palustre, S. squarrosum, S. subsecundum,
     S. teres, S. warnstorfii
5   Pools
     S. auriculatum, S. compactum, S. cuspidatum, S. riparium

Sphagna may be restricted in habitat or wide-ranging, and their habitat may vary regionally. Within one area, the species are good interpreters of the habitat in which they grow.

Table 4.15 Blanket bog characteristics (after Lindsay 1995)

1   Landscape is cloaked with peat, the non-peat parts showing as islands or corridors
2   Peat varies from a few centimetres to 7 to 8 metres deep
3   Peat basically rain-fed
4   Ground has (generally) become wetter as peat spreads
5   The peat shape is mostly that of the ground below
6   Separate hydrological units can be detected, although many are fused together
7   Surface patterning is usually prominent
8   Erosion features are common

Table 4.16 Bog forms and their vegetation (after Lindsay 1995)

Species listed are constants, often dominants, often distinct vegetation types

LAND

1   Peat mounds: (a) *Racomitrium*; (b) hypnoid mosses; (c) *Calluna/Empetrum*
2   Erosion hags: (a) *Racomitrium*; (b) *Calluna*/mixed dwarf shrub; hypnoid mosses; *Calluna*; bare peat/dwarf shrubs/*Cladonia*
3   Tall hummocks: (a) *Sphagnum*; (b) *Racomitrium*
4   High ridge: (a) undamaged, 5 subcommunities, mostly *Sphagnum*-mixed; (b) damaged, 5 subcommunities, mostly *Calluna*- or *Eriophorum*-mixed.
5   Low ridge: (a) oceanic, 3 subcommunities, *Sphagnum*- or *Schoenus*; (b) damaged, 4 subcommunities; (c) natural, 2 subcommunities, *Sphagnum* spp.
6   Water's edge: (a) *Narthecium–Sphagnum*; (b) *Sphagnum pulchrum*

WATER
Hollows

1   *Sphagnum*, 5 subcommunities
2   Mud-bottom, 5 subcommunities, *Eriophorum angustifolium*, *Rhynchospora* spp., etc.
3   Erosion gullies, 2 subcommunities.
Pools
4   Drought-sensitive, 4 subcommunities, *Sphagnum*, *Utricularia*, *Menyanthes*
5   Permanent, 2 subcommunities, *Menyanthes*, *Sphagnum*

Table 4.17 Areas of peat soils (mainly) over 1 m deep (Lindsay 1995)

	Fen (ha)	Bog (all) (ha)	Blanket bog (ha)	Total (ha)
England	131 672	252 532	214 138	384 204
Scotland	1215[a]	1 094 743	1 056 198	1 095 958
Wales	2867[a]	162 941	158 770	165 808
Total measured	135 754	1 510 216	1 429 106	1 645 970

[a] Incomplete.

means that rainwater, low in pH and with few dissolved substances, becomes even lower in pH and solutes. Even in fen, sphagna water is hardly over pH 4. This makes a very harsh habitat for the flowering and other plants also growing in the bog, made worse by the exudates released from the sphagnum. In consequence, while (short) fens can be very species-rich indeed, bog species are few, and many have adaptations. The heather family,

Table 4.18 Types of damaged bog (after Lindsay 1995)

Using Primary 'natural' bog as a standard.

1   Primary bogs
    Moribund: burned, dried; micro-erosion; wooded/active; eroding (deep gullies); drained; wooded/moribund
2   Secondary bogs
    Re-vegetating: domestic peat cutting; commercial peat cutting; eroded
    Moribund: block milling; surface milling; sheet erosion
3   Converted to agriculture
    Opencast mining

Table 4.19 Changes in rain-fed bog on drying (after Lindsay 1995)

1   Pre-drainage. Water table high, *Sphagnum* hummocks above water level, *Sphagnum* hollows below, acrotelm (active upper peat layer) thick
2   Water table below ground level. *Sphagnum* species change, acrotelm thinner
3   Water table 30 cm or so below ground level. *Sphagnum* lost, dwarf shrub and tree present
4   Water table over 50 cm down, acrotelm lost, surface mainly dry peat, but dwarf shrub and tree developing

some grasses, pine and birch, have mycorrhiza (fungi that help the roots absorb minerals, and themselves use organic substances from the roots). Some are insectivorous, like the sundews. Others, such as bog myrtle, have nitrogen-fixing actinomycetes (allied to fungi) on their roots. The plants typically grow slowly and do not respond vigorously to added nutrients. Nitrogen may be limiting (although, as in much of the Netherlands, there is substantial nitrogen pollution from the air) (Page and Rieley 1990; Proctor, in Wheeler *et al.* 1995).

Rain-dependent bogs are mostly 50–70°N in Eurasia and western North America, and 40–55°N in eastern North America. They can occur only where precipitation annually exceeds evaporation. Peat accumulation raises the surface above the surrounding drainage, and, obviously, depends on organic matter growth exceeding decay. Raised bogs are the most widespread, typically developing in infilled lake basins, wide flood plains or coastal flats. The gently domed peat masses are often 5–10 m above the water table around. The conditions for bog initiation are likely to be more rigid than those for continuation. They were extensive in the Salop–Cheshire–Lancashire plains, the Trent–Ouse plain south of the Humber, Morecambe Bay, Solway firth, the Forth valley, Somerset levels, west Fenland, etc. Rain-dependent bogs are not developing in the valley bogs of Dorset and the New Forest (see later discussion, however).

In the bog, the upper 3–5 cm or so is still-living sphagnum, and active chemically. It is termed the 'acrotelm': the active acrotelm. As the mosses grow, their weight increases and the material below, mainly dead sphagnum, becomes compacted. This lower layer, going down to the bottom of the bog, is termed the 'catotelm': the compact catotelm. The peat below becomes firmer. Blanket bog peat ranges from a few centimetres up to about 8 m deep, raised bog grows even deeper.

Even with rain as the only water source, (non-pollution) chemical variations occur. Sea salt brings solutes, and, for example, *Anagallis tenella* in the west of Ireland. In large bog

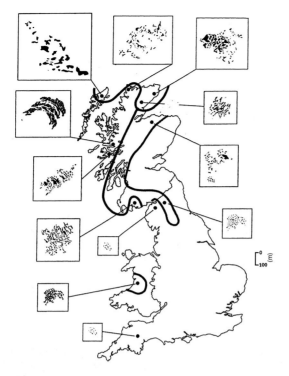

Figure 4.3 Bog surface patterns (extreme types) (from Lindsay 1995).

pools, waves cause erosion and aeration, and in consequence, the peat erodes, giving more mineral turnover and, for example, *Narthecium ossifragum*, *Empetrum nigrum* and *Myrica gale*. Run-off streams bring a flow of minerals in their path, and, for example, *Narthecium ossifragum* on their edges (which, downstream, become flushes). Where peat is shallow enough to have roots reaching the mineral soil below, there may be, for example, *Carex rostrata*, *Menyanthes trifoliata*. Bird droppings can bring *Carex nigra*, *C. acuta*, *C. echinata*, *Hydrocotyle vulgaris*, *Juncus effusus*, in extremes even *Urtica dioica* and *Polygonum hydropiper* (Barkman, in Verhoeven 1992).

The hummock and hollow pattern is a characteristic water pattern, the pools generally being larger on flat surfaces such as the top of the dome of a raised bog, or the part of blanket bog on flat land. In detail, the pool pattern varies over Britain (Figure 4.3), a fascinating and still not interpreted phenomenon. Part is because the main peat–forming species have characteristic growth forms – the habit shapes the habitat. Part is that the wetter the climate, the longer the pools. In fact, the features can be listed as not just hummock and hollow but also

peat mounds	sphagnum hollows
laggs	erosion gullies
hummocks	mud-bottom hollows
high ridges	drought-sensitive pools
low ridges	permanent pools

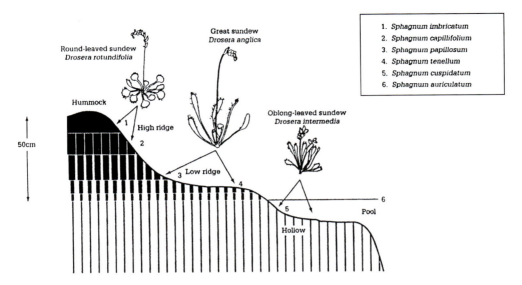

Figure 4.4 Distribution of hummock and hollow and the species of *Drosera* and *Sphagnum* characteristic of each (from Lindsay 1995).

Figure 4.5 Blanket bog.

and the combination of those that are recurrent under the same conditions forms the features of the bog (Lindsay 1995). The wettest centre of the bog has the most sphagnum. The other species, the cotton grasses, low shrubs, sedges and grasses, increase, as drier ground increases.

Species vary; Figure 4.4 shows the position, in relation to water level, of the three sundews. Heather grows in drier places than heath (*Erica tetralix*), and so on. The species lists for different bogs may be similar, but the pattern varies, the pattern due to ridges (higher and lower) to hummocks (higher and lower, larger and smaller), and to pools.

Figure 4.6 Bog in foreground, acid heath, etc., behind.

Blanket bog peat (Figure 4.5) has been dated (as the start of widespread peat accumulation) to about (Tallis, in Wheeler *et al.* 1995):

700 BC Berwyns
1000   South Wales (1600), mid-Wales (1900), Northern Pennines (1500)
2000   Orkney, Shetland (1800), Antrim plateau (1800), Sperins (2200), Western Ireland (2500)
3000   Galloway (3000), mid-Pennines (3000), Outer Hebrides (3500)
4000   Caithness (3900), Dartmoor (4000), Black Mountain (3900), Rannock Moor (4100)
5000   Southern Pennines (5200).

Bogs, so uniform at first sight, are in fact varied, with many combination of different species in only slightly different conditions. Even very small changes in water level lead to changes in species composition of sphagnum.

There are several dozen *Sphagnum* species, even in Britain, and they vary in habitat. There are fen sphagna, capable of colonising mineral-formed peat. There are bog sphagna. Within that there are bog-pool sphagna and bog-hummock sphagna. There are sphagna commonly capable of building peat, and sphagna that are not. One sphagnum species may create a habitat suitable for another (e.g. fen species allowing overgrowth by bog species).

The genus *Sphagnum* is the most successful in terms of area covered, and period of dominance, of any of the mosses or other Bryophytes. Other mosses (e.g. *Racomitrum lanuginosum*) and lichens (e.g. *Cladonia* spp.) grow on the bog in the same way as flowering plants. Sphagnum forms the bog, and forms most of its peat.

Bog can start, given suitable other conditions, on any soil or rock type. It is rare on limestone, because if water is held above limestone (by sphagnum) the limestone dissolves, the lime-rich water soaks into the peat as well as into the rock below, and bog does not form. Yet it can and has done so, for example in the Pennines, where peat sits thick above limestone. Bog development above peat or mineral soil happens when rainwater can be kept on the surface. Rainwater lenses develop, necessarily nutrient-poor, and sphagnum colonise. Once there, the sphagnum hold more water themselves, water from rain, and so the process continues, fen sphagna then bog sphagna. Alternatively, an existing bog, whether or not it started over another wetland (i.e. raised bog) can creep along, and creep over new areas. The existing bog holds water – so new bog can be formed at its edges – so it creeps. It may creep over land. It may creep over water (so starting as quaking bog or quagmire).

Bog does not now develop on well-grazed land, and it may be that, after the last Ice Age, elk and auroch grazing prevented bog formation in parts. Also, beaver were abundant (still locally present in Wales in the twelfth century AD, as described by Gerald of Wales, and later, in Scotland). Beaver dams cause flooding around them. In the Somerset levels, beaver on the river Bure would spread calcium-rich water far and wide. Flooding by nutrient-rich (i.e. non-rain) water renders bog development impossible (Coles, in Wheeler *et al.* 1995) – as did, in later centuries, controlled flooding of fens and marshes. Other animals also lead to much diversity in wetland type.

On Dartmoor, Proctor (1989) describes, on an intact bog, the sphagnum carpet (mostly *S. papillosum*) with heather, *Scirpus caespitosus*, etc. There was little heather in the wet central parts, and *Carex rostrata* in the lowest and wettest. The soils present, in two places, from top (recent) to bottom (oldest) were:

Site One	Site Two
Sphagnum (*Carex*)	*Sphagnum*
Sphagnum Carex (*Polytrichum*)	*Carex* (*Sphagnum*)
Carex	*Carex* (*Polytrichum, Menyanthes, Calliergons*)
Carex	*Carex* (*Potentilla palustris, Polytrichum, Salix?*)
Carex, Menyanthes	Amorphous, sandy
Carex	Stone
Polytrichum, Carex	Gravel, *Polytrichum*
Carex (*Polytrichum*, coarse sand)	Gravel, *Polytrichum*, sand
	Gravel, sand
	Carex (*Polytrichum*)
	Gravel, sand

This shows the full variety of habitats here, and contrasts with the summary in Basic Charts H and I.

Few things can stop the creeping of bog: change of climate was the most common in the past, apart from insurmountable obstacles (the sea, steep slopes, etc.). Now, human impact stops it. Impact tends to start at the edges, which are most accessible and driest, and to creep inwards: or, now, jump inwards.

While blanket bog can cover half a county, and raised bog can extend to hectares, valley bogs tend to be smaller, over lower slopes of a shallow valley. They are well exemplified in the New Forest, where the hills are heath. Bog now starts close (e.g. 20 m) to the river and expands round the valley (Figures 4.7, 4.26 and Basic Figure 3). They receive – and, being bog peat, keep – solute-poor run-off from the heath around as well as from rain. Nearer the stream nutrients build up, partly from the stream (because stream banks flood

Figure 4.7 Valley bog, carr along stream at base of valley, far. Bog next to carr, rising to heath in foreground. Note alder saplings kept down by grazing.

Figure 4.8 Moorland, heather and grass, local bracken and trees, rush (*Juncus effusus*) in damper valleys, Pennines.

and bog could not form) and partly from moving run-off. There may be carr along the stream bed (Figure 4.26 shows the loss of fringe reedswamp with both draining and grazing).

Drying bog peat can be colonised by species of drier habitat, and become moor (heather, grass, rush, etc., see Figures 4.8–4.10), which may cover large areas.

While a marsh dies back yearly, and has a one-year turnover, bog peat embodies energy and has a turnover of perhaps several centuries (Kangas, in Lugo *et al.* 1990).

The extent of deep peat and raised bog in Britain is shown in Table 4.16, Figures 3.2 and 4.3. The blanket bog and fen distributions can be deduced from these two. Blanket bog is in the north and west, much fen peat is less than 1 m thick, and not shown.

Figure 4.9 Moorland, grass, Pennines.

Figure 4.10 Moorland, sedge (wet foreground), rush (damp) and grass.

Drying peat – dried by change in climate as well as more impact – allows tree colonisation, and the cracks and gullies and general drying and decomposition all increase mineral status, so lead to increased species of both drier and more nutrient-requiring habitats (see Chapters 6 and 7).

The Gwent levels are an example of bog stopped by flooding (Smith and Morgan 1989):

~1900 BC	Estuarine deposits ended the bog
~3200	Bog
	Willow-alder
~9700	Fen grassland
	Cladium
	Phragmites
~5500	*Phragmites* on estuarine clay
	(probably succeeding saltmarsh)

The encroachment by bog on fen starts as sphagnum patches, with fen plants still rooted in the fen soil (e.g. Figure 4.19). Bog tends to start on higher ground (e.g. Tansley 1911) where it has least contact with groundwater. Then as the sphagnum builds upwards, short-rooted fen plants no longer have access to the mineral-rich soil below, and do not continue. Gradually the deep-rooted ones also cannot reach the fertile soil: even those species whose bases have grown up to keep pace with the new surface (Haslam 1978) are increasingly separated from the fen peat. Finally, only the trees are left. These, of course, cannot change in relation to ground level, they grow where they are, so they retain root contact with the fen soil, but eventually die. Flooding with surface water, say, a relative increase in sea level, quickly kills the sphagnum and its associates, and lake or fen conditions return (and see Basic Chart K and Table 2.1).

Scientific literature, as Mitsch and Gosselink (1993) point out, is full of successional pictures and diagrams. Patterns are not always that simple, as in the Dartmoor example above, with nutrient-poor and cyclical patterns of *Carex* (wetter) and mosses (drier), until and only finally, sphagnum bog is found. Zonations now found spatially may not be part of a succession. In fringing reedswamp, sediment may indeed build to fen conditions. Or it may be, in terms of a century or so, stable (Spence, in Burnett 1964). Or again, erosive conditions may develop and the lake and its sediment vanish. The same applies to peat-forming conditions. They apply locally, not necessarily to somewhere even 2 km away. There are vast peat sheets. There is also local variation in the ability to form peat (Tallis, in Wheeler *et al.* 1995).

Marsh

Flood plains (Figures 1.4, 2.4, 4.27, 4.28)

Rivers that transport, or (even in the late Glacial period) used to transport much sediment, create flood plains. In these there are one, two or many (Figure 4.28) channels that carry water in dry periods, and a greater width over which the water spreads in flood and fills with sediment or peat (Basic Figures 1, 4 and 5). Active flood plains are intermediate between river, lake and dry land. Water may be running or standing, present for all of the year or draining down or out and dry, except for great sediment-carrying floods. Although, overall, it is a *plain*, in detail there may be much variation, with oxbow and other lakes, and terraces from past change in land level and speed of flow, as well as present erosion and deposition. Waters are connected or separated from the main channel. Sedimentation, erosion and current are all three important, and all liable to change with the

Figure 4.11 Bog – birch wood, *Sphagnum – Potamogeton polygonifolius* in foreground.

centuries. The British flood plains were often drained (in part – not like now) before *Domesday Book* times.

There is an intimate chemical and biological relationship between a river and its flood plain. If the links are disrupted – as common, even usual in Britain – diversity of form, flora and fauna, and productivity all decline (Junk and Welcomme, in Patten 1990). The loss of this link harms the purifying power and chemical water stability power of the system (see Chapter 8).

In Britain, such plains tend to be small (Figures 1.4a) and seldom over about 4 km (2 miles) wide. The large plains, like the Fenland and the Somerset levels, facing the coasts were much affected by relative sea level, altering down the centuries and building peat and marine clay more than river sediments. These could occur upstream in the same rivers, e.g. River Great Ouse for the Fenland. Both of these types, the river flood plain, and the freshwater-marine wetlands grade into the marshes of the coast, Romney Marsh, Pevensey levels, North Kent marshes, Gwent levels, etc., which are predominantly mineral soil and

Figure 4.12 Bog (dry) – pine wood.

marine clays. Marine areas tend to be flatter, the sea having a more uniform action than the river.

Flood plains, for their history, have generally received more research from archaeologists (e.g. Williams 1970; Haslam 1997; Oxford Department of External Studies 1981) than from ecologists. Unlike peat, sediment does not store the vegetation past; silt and gravel are not made of vegetation, and do not (usually) store the pollen rain of the time they were laid down. The flood plains were mostly converted to grassland (marsh hay, water meadow, etc.). Only in the twentieth century have they gone to arable in a major way, after much drainage.

The deforestation in Europe in the Middle Ages accelerated erosion and increased sedimentation in flood plains. The flood plains rose while the water level fell. In the nineteenth century, the river courses were much altered, accelerating storm flows, increasing channel erosion. The groundwater table dropped further. In the twentieth century the plains were, as noted above, increasingly separated from the channels. The flood plains were then not flooded from the river, but from slowly rising groundwater (Junk and Welcomme, in Patten 1990).

Flood plain forests (Figures 2.3, 4.13–4.15, 4.28)

In Britain, the wetland woods are not generally on flood plains, but more on fen, bog and wider marsh. The willow-type in Basic Chart K is still found, even if the ancient mixed

Figure 4.13 Alder carr.

Figure 4.14 Sallow – mixed carr.

Figure 4.15 Willow – mixed carr.

forest is not. The degree of flooding and the drainability (how long waterlogged after flooding) are very important in determining the species (see also Chapter 6). In Europe:

Salix spp. often tolerate flooding well;
Salix alba is characteristic of gravelly soil and can tolerate floods of 4–8 m deep, for 190–300 days per year (on average);
Quercus robur and *Ulmus minor*, sand-loam and loam, 80–95 days per year flood;
Populus alba, sand-loam and loam;
Fraximus excelsior, 40 days per year flood on average (more in wet years);
Alnus glutinosa, waterlogged soils.

(Junk and Welcomme, in Patten 1996)

Pollards and other riverside trees (Figure 4.16)

Britain is still rich in riverside trees. These may be hedges beside ditches and brooks, copses or bands of trees by streams, or isolated trees along one or both banks. These isolated trees are often pollarded, and willows are the commonest pollard. Pollards are trees 'coppiced' above grazing level (often about 8 feet, rather over 2 m). Coppiced every 5–10 years or so, the cut withies were available for fencing and craftwork (see Chapter 2) and formed a valuable riverside harvest. The value these days is small, but fortunately for conservation purposes, the Environment Agency is now likely to spare, and manage, pollards. Pollards were depicted a millennium ago. (Such a suitable way of combining a grass crop and a wood crop!)

River banks

These – depending on definition – are (1) the edges, fringes, of the river; (2) the sloping bank up away from the river, where the lower part – if not all – is intermittently flooded and waterlogged; and (3) the top of the slope and the land stretching away from it (for 5–50 m, depending on taste). The fringes may (or may not) bear reedswamp or other

Figure 4.16 Pollarded willows along river, and adult willow standing alone (contrast Figure 4.15) on wet grassland.

emerged aquatic vegetation, usually in a band 0.3–2 m wide, and is here considered as river (see Haslam 1979, 1982, 1987, 1997; Haslam and Wolseley 1981). The land away from the river may or may not be wetland but is here described under flood plain, above (and see Chapters 1 and 3 and, as Buffer strip, Chapter 8).

Undrained rivers tend to have very low banks (flood plain), or to have, e.g. cliffs. Drained rivers often have banks dug during draining. When ungrazed, these frequently bear tall-herb vegetation (e.g. *Urtica dioica, Filipendula ulmaria*, brambles, rough grasses, generally species-poor). Sometimes there is scrub, sometimes managed grassland (e.g. grazed). It is a useful habitat as being Greenway, snaking across the country and often across the town, valuable for the habitat of animals, both large and small – moorhen, reed warbler, invertebrates.

An unfortunate new 'use' of these is for the spread of three invasive aliens, *Fallopia japonica, Impatiens glandulifera* and *Heracleum montegazzianum*, which are spreading quickly, and taking the place of native species. *H. montegazzianum* is like a giant rhubarb. It has poisonous sap (with furmacumarins) that has photosensitising properties and leads to (phyto) dermatitis on contact with skin. Blisters and hyperpigmentation can result, and the skin's ability to filter ultraviolet A rays can be permanently impaired so the trouble can recur, long-term, in sunlight. Eradication of the plant is difficult, but spraying with, for example, glyphosate or grazing with sheep, early in the season, for two years, may be satisfactory.

Gravel and sand pits (Figures 2.4 and 2.5, Basic Figures 11 and 18)

Gravel and sand pits are surprisingly frequent along many large rivers, where the upper more recent siltier deposits overlie gravel (or, less often sand) swept down in earlier times

Figure 4.17 Carex paniculata tussock sedge–swamp, alder colonising tussock above water level.

when the river flows were much fiercer than they are now. As the riverside land is drained, it is only pits that are flooded. At the bottom there can be shallow edges (and islands for birds). Shallow slopes will bear reedswamps (usually *Typha* spp. or *Phragmites australis*), and sallow carr often develops on damp areas behind. There may or may not be a connection to the river – which will bring in pollution as well as water.

Reedswamp (see Basic Figures 2–4, Basic Charts E–M,P,Q, and Figures 4.18–4.21)

Reedswamp, like bog, is a primary invader. It invades water, bog invades land. It occurs in any flooded habitat except bog (but see later). It is narrow in habit, with tall mono-cotyledons, and in usually monodominant stands, although mixed stands and associate species do occur. Each of the dominants has a different habitat range, although there is considerable overlap. The NVC lists 16 dominants, four for sedge swamp, plus one extra, the horsetail, *Equisetum fluviatile* (Table 4.7). Of these, only *Carex rostrata* dominates in bog pools. It also dominates, and spreads, on flood plains of oligotrophic rivers, shores of oligotrophic lakes, etc.

Reedswamp requires soils that it can colonise: soils neither very coarse, nor unstable. This eliminates many river shores, and lakes in, say, the Scottish Highlands. *Equisetum fluviatile* and *Eleocharis palustris* can grow in oligotrophic waters, but do not dominate in dystrophic ones. *Sparganium erectum* ranges from nutrient-low to nutrient-rich but is commonest on nutrient-medium river edges. *Cladium mariscus* is usually in calcium-dominated or calcium-rich medium fen. *Phragmites australis* as dominant (rather than associate) stretches from bogs with added nutrients (e.g. much run-off) and nutrient-poor

Figure 4.18 Reedbed.

lochs via the calcium-rich medium fen (not calcium-dominated) to the most eutrophic found. *Typha* spp. come in the nutrient-rich end of this, and, as noted earlier, *Glyceria maxima* stands are in silt-rich habitats.

Table 6.1 shows depth ranges, here taken from the NVC, for the abundance or dominance of species. *Scirpus lacustris* is in the deepest water.

There is also a range for water movement. On riversides, at one extreme, *Phalaris arundinacea* grows in fierce flow, with great vertical fluctuations, and on coarse substrate (such as the large gravel bars in the river Rhône). It also needs to be dry at least in late summer. *Carex rostrata* and *Sparganium erectum* tolerate considerable movement, while *Cladium mariscus* is fairly intolerant. *Phragmites* can tolerate lake waves much better than flowing rivers.

Peat formation in reedswamps depends on both habitat and species. *Phragmites* is the most widespread and common peat-former: yet it cannot do it in nutrient-poor lochs where the small reed stands have much wave action and cannot retain (the paltry amount of) leaves, stems, etc. *Scirpus lacustris* is usually in deep or moving water and is not a main peat builder. *Phalaris arundinacea* is normally in places either too unstable or too dry, or both, for fen peat. *Cladium mariscus*, in its limited range of nutrients, and in its general habitat range, does build. *Carex elata* is very local, but in flooded sites is a peat-former.

Fens (see Chapter 3, Basic Charts A–L, Basic Figures 1–22, Figures 4.13–4.14, 4.18–4.22, 4.25, and Chapters 6, 7, 10)

Fens develop beneath surface water, rather than, as bogs, from rain. Obviously there are intermediates: valley bogs, blanket bog basins now with mineral run-off from rocks as well as bog, and so on, but there is no doubt of the difference between the typical ones.

Figure 4.19 *Sphagnum* reedland with two reed clones. *Sphagnum* colonising fen where surface rainwater dominates, while reeds are well rooted in fen peat. (A similar community can, however, occur in bogs where some mineral inflow allows good reed growth.) The reed in the foreground is a clone with sparser, wider, taller shoots; that beyond is denser, narrower and shorter.

Reedswamp is the community invading open water, and, in a typical succession, or zonation, it turns the reedswamp to a fen habitat, via peat formation, so enabling species with drier requirements to invade and replace the reedswamp. This replacement is carr, in the absence of prevention by, for example, human impact, animal impact, or climate or sea level change.

Four types of fen are recognised here. Poor fens have overall poor concentrations of nutrients, and have not just *Sphagnum* but *Sphagnum*-plus-fen-species, such as *Sphagnum* reedland (Figure 4.19), small sedge-fen with bog species such as *Eriophorum angustifolium*, *Sphagnum*, *Carex nigra* as well as rather richer species such as *Galium palustre*, *Potentilla palustris* and *Succisa pratensis*. They tend to be small in area, as in mosaics with other peatland types, or by river shores.

Figure 4.20 Sward–sedge swamp, dry enough to allow woody plant invasion after cutting.

Figure 4.21 Species-rich sward–sedge swamp.

Table 4.20 Comparison of the more-silted Yare valley and the less-silted Bure valley communities, Broadland (Tansley 1911)

Nutrient status has increased over the intervening about 90 years, with drying (and pollution)			
BURE		YARE	
Dominants		Large societies	
Phragmites australis		*Glyceria aquatica*	
Cladium mariscus		*Phalaris arundinacea*	
Juncus subnodulosus		Small societies	
Large societies		*Poa trivialis*	
Molinia caerulea		Accompanying plants	
Small societies		*Lychnis flos-cuculi*	va
Carex filiformis		*Filipendula ulmaria*	va
(*Sphagnum*)		*Galium palustre*	va
Eriophorum angustifolium		*Valeriana officinalis*	va
Accompanying plants		*Myosotis scorpioides*	va
Rhamnus frangula	f	*Ligustrum vulgare*	f
Myrica gale	f	*Thalictrum flavum*	f
Ligustrum vulgare	o	*Eriophorum angustifolium*	r
Liparis loeselii	r	*Cladium mariscus*	r
Drosera intermedia	r	*Rhamnus frangula*	o
Pyrola rotundifolia	r	*Myrica gale*	o

va, very abundant; f, frequent; o, occasional; r, rare.

Sphagnum, in Broadland, occurs both on deep fen peat (the typical raised bog position) and on the margins beside mineral soil of low base status. Cation and pH levels are much less in *Sphagnum*. *Sphagnum* lessens water fluctuations (e.g. 5 cm in *Sphagnum* and 30 cm outside this) (George 1992).

Calcium-dominated fens may be even lower in non-calcium nutrients (see Chapter 7), but the very high lime content makes the flora, and therefore the fauna, different. *Schoenus nigricans* is the most characteristic dominant, it can grow in quagmire, building peat between tussocks, this peat growing up with the tussocks (see Chapter 10, Figure 10.8). Quagmire has almost gone: there was much a century ago (e.g. Ashfield 1861, 1862; Bunbury 1889; Parmenter 1995). Quagmire was more likely to be in abandoned peat diggings than from primary invasion of a hitherto open water lake. *Cladium mariscus* is also characteristic and in wetter times was on the firmer soil. *Cladium* extends also into calcium-influenced fen. In the more open *Schoenus* community, the associated species include sundews, *Sphagnum*, butterwort and grass-of-Parnassus, as well as the calcicoles *Briza media*, *Valeriana dioica* and *Epipactis palustris*. This grades into the richer (*Carex davalliana*-type) of the poor fens, and the calcium-influenced medium fen. Birch is the typical tree coloniser. If abstraction removes calcium-rich incoming groundwater, and the fen is then supplied by rain and acid run-off, poor fen will result. If pollution brings much fertiliser, etc., calcium-influenced medium fen will result. On drying, and yet more on drying and disturbance, the calcium hold over nutrients is relaxed, and nutrient status rises sharply: even to that of rich fen, with the rich nettle-mix tall herb communities. Keeping soil damp and not disturbed leads to birch wood. Damp without trees bears a more *Juncus subnodulosus–Cirsium dissectum* type of vegetation.

Calcium-dominated fen is necessarily local. It is of much scientific and conservation interest with the unusual habitat type, and very diverse vegetation. Dutch calcium-

Figure 4.22 Species-rich sedge–grass–rush fen.

dominated fen is less calcium-rich, as the water comes not from limestone but from lime-rich alluvium. They have *Carex davalliana*-type small sedge communities.

Medium fen can be based on rather poor but silty water, or on nutrient-medium not nutrient-rich peat, or on nutrient-medium and calcium-rich peat. Esthwaite fen in the Lake District is a good example of the first type. It was researched by the pioneer wetland botanist, W. H. Pearsall (Pearsall 1917, 1920; Tansley 1939). Reedswamp (*Phragmites, Scirpus*) is succeeded by *Carex* spp., then carr of mainly sallow (*Salix cinerea*, etc.).

Half a century on, the reedbeds are further into the lake, because more silt has been laid down, turning deeper, stonier and unsuitable habitat to shallower, silty and suitable. Behind the new reedbeds, *Carex* has invaded and replaced, as dominant, the reedbed, and behind that again carr has expanded into the former sedge fen. This is a classic case where zonation and succession are the same. And is an example of silted and peaty medium fen.

Calcium-rich medium fen is also represented by a classic – Wicken Fen in Cambridge-shire – on which the pioneer fen worker, H. Godwin, first published in 1929. *Cladium* is the dominant, representing the large-sedge, *Magnocaricetum*, community. The present reedbeds give the impression that *Phragmites* was an earlier dominant there, but in fact they date only from the 1960s, and were planted on abandoned arable, where nutrients had of course been increased (both by drying and disturbance and by fertiliser). It is all too obvious to the conservers that the fen is dry enough for carr! It is a very mixed carr, dominated by *Frangula alnus*, but including many others, both nutrient-poor (e.g. birch), dry (e.g. *Quercus robur*), and waterlogged (e.g. *Alnus glutinosa*). The diversity is presumably partly because it is ancient, and partly because of the intermediate nutrient status. Areas mown annually bear species-rich (medium) litter. In the mid-nineteenth century, *Schoenus nigricans* was abundant. There were then open peat cuttings to colonise. Also, with plenty of uncontaminated calcium-rich peat, moving rain/run-off water would become more calcium-rich.

In rich fen, *Phragmites* is the main reedswamp dominant and peat-former. The line dividing rich from medium fen is arbitrary. It falls somewhere within the large-sedge, *Magnocaricetum*, format. *Cladium* is undoubtedly medium, *Carex riparia* undoubtedly rich. *C. appropinquata* lies close to *Cladium* but is less calcium-rich, *C. paniculata* is between that and *C. riparia*. The major large Carices can all invade open water, but (except for *C. rostrata*) are usually present inland of reedswamp. Mechanisms of succession are discussed in Chapter 10. *C. paniculata* fen is an important Broadland community, being, in its turn, invaded by alder carr. In more nutrient-rich fens, carr, usually sallow (*Salix cinerea*) before alder, invades reedswamp directly. *Glyceria maxima* comes in as the most nutrient-rich reedswamp species. *Typha* spp. usually occur between this and *Phragmites*: *Phragmites*, of course, tolerates the nutrient-rich habitats, but may go under to, or be prevented by, *Glyceria maxima*. *Glyceria maxima* and *Typha* spp. appear to tolerate low aeration better than *Phragmites*. The vegetation of the fens of the two major Broadland rivers is shown in Table 4.20. The separation is distinct (though the two overlap). One has the less nutrient-rich large sedge community, the other, where the peat is more silted (so more nutrient-rich), has *Glyceria maxima*.

Tall herb and short herb communities (see Basic Charts K–Q and Figures 4.23–4.26)

Tall herb and short herb communities are those that are neither reedswamp, sedge or wood, nor grass or arable communities of fens and marshes. As they barely appear in peat columns, they are not long-lived natural communities. Most have developed under human impact, via mowing, grazing, surface removal (peat-digging), abandonment, burning, trampling, and selective cutting in many different combinations and intensities. The short herb communities and wetter tall ones are often species-rich and valuable. They are shown in Basic Charts K–N and P. The tall herb communities are generally the more nutrient-rich. Short herb communities range from poor fen to, but infrequently, the richest ones. Tall herb communities occur equally on marsh (although not necessarily with the same communities).

The associate species often mark the habitat conditions more closely than do wide-ranging dominants such as *Phragmites* or even *Cladium*.

- Short herb types at the nutrient-poor end include the purple moor grass (*Molinia caerulea*) communities, replaced in slightly more nutrient-rich places by the less tus-socky, *Calamagrostis* spp.

Figure 4.23 Species-rich grazed bog–pasture.

Figure 4.? ...ssland (see Figure 2.3).

- *A.* ...nating grasses in the richest
  fen
- *Jun.* ...*pendula ulmaria–Angelica*
  sylves. ...r than *C. canescens*.
- The *?* ...bably the most species-rich
  Broads ...s.

Even in n?... ...up with the care required to
overcome the ...thus invades. This is reducing
the size of this ...ell 1995).
Fens that wer?... ...nd grazed for 70 years or more
are often not dis... ...nose abandoned more recently, e.g.
the 1940s, show i... ...sion to fen (George 1992).

Figure 4.25 Tall-herb fen.

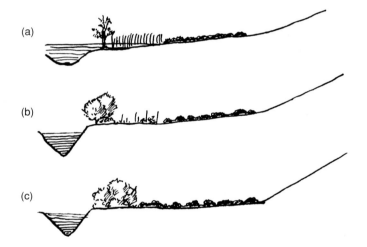

Figure 4.26 New Forest valley bog patterns in the twentieth century. (a) Undrained, carr too
    wet and in poor condition, reedswamp in good condition, too wet and soft for
    grazing, bog on higher (firmer) ground only; (b) intermediate; (c) drained and
    much dried, carr develops well, reedswamp is dry enough for grazing and is slowly
    lost, replaced by bog. (Note: channelling exaggerated, and misplaced, for visual
    effect.)

In a few remote places (e.g. south of Houston Great Broad) the fens may date back to
the vegetation at the end of the second (Romano-British) marine incursion. They have a
much larger number of woody species than similar areas that have developed carr only in
the past century (*cf.* Wicken Fen). The new ones have not had time to develop such
diversity (George 1992).

Most vegetation, however, was managed (see also Chapters 2 and 3), and there was
little woodland: reedbeds, sedge beds, marsh hay, litter, rough pasture, sheaf, peat-cuttings,
and then re-exploited after infilling. A mosaic of communities occurred (George 1992).

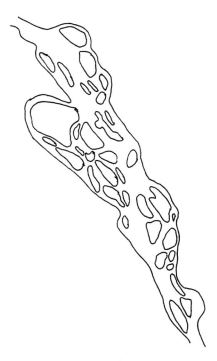

Figure 4.27 Part of the Rhône flood plain, *ca*.1860, showing braiding and islands (modified from Bravard 1987).

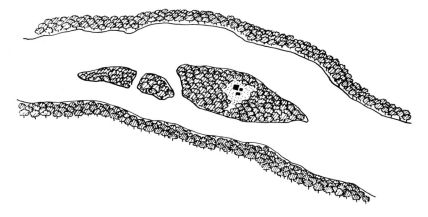

Figure 4.28 Wooded river islands, Danube, 1990s. Note the pattern follows that of (the more natural) Figure 4.27 and differs from that of Figure 2.5 (where the lakes are man-made and not river related).

Grassland (Figures 1.4, 2.3, Chapters 1, 2, 3, 5, 13)

Although there is no confusion between short grazed pasture and tall herb communities over 2 m high, other separations are questionable. At what point is rush-pasture a poor fen rather than a grassland? At what point does (tall) marsh hay have enough contribution from other species to be tall herb? Rough grazing can be as species-rich as short herb. Grasslands vary with water regime and nutrient regime (internal and added). They vary with seeding, and intensity of grazing. They vary with management in structure as well as in species (Figure 2.3) and this can be quite as important to the associated species, especially the animals. They may be large (Figure 1.4), covering part or all of a large fen or marsh, or small, just along a brook.

Abandoned land gets more nutrient-rich as it gets drier and the soil becomes more oxidised. In pasture, however, trampling compacts the soil, enhancing damp and decreasing mineralisation. In the one habitat for water and intrinsic nutrients there may be the most nutrient-rich *Epilobium hirsutum–Urtica dioica–Galium aparine* on one side of a fence, and on the other, pasture with *Eleocharis palustris, Lychnis flos-cuculi* and even *Eriophorum angustifolium*. This is recognised in the NVC by describing grass communities as 'Mesotrophic Grassland' (MG).

Woodland (Figures 4.11–4.15; see also Chapters 3 and 10)

Birch colonises bog (with pine) and may also dominate on poor and calcium-dominated fen, and extend, as a sparser species, into richer places. Alder invades fen and marsh of all types, either directly or after sallow as the fen dries. It is also common along riversides, alone, in mixed stands, or in a valley wood. Sallow invades fen and marsh which are wetter and more stagnant than alder can tolerate. It also occurs along riversides, and (with a spread of species) into lower nutrient communities. Willows are cultivated in osier beds, grow in riverside woods and as isolated trees, etc. Ash colonises drier sites, and oak even more so. Oak does not dominate in woods properly called 'wetland woods'.

Alder woods of the Breck fens in the 1950s had been alder woods in the 1830s (Tithe Maps). By the 1990s, a few 1950s-invading alder copses had become genuine copses or small woods. Open fen of the 1830s had therefore taken a minimum of a century and a half – managed for over half of that – to become alder woods. Sallow carr was similar. Few 1830s open fens had sallow dominant in the 1950s, more had developed in the 1990s. Abandonment and drying help woody invasion (but see Chapter 10).

5 The animals

All along the backwater
Through the rushes tall
Ducks are a-dabbling
Up tails all.

Said the first swallow . . . 'The call of the lush meadow-grass, wet orchards, warm insect-haunted ponds, of browsing cattle, of hay-making.'

(Kenneth Grahame, *The Wind in the Willows*)

A herd of beeves, fair oxen and fair kine
From a fat meadow ground.

(Milton, *Paradise Lost*)

It is no longer enough to rely on luck, natural processes and the spin-offs from management for other groups to supply the needs of invertebrates.

(Kirby 1992)

One must question what value there is in monitoring invertebrates to demonstrate changes that are obvious by casual inspection or botanically, until there are detailed studies that correlate changes in the composition of the indicator species with comparable changes in the conservation value of the whole invertebrate fauna.

(Drake *et al.* 1989)

Introduction

Wetland animals live in, on, or around vegetation, using it for food (directly, or via animals), shelter (whether trees or decomposing grass), the opportunity to hunt prey (e.g. spiders' webs) or to hide from predators (e.g. otters). Vegetation habit and form, its structure and species, provide the habitat available to animals. Vegetation chemistry influences and determines the animals feeding on it. Some animals can feed on a wide variety of plants (e.g. sheep), others are restricted to few, or even one species (e.g. reed bugs). Bog and fen fauna differ, because their flora differs.

Water influences animals directly, as some species require water for part or all of their life cycle (e.g. the young of mosquitoes and dragonflies), or indirectly, when animals require a specified plant species (e.g. Large Copper butterfly), or a specific structure (e.g. reedbeds for bitterns).

Many wetlands have high diversity, providing a variety of food plants: and therefore animals. Although diversity decreases in nutrient-poor habitats, e.g. bogs, there are probably more wetland than land or water animals (Gopal, in Patten 1990).

If the vegetation is right, and the disturbance low enough, the animals will be right (good and appropriate for that habitat).

This is true, as long as not misinterpreted. First, animals living outside as well as inside a wetland must have the outside habitat right also (migrants must have good habitat in all phases, otters on the bank must not be killed by pollution in the river, etc.).

Second, 'right' vegetation may mean species or structure or both, a community with 'frequent' alder and sallow will not bear woodland birds if the 'frequent' woody plants are not over 50 cm high, being mown annually in a tall herb community. In badly degraded and eroded bog in the Isle of Lewis the breeding waterfowl were as good as in satisfactory bog (Stroud *et al.* 1988). Plant species were impoverished, but structure was satisfactory. Invertebrates would be poor, though.

Third, the animals must happen to arrive, or to have ancestors in the site. Invertebrates, unlike, say, swans and deer, are not roaming the country, fit and well. This is demonstrated in the between-site variation in Broadland fens: 70 species in one fen were absent at another a few miles off, 84 beetles were in a fen in one valley, 28 in another (George 1992). Generalisations from spot sites are thus unwise.

Wetland animals are those that depend on a wetland habitat for food, shelter, reproduction and minor resources at any time in their life cycle. Bogs and fens may be isolated in their landscape like islands, they bear species restricted to these habitats, and they bear generalists. Many animals move in and out, particularly the larger and more mobile ones (e.g. deer, migrant waterfowl) while many others spend all their lives within the wetland.

Invertebrates

General principles

Insect hunters flourished from the nineteenth century. The undergraduate Charles Darwin, hunting in Wicken Fen, was one of them. Good lists – for some wetlands – go back to that era. Lists are still compiled that show the changes, and more wetlands have been mapped. Examples include, for Wicken Fen, guides and check-lists; for Broadland (e.g. George 1992); for Welsh peatlands, Countryside Council for Wales publications (e.g. Holmes *et al.* 1991a,b; Ismay 1978).

There are a few ecological tables, such as Table 5.1 for good raised bogs, and Table 5.2 for some species in different habitats in some Cumbrian bogs. The bog community in Table 5.1 is as much characterised by the sparseness of some species (e.g. snails, snail-killing flies, millipedes, woodlice and harvestmen – except for the millipede *Omonatoiulus sabiolus* and the harvestman *Mitopus mario*) as by the presence of other species. Many of these others are common species with wide ecological tolerance, some, like bees and hover flies, being non-resident.

Edge effects affect wider strips for invertebrates than for plants, so the edge of the 'proper' vegetation is still transitional for invertebrates, making the 'proper' fauna occupy a smaller area. This is especially important for small and isolated sites that may lose their 'proper' fauna while retaining 'proper' vegetation.

In general, habitats open and species-rich in plants are species-rich in invertebrates, e.g. those with much rush, short *Carex, Schoenus nigricans*. Tall dense *Cladium mariscus* has a poor fauna compared to a sparse *Cladium*-mixed community: the former has uniform, thick structure, thick litter and heavy shade.

Table 5.1 Typical invertebrate species of good raised bogs (Drake *et al.* 1989)

Pyrrhosoma nymphula (Sulzer, 1776)	common	*Hybomitra montana* (Mg)	local
Sympetrum danae (Sulzer, 1776)	common	*Phyllodromia melanocephala* (Fabricius)	common
Salda morio (Zetterstedt)	nr	*Dolichopus atratus* Meigen	local
Saldula saltatoria (Linnaeus)	common	*Dolichopus lepidus* Staeger	nr
Chartoscirta cocksi (Curtis)	local	*Dolichopus vitripennis* Meigen	local
Cosmotettix panzeri (Flor)	notable/nb	*Hercostomus aerosus* (Fallen)	common
Macustus grisescens (Zetterstedt)	common	*Campsicnemus compeditus* Loew	notable/nb
Tyrphodelphax distinctus (Flor)	local	*Campsicnemus loripes* (Haliday)	common
Paradelphacodes paludosus (Flor)	na	*Helophilus pendulus* (Linnaeus)	common
Pterostichus diligens (Sturm, 1824)	common	*Sericomyia lappona* (L)	local
Agonum ericeti (Panzer, 1809)	notable/nb	*Sericomyia silentis* (Harris)	common
Cymindis vaporariorum (L. 1758)	notable/nb	*Opomyza germinationis* (Linnaeus)	common
Agabus bipustulatus (L. 1767)	common	*Scaptomyza pallida* (Zetterstedt)	common
Anacaena globulus (Paykull, 1798)	common	*Myrmica ruginodis* Nylander	common
Plateumaris discolor (Panzer, 1795)	local	*Myrmica scabrinodis* Nylander	local
Limnephilus luridus Curtis, 1834	local	*Myrmica sulcinodis* Nylander	local
Coenonympha tullia Muller	notable/nb	*Formica lemani* Bondroit	common
Pedicia rivosa (Linnaeus)	local	*Mitopus morio*	common
Limnophila pulchella (Meigen)	notable/nb	*Pardosa pullata* (Clerck)	common
Limnophila squalens (Zetterstedt)	local	*Pirata piraticus* (Clerck)	local
Limnophila meigeni Verrall	local	*Pirata hygrophilus* Thorell	common
Erioptera gemina Tjeder	local	*Pirata uliginosus* (Thorell)	local
Molophilus occultus de Meijere	nr	*Antistea elegans* (Blackwall)	local

na, notable "a"; nb, notable "b"

Table 5.2 Common leaf hoppers at each habitat on some Cumbrian bogs (Drake *et al.* 1989)

	Status	Exposed bog	Bog near trees	Wet heath	Dry heath	Wet wood	Moist wood	Dry wood
Tyrphodelphax distinctus	L	2	1	1				
Paradelphocodes paludosus	N	2		1				
Cosmotettix panzeri	N	1		1				
Sorhoanus xanthoneurus	N	1	1	1				
Aphrodes bifasciatus		1		1				
Ulopa reticulata				1	1	1		
Neophilaenus lineatus		1(2)		1	1	1		
Stroggylocephalus livens	N	1	1	1		1	1	
Macustus grisescens		1	1	1	1	1(2)	1	
Philaenus spumarius		1	1			2	1	1
Oncopsis flavicollis			1	1		1		1
Thamnotettix confinis						1	1	1
Speudotettix subfusculus						1	1	1

N, national importance; L, local importance.

The following section (up to 'Wet Grassland') comes mainly from the authoritative work of Kirby (1992).

Invertebrate needs include the following:

1 annual or biannual life cycles are common: therefore, the required conditions for breeding must occur each year;

2   long-term resting stages are usually absent: so again invertebrates are without the resilience of those plants that can survive bad seasons;

3   complex life cycles are common, and each stage may require a different habitat (e.g. dragonflies: shallow water, emergent plants in it, and tall grass and scrub);

4   highly specialised species are common, and this is often restrictive, e.g. the swallowtail butterfly lives on milk parsley (*Peucedanum palustre*) only;

5   many require a specialised structure, e.g. a tussock of grass, which therefore must be present for survival. Management for a short sward (e.g. Figure 2.3) would not do;

6   many cannot move far; flightlessness is frequent in those habitats most threatened – the ancient, stable and isolated;

7   they are cold-blooded, and different species are adapted to different temperatures: so need to find these;

8   their micro-climate may be as affected by minute ridges as grassland is by mountains; opening a fen drove can be equivalent to an Ice Age;

9   over-small populations die out; extinction continues, and there is too little immigration to balance it (see 6 above).

Some general principles can be laid down:

1   even tiny open spaces of bare ground (e.g. bird track) can be crucial;

2   more than one habitat may be required;

3   in transitions, gradual changes are preferable to abrupt ones;

4   a mosaic of vegetation structure is valuable;

5   little variants should not be destroyed without evaluation, seepage areas too small for specialised plants may be of great importance to invertebrates;

6   more complexity and variety of structure means more invertebrate habitats, and structure is important at all levels: a few centimetres of bare ground may mean a warmth-loving invertebrate can live on a food plant; wood differs to reedswamp;

7   complex structure can come from complex plants (e.g. trees), from seasonal changes, from impact (mowing, grazing, etc.);

8   sites with many common plants, a stable history and a varied structure may be better for fauna than ones with rare plants and a more varied history;

9   large populations of a plant are more likely to support a rich invertebrate fauna: they are easier to find, and large populations are more resilient;

10   plant-eaters may need particular forms of a species, e.g. a rosette; the more the growth forms and habitat tolerance of a plant, the larger its potential invertebrate fauna;

11   all plant stages are needed, flowers, bark, etc., for different invertebrates;

12   stressed plants are weaker, and more easily attacked by invertebrates;

13   large sites have less edge effects. They are more likely to contain habitat variety.

Because of drainage and other impact, many aquatic and wetland invertebrates are very rare and restricted. Ancient wetlands, with many interlocking habitats, are of great value (e.g. Somerset Levels). The more the habitats, the better the buffer against disruption.

Wood and trees

Wood is the richest British habitat for invertebrates (and see Table 5.3). There are more of species such as craneflies in shaded than in open wetland. Dead wood is a distinctive

Table 5.3 Invertebrate species supported by some trees and shrubs (selected from Merritt 1994)

Species	Wildlife value	Number of insect spp. supported
Alder (*Alnus glutinosa*)	Moderate, cones eaten by siskin and redpoll	90
Alder buckthorn (*Frangula alnus*)	Moderate, berries, food plant of brimstone butterfly	
Ash (*Fraxinus excelsior*)	Moderate	41
Bird cherry (*Prunus padus*)	Moderate	
Birch spp. (*Betula pendula* and *B. pubescens*)	High, seeds and buds eaten by birds	229
Blackthorn (*Prunus spinosa*)	High, berries, nest sites for small birds	109
Buckthorn (*Rhamnus catharticus*)	Moderate, food plant of brimstone butterfly	27
Dog-rose (*Rosa canina*)	Moderate	~100
Dogwood (*Cornus alba*)	Moderate	18
Elder (*Sambucus nigra*)	Moderate, berries	19
Field maple (*Acer campestre*)	Moderate	~40
Guelder rose (*Viburnum opulus*)	Moderate, berries	17
Hawthorn (*Crataegus monogyna*)	Very high, berries, nest sites for small birds	230
Hazel (*Corylus avellana*)	High, nuts	107
Holly (*Ilex aquifolium*)	Moderate, berries, roost site for birds in winter	13
Hornbeam (*Carpinus betula*)	Limited, nuts	28
Norway spruce (*Picea abies*)	Moderate, cones, winter roost site for birds	37
Pedunculate oak (*Quercus robur*)	Very high, nuts, high densities of moth caterpillars; mature trees offer nest sites for hole-nesting birds	~250
Poplars (*Populus* spp.)	Moderate	97
Aspen (*Populus tremula*)	Moderate	
Black poplar (*Populus nigra*)	High	
Scots pine (*Pinus sylvestris*)	Moderate, cones, roost site for birds in winter	91
Willows (*Salix* spp.)	Very high	266
Crack willow (*Salix fragilis*)	High, older specimens often develop holes used by nesting birds	
Goat willow (*Salix caprea*)	High	
Osier (*Salix viminalis*)	High, nest sites for coots and grebes	
Common sallow (*Salix cinerea*)	High	
White willow (*Salix alba*)	High	

habitat in which many species can breed. Ancient woods are of extreme importance, because of their continuity over time. Different states form different habitats and support different fauna, e.g. pollards: old gnarled trunk, newer shoot bases, young withies, old withies ready for cutting. Sap runs support an interesting and varied fauna, chiefly beetles and flies. Management should mimic that of the past.

Marginal trees by water may contain aquatic and dryland species: rich, particularly on alder and willow. Fringing tree-lines are good.

Dead wood should be left wherever possible. Different parts, from twig to trunk, have their own distinctive population. Woodland provides shelter for flying insects emerged from water. Mayflies and other feeble fliers benefit considerably. Spongeflies spend much of their adult life high in trees. Some insects lay eggs in branches and leaves overhanging water, others use crevices in the bark. Dead wood and other vegetation falling into water form a good niche.

Shaded ponds are a poor habitat, but the species include specialists not found elsewhere. A shaded woodland stream will be cold and have a different fauna to an open one warming in summer.

Lowland peatlands

All the successional stages have interesting invertebrate communities. Small-scale peat-digging can create and maintain the early successional pools, bare peat, etc. Grazing gives diverse peatland vegetation, but is not good for all peatland types. It damages, for example, *Sphagnum*, but light grazing gives structural diversity. Mowing every 1–3 years can give a very rich fauna. Some species eat the plant before the cut or after regrowth, others develop in wet peat. It is important to retain the traditional management, particularly on ancient sites.

In the lowland bogs of south west Scotland, Curtis (1977) found sites with few spider species were influenced by burning, grazing, flooding, perhaps drainage and afforestation, and perhaps exposure. Except for the flooding, which has a direct influence on the spiders, these factors act indirectly, the vegetation and its structure being changed. Most of the bogs are threatened remnants, and of them it could be said that these wetland sites may appear of little use to the general human populace in terms of scenic value, recreational asset, building or agricultural use, etc., but they do retain communities that are an essential component of the ecosystem from which we cannot divorce ourselves.

Reedbeds (Phragmites australis)

All stages in the development and disappearance of reedbed (or swamp) support important invertebrate communities. Some feed on reed, some eat or parasitise those. Others live inside shoots, old or young. Many live in leaf litter. Some are just adapted to the general structure. Permanent water has the least interesting fauna, although it includes some specialists, but most of the species occur as much or more in drier places. These drier places (see Basic Charts P and Q) are, of course, man-made, where tree invasion has been prevented.

Long rotation cutting, giving many stages, is good, as is litter left in piles. For commercial cutting the bed should not be left more than two years. This is too short for the development of some species, and creates an over-uniform habitat.

Water margins

Fringes of water bodies usually have great interest, so should be managed for this interest. Watercourses with gentle slopes above water level support good vegetation and invertebrates. Oligotrophic lakes with reedswamp and stony margins support the most distinctive fauna. The richest invertebrate communities are usually associated with the densest vegetation. Varied structure is good, and so are factors creating this, e.g. trampling, grazing.

Temporary pools support important populations, including species succumbing to competition or being eaten in permanent pools. After a few days a temporary pool will start having common invertebrates from water bodies, and if the pools persist longer, the earlier specialists will be mostly eaten or replaced.

Hollows in grassland are more likely to be of conservation interest for fauna than flora. Hoof prints are good habitat, patchy, drying for part of the year, and occurring in different places at different times.

(Very) wet heathland

With its different plants, wet heathland has a different, yet distinctively wetland, fauna. As elsewhere, a mosaic of structurally diverse communities is best: low sphagnum and sundew, tall sedge, grass, rush, low shrubs (e.g. bog myrtle), sallow carr. Diversity has decreased as patterned management has decreased. Purple moor grass, *Molinia caerulea*, is of much invertebrate interest when in sparse tussocks, less so in rank dominant stands. Some birch is good. Pool margins and sphagnum fringes can support a diverse fauna, including rare species. Weed-choked pools should be left, not cleared.

Small marshes

Even small marshes, remnants, small areas of impeded drainage, etc., may have diverse invertebrates, whether or not the vegetation is of conservation value. They should be retained.

Seepages and small springs

Seepages and small springs are created by groundwater, so are buffered against changes in weather. They commonly have perennial flow, and their temperature is reasonably constant (until this is more influenced by the land it flows over). These may date back millennia, if due to land form and rock type (and have not been drained or abstracted). Some species are below-ground ones, occasionally washed up. Some are surface but cool-water dwellers. Species of thin films of water, oxygenated surface mud or permanently wet moss do well. Many core species are linked, partly or wholly to seepage. Tiny ones can be of much interest, a seepage line, even more.

Wet grassland

This may be perennially wet (high water table or impeded drainage), winter-wet or with other patterns. Those winter-flooded and summer-dry have a rather harsh habitat. Washlands summer-flooded and winter-dry have more diverse invertebrates (but may be poor in plant species). Mud and decaying vegetation are important in early stages. Structure is more important than floral composition (Figures 2.3 and 5.1, Basic Figures 2–6, 7 and 16).

Variation in wetness produces many niches and therefore many species. Such variation also protects populations from changes in water regime, some of each habitat remaining. Slight ridges or hollows can make a great difference to wetness and duration of flooding. Damp hollows and temporary pools may be important. So may variations in soil type, peat communities differing from those on clay or silt.

Bare ground warms and dries in the sun. Solitary bees, wasps, etc., dig burrows, other insects lay eggs or hunt. Tall grassland generally supports more species (although tall rank abandoned grassland is unsuitable because of litter and changing composition). It is good to have a range of growth forms growing, flowering, setting seed, bare patches and dense tussocks (which form shelter and hibernation sites).

Flood plains

Flood plain Sites of Special Scientific Interest (SSSIs) bear remnants of a fauna that must previously have been widespread, and is now fragmented and damaged. Wet mixed woodland is very rich, one now bearing over 2000 invertebrates (out of the 30 000 species in Britain). Predators, scroungers and species of the bottoms of flooded trees are important, with populations varying with flooding. In fen, species vary with the structure and habitat, and 'wild' fens are very species-rich. Hay meadow has a valuable but sparse community. Over half the species of British arthropods (beetles) are found in flood plains. Spiders and ground heather vary with the vegetation type.

Invertebrates are lost with the loss of a functioning flood plain, i.e. one flooded by the river. Increasing water regulation causes loss (Flood plain symposium 1995).

Agriculture

Cultivated wetlands may also show interesting patterns. In drained wetland pasture in western France, Georges (1994) found that ground beetles, carabids, are linked to wetness. Wet habitat species are less in drier areas, and of course decrease as drainage increases. Drainage is associated with loss of old hedgerows. *Carabus* spp. are most in woodlike habitats and ancient grassland: *C. granulatus* dominated in wetter places, and disappeared with water regulation, *C. auratus* occurred only in wet and reduced soils, *C. violaceus* in changed-use areas by plains or old polders. Between the arable just outside the flood plain and the drier parts of the plain, there is a partial overlap of species.

The main land-use patterns can be characterised by species assemblages composed of dominant species that influence population diversity. The carabids have highly intricate habitat relationships.

Fish

Fish are not described here, as in wetlands as here defined they are present only in margins and deeper flooded reedswamps usually beside open water, so are better considered as aquatics. Fish use reedswamps and equivalent vegetation for shelter protection and food (plant or animal).

Birds

As with invertebrates, early work was distributional, taxonomic and behavioural. Ecology has become necessary only because numbers are in such decline. Several groups are resident all year in Britain, more use different and separate wetlands in different seasons (Table 5.4). Migrant birds usually return to the same areas every year, whether for feeding or for nesting. The observed shifts in space (regionally) and in time (seasonally) usually depend on current conditions – food supply, nesting habitat, etc. Different water levels are wanted for different species (e.g. Table 6.2).

To get a proper variety of waterfowl, there must therefore be a variety of waters, together with open marsh and the other damp and drier habitats. The maximum diversity of habitats leads to the maximum diversity of animals and plants in it.

Why is intermittently flooded grassland or short herb habitat so attractive to waders? In flood, soil invertebrates come to the surface to try to avoid being drowned, so there is an exceptionally high density easily available to birds hunting for them. Then small fish, frogs, etc., may come on to the flood for food – and become food. Flooded fen or marsh is usually nutrient-rich, so phytoplankton and then zooplankton appear (there is high light, high nutrients, and development is limited only by temperature). The zooplankton are good food for ducks.

Flooding wet grassland, therefore, gives a high short-term boost of food. The supply gives out after a few weeks (except for the plankton), and the fowl may move on.

Waterfowl have two particularly vulnerable seasons: nesting and moulting. For nesting, there are the hazards of (1) finding a place with structure suitable for a nest, making the nest, and keeping it safe (e.g. no flood, plough, or trampling); (2) survival of the eggs through to hatching; and (3) survival of the young to maturity. This is not only protection (from, for example, predators and adverse weather) but feeding: and the chicks may eat different food to the parents, needing different habitats.

Waterfowl may feed on farmland near reserves, leading to a conflict of interest between the bird preservers (wanting more birds) and the farmers (wanting full crops). Complaints need watching, an unpleasant sight does not necessarily mean severe crop damage. Most migrants arrive in Britain too late to do much damage. Winter feeding can clean fields of debris, waste potatoes, grain, etc. Grazing and trampling in winter and spring do not usually reduce the yield of silage grass, winter wheat and spring barley. Spring wildfowl, however, can be more difficult, competing with livestock for the first new grass.

Large flocks of birds may have locally serious effects. Much food is needed. Their faeces give areas of very high nutrients, with much organic nutrient, which can last for decades, causing changes in vegetation, so in habitat, and perhaps in bird populations. Additionally, the sheer weight of flocks, even of as small a bird as starling, can flatten non-woody vegetation. This damages crops, possibly changing the 'wild' communities. Recovery is slow.

Full bird lists for individual wetlands exist, for example for Woodwalton Fen, Harold (1990) lists eight on-water birds (including wigeon and mallard), seven near-water birds (including grey heron and sedge warbler) and 49 birds that could also occur in dryland vegetation (wood, tall or short herb communities). It must be remembered that most wet fens and marshes are intersected by dykes, so on-water birds can live there, and go into the wetland as wanted. Wicken Fen birds are described, and references given, in Finlay (1997). For Broadland, George (1992) gives much detail, including:

Alder carr:	Marsh-, willow- and long-tailed tits, great and lesser spotted woodpeckers, blackcap, chiff chaff, red poll, sparrow hawk.
Sallow carr:	Colt's warbler (new invader).
Open fen:	Reed- sedge- and grasshopper-warblers, reed bunting; in reedbeds, reed warbler.
Open fen or scattered bushes:	Savi's warbler, bittern, bearded tit, marsh harrier.

Table 5.4 Birds of different habitats (Andrews Ward, in Haslam 1994)

In the following lists, species marked S are present only as breeding birds (assume the period May to August) and those marked W are present only in winter (assume November to February). Outside these periods these species may not be present. Birds not marked S or W are present all year.

Blanket bog
The habitat is assumed to contain heather

Peregrine S	Snipe S	Cuckoo S	Meadow pipit S
Red grouse	Curlew S	Skylark S	Grey wagtail S
Golden plover S	Common sandpiper S	Wheatear S	

Raised bog and fen
The following list includes only species dependent on the open surface of the primary active bog or unwooded fen. Birds using the wooded rand slope of the bog or occurring in wooded fen are listed below

Mallard	Snipe	Skylark	Reed bunting
Teal W	Curlew	Meadow pipit	
Moorhen	Cuckoo S	Pied wagtail	

Reed beds
The habitat comprises predominantly reed but with other fen vegetation at the margins, excluding extensive scrub or tree cover

Mallard	Water rail	Reed warbler S
Moorhen	Cuckoo S	Sedge warbler S

Woodland and scrub on raised bog rands, and fen carr

Pheasant	Great tit	Song thrush	Dunnock
Woodpigeon	Coal tit	Mistle thrush	Tree pipit S
Cuckoo S	Willow tit	Redwing W	Linnet
Tawny owl	Long-tailed tit	Robin	Redpoll
Carrion crow	Treecreeper	Willow warbler S	Bullfinch
Magpie	Wren	Sedge warbler W	Chaffinch
Jay	Blackbird	Whitethroat S	Yellowhammer
Blue tit	Fieldfare W	Goldcrest	

Lowland wet grassland
The habitat is assumed to be subject to short duration winter flooding and have a high summer water table with freedom from livestock grazing during the bird breeding season. Permanent standing water is present in dykes. There is no significant scrub or tree cover. *Note*: Snipe will be present in summer only on peat substrates

Little grebe W	Moorhen	Skylark	Reed warbler S
Heron W	Coot	Carrion crow W	Meadow pipit
Mute swan	Lapwing	Rook W	Yellow wagtail S
Mallard	Redshank	Magpie W	Pied wagtail W
Teal W	Common gull W	Jackdaw W	Starling W
Wigeon W	Black-headed gull W	Fieldfare W	Reed bunting
Kestrel W	Kingfisher W	Redwing W	
Pheasant	Woodpigeon W	Sedge warbler S	

Bird populations are related to the physiognomic category of vegetation. Generally, the structure (wood, tall herb, grazed, etc.) and the nutrient status (bog, fen, etc.) are important rather than the exact plant species involved. There are, of course, exceptions to this: *Phragmites* reed stands differ to, say, *Typha* ones. Structure affects the bird habitat for nesting, loafing and hunting. The importance of structure rather than composition of degraded blanket bog on Lewis for waterfowl (Stroud *et al.* 1988) is noted above.

Birds occupy larger territories than invertebrates, so a bird confined to one wetland habitat must have a large territory. Migrant birds need good habitats at each end, and passage sites in between. Birds may use a wetland part-time only, large flocks feeding on more productive fields, lakes, etc. (Gopal, in Patten, 1990). Many birds in the south-west Highlands of Scotland depend on blanket bogs, lochans and marshes for feeding and nesting, while ranging widely (Davies 1977).

Altering the vegetation type, however, for instance by drying reedswamp and allowing carr to invade, is as devastating to birds as to other groups. Disturbance is usually more disastrous to birds than to plants and invertebrates. Bird conservationists are particularly liable to the idea that habitat benefiting birds is ethically good, regardless of other groups, e.g. bulldozing a saltmarsh to make pools and islands.

Wet grassland

The conversion of wet habitat to (wet) grassland took a good thousand years. Has there been adaptation of waterfowl to the wet grassland? Have strains suited to wet grassland been selected from a previously varied population?

In wet grassland, eight waders commonly breed. Six are in decline. (Oyster catcher and curlew are doing well.)

Oyster catcher (*Haematopus ostralagus*) is mainly (not entirely) migratory. Only recently has it spread its breeding range from the coast to wetlands. It probes grass and arable for earthworms and insect larvae. Nesting is typically in open or short vegetation. Numbers have increased since the move inland: intensive agriculture has led to an increased biomass of soil invertebrates.

Lapwing (*Vanellus vanellus*) is mainly migratory, and likes places of unbroken views in all directions (in Britain, typically below 500 m). Although it is, essentially, a land bird, it prefers to forage, for invertebrates, on or just below the surface, so goes to moist to saturated bare ground or short vegetation. Nests are in fairly open ground often slightly raised (about 8 cm high in denser stands and about 15 cm high in open ones), with wetter and feeding grounds close by. Its traditional habitat was the steppes, but it expanded into agricultural areas, grass and arable.

Increased farming, however, means more disturbance. As lapwing can re-lay up to eight times, it tolerates some disturbance. The young often feed in harvest stubble, pools and wet areas. Although declining in Britain, lapwing is increasing in parts of Europe.

Ruff (*Philomachus pugnax*) is migratory. It nests in damp grass, wet heaths, moors, etc., preferably with access to short grass and shallow permanent or flood water. Its food is mainly invertebrates found by probing soft soil or mud at the surface, or underwater. Ruff abandons fully drained, arable land. Breeding birds want their several habitats close (e.g. 5 ha). The rest of the year food, rest and roost can be 20 km apart.

Ruff is in drastic decline, and is increasingly restricted to bird reserves. It is vulnerable to even small habitat change, including change in drainage and intensity of farming.

Dunlin (*Calidris alpina*) is found in wet moors and rough grazing as well as at the coast. It probes the surface for small invertebrates. The nests are in moist and concealed places. The breeding population is declining with drainage and conversion to arable.

Snipe (*Gallinago gallinago*) is partly or fully migratory. It occurs on damp or wet lowlands with access to waters, preferably peaty but not too acid. It needs a soft invertebrate-rich soil, where it probes deeply. It prefers to have lookout posts (at least 1 m high), among medium vegetation: tussock sedges, rushes, coarse grass, and shorter plants. In winter, there is a greater variety of food in wet places. Snipe, like ruff, is very vulnerable to intensification of agriculture. The breeding season is unusually long, so the soil must be kept wet enough to be soft late in the season.

Black-tailed godwit (*Limosa limosa*) is also migratory. It used to inhabit traditional wetlands and now lives in moderately high and dense grass, in damp places near water. It can nest on arable, and it nests in flocks, which is bad when the habitat becomes small and fragmented, not supporting many pairs. It feeds most on grassland in spring, then moves to estuaries. It probes soil or mud, and pecks in shallow pools for invertebrates. Outside the breeding season it eats some vegetation. It declines with habitat loss, and with disruption by, for example, trampling or mowing.

Curlew (*Numenius arquata*) is mostly migratory. It traditionally inhabits moorland, rough grass boggy hollows, riverine flood plains, wet meadows, and agricultural pastures, and even some arable. It tolerates scattered trees, but not carr. Curlew mainly eats invertebrates. It pecks and probes. In late summer it also eats berries. It nests on the ground, often on tussocks or low hummocks – so is in decline. There are local increases in Europe, where it invades agricultural land. When breeding, curlew want good access to moist ground for foraging. It may fly 1–5 km to feed. The nests are in tussocky or taller vegetation with lookout perches on shrubs, posts, etc.

Redshank (*Tringa tringa*) is mainly migratory, and Britain is a major breeding area. It lives in moist or wet grass and marsh. Redshank frequently feed on muddy water margins, where they probe and peck for earthworms and invertebrates (including larvae). It is one of the more vulnerable species, and is declining inland with drainage and intensification. (There is less decline in coastal populations.)

Britain has one-quarter of the European Union populations of birds (though less than the Netherlands). In order of numbers of breeding pairs, lapwing come first with 181 500, followed by oyster catcher, curlew, redshank, snipe, dunlin, black-tailed godwit and finally ruff, only 10 pairs. But these numbers are of course trivial compared with the eighteenth century and before (section adapted from Dunn 1994).

River banks can have interesting birds. Reed bunting has recently increased in habitats drier than reedbeds, in low scrub and tall herb and hedge scrub along drained rivers, deep-banked, and fenced against arable. Reed bunting invades this increasing habitat provided it is allowed its food. It needs seeds: and so not autumn removal of seed heads.

Many birds of grass, wood, tall herb and the river fringe will use the greenway of the river corridor, which can be inhabited in place, or used to spread populations along river lengths (Ward 1991).

Mammals

Compared to the information obtainable for birds and invertebrates, that available for other groups of animals is pitifully small. No British (or European) mammals are restricted, as species, to wetlands, although some favour wetlands. Many smaller individuals, however, may spend their entire life in a wetland: given the animal's chosen territory falls entirely within the wetland. Species include:

cattle (*Bos primigenius*)
coypu (*Myocastor coypus*), introduced and eradicated
deer: red (*Cervus elaphus*); fallow (*Dama dama*); roe (*Capreolus capreolus*); Chinese water
   (*Hydropotes inermis*), introduced; muntjac (*Muntiacus reevesi*), introduced; sika (*Cervus
   nippon*), introduced
dog (*Canis familaris*)
fox (*Vulpes vulpes*)
hedgehog (*Erinaceus europaeus*)
horse (*Equus caballus*)
mink (*Mustela lutreola*), introduced, decreasing as otters return
mole (*Talpa europaea*)
mouse (*Apodemus sylvaticus*)
otter* (*Lutra lutra*), increasing after severe decline from habitat loss and organochlorine,
   etc., pesticides
rabbit (*Oryctolagus cuniculus*)
rat, brown (*Rattus norvegicus*)
sheep (*Ovis aries*)
shrew*, water (*Neomys fodiens*)
stoat (*Mustela erminea*)
vole (rat)*, water (*Arvicola terrestris*)
weasel (*Mustela nivalis*).
(* also aquatic; the other species are basically land animals)

In considering the effect of mammals, the extinct ones also must be taken into account: beaver, and large grazers such as elk and auroch (see Chapter 3). Beaver would cause, over such flat land, very wide and shallow ponds, and would cause them at frequent intervals, frequent both along most streams, and frequent over the decades at any one place. Flood means more nutrients, flood means drowning sphagnum. On both counts, beaver would prevent the growth of bog, and encourage that of fen and marsh, wherever their pools could reach – which is not very high in level above the stream.

Before there were large herds of domesticated livestock, there were wild grazing mammals. As the buffalo records in nineteenth-century North America show, wild herds can be just as effective in producing and maintaining grassland as can domestic ones. Grassland, however, can be produced only where the ground is dry enough for the forage grasses to become established and grow well (see Chapter 4), as well as having the animals potentially present, and the ground being not only dry but also firm (not a thin skin above water; see Chapter 3).

Therefore, from as soon as wetlands were formed after the Ice Age there could potentially have been drastic effects by mammals – aided by the grazing of geese and other waterfowl – on fens and marshes. They could well have prevented bog formation in a lot of places. Once bog has formed, however (and formed above beaver-level), grazing alone will not produce a good grass sward, the bog is too wet and too dystrophic. At the present time both drainage and fertiliser are needed to get even rather poor grassland. Drainage mimics the drying of the Boreal and sub-Boreal eras (see Chapter 3). Large flocks of birds can mimic fertiliser only locally. This, on a bog, is too local and uncommon to have any widespread effect.

Rough grazing is possible on any bog or other wetland firm enough to support the relevant animal (firmer for red deer than for water rat!). Such grazing decreases or prevents tree invasion. Figure 4.7 shows badly eaten short alders in a New Forest valley bog. Ponies and cattle have not succeeded in preventing their appearance, but, while grazing stays at the present intensity, are maintaining bog and preventing wood. A relaxation of grazing for a few decades would allow growth. Also in the New Forest valley bogs, around 1900 there was, as now, carr along the streams, but there was then an outside band of reedswamp before the bog. There was, then as now, grazing by ponies and cattle. There was one major difference. The streams had not been drained, either in the forest or downstream. The wet was such that the carr was degenerating, instead of flourishing on dry ground, as now. And the reedswamp was in soft peat some 2 m deep, in which no horse or cow could stand. It therefore remained ungrazed (Tansley 1911). With the subsequent drying, the reedswamp could be reached, and was grazed. By the 1970s the *Phragmites* populations were typically in bog, were depauperate with grazing, and were lost from much of the area. By the mid-1990s there was indeed little left. Thus grazing causes change. When wet excluded grazing, there was different vegetation to that prevailing when grazing was present (reedswamp species tolerate only light grazing).

Similarly, in raised and blanket bogs, even light grazing can be sufficient to keep out grazing-sensitive species, and a different flora may be found on places too wet for such animals; *Carex rostrata*, for instance, and indeed *Phragmites* where even a little mineral run-off accumulates. The effect of drying plus even very light grazing can be great.

As human impact increases, flood plains, fen edges (and to a lesser extent bog edges) are dried and converted to grazing: rough grazing, then unimproved grass, finally improved grass, and perhaps ley and arable.

When forage quality improves, stocking density increases, and the exact species becomes more important (Figure 5.1). Invertebrates respond to even small differences in structure, varying in distribution and abundance. Even hoof prints can be significant niches. Sheep graze to a low, thick, sward and, being light, can walk on the wettest ground. Cattle and

(a)   (b)

(c)   (d)

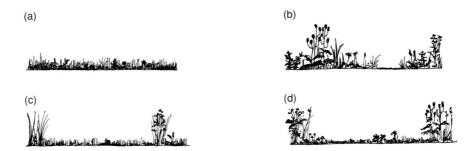

Figure 5.1 Grassland structure resulting from different grazing mammals: (a) sheep; (b) cattle; (c) horses; (d) rabbits (after Kirby 1992). (*Note*: Livestock may be mixed.) There were also goose-meadows near water (see description in text).

horses are ranker eaters, and are more likely to have uneven structure and sunk hoof prints. Horses may leave more structural variety: better for invertebrates (e.g. Sheppard 1988). It must be remembered that, before the car, horses were the main means of transport, being used for riders, carriages and, alternatively with oxen, ploughing.

It must also be remembered, however, that in earlier centuries – as is slightly reappearing now – there was, especially near villages and farmhouses, mixed grazing. Horses needed regularly could be put in the same field with cows, indeed with sheep also.

The eighteenth-century change in the Scottish Highlands from cattle to sheep must have had a significant effect on the moor and bog vegetation. The English wetlands described by Defoe (1724–7) mostly had cattle.

The traditional way of keeping a low stocking rate through the season gives better habitat diversity than short but intensive periods of grazing, which, at the worst, can leave fields bare, and, at the best, short, uniform and trampled. This is also more likely to lead to stream pollution: if the livestock are placed near the waters. Sward species composition varies with the date of grazing (supposing the sward is not much managed and of low diversity). Different species grow at different times in summer, so the species that are up and available to be grazed and damaged vary. Next, plant habit. Species like daisy (*Bellis perennis*) and plantain (*Plantago* spp.) grow short and flat, not easily grazed, and do well under short grazing, but poorly when plants are left tall enough to shade them out. Only by (proper) management does wet grassland become species-rich. Abandoned, it becomes tall and rank, so bereft of the short plants, and species-poor.

The red deer is the usual species of the Scottish Highlands but, as cited in Chapter 1, eighteenth-century species (at least to Inverness) used to be more varied. Deer entering 'wild' lowland wetland are more likely to come for the cover afforded by carr or tall, non-woody vegetation. Grazing is present, but may also be easier outside the wetland. Deer territory is large. Deer tracks can break up tall vegetation, making tracks in, for example, reedswamp and bogs. These create habitat diversity, allowing other plants and indeed smaller animals to enter the niches they make.

Smaller herbivores include the water vole, which needs good high vegetation for 20 m or so back from open waters to give cover from predators, and a bank to burrow in. Water rats have greatly declined with the drying, 'tidying and cleaning' of stream banks and similar, giving loss of habitat, made still worse when engineering works lead to floods in the burrows; at the same time, the number of mink has increased.

Mink were introduced for breeding for fur, and, especially until the recent stricter controls, were liable to escape, and there are now wild populations. They seem to be decreasing where otters are returning. This is an interesting example of presumed competition. Otters cannot tolerate organochlorine and other pesticides in their food, nor the cleaning and tidying of banks that destroys their habitat. Otters declined and mink spread. A habitat-controlled change? Otters were a pest species until recently (e.g. Haslam 1991), both for taking fish (i.e. that fishermen might have caught) and for raiding stores of, for example, grain. In Broadland, however, surely decreasing from the eighteenth century, they were still common in the 1950s until pesticides got unacceptably into their food (George 1992). This was followed by habitat loss, severe by the 1970s. Introductions in the 1990s are leading to successful recolonisations over wide areas. Otters need long areas of stream (up to 40 km for males, 10 km for females) with at least intermittent tall bankside vegetation (marsh, carr, brambles, tall herb community or even artificial holts). Cover should extend at least 50 m from the river. Disturbance is bad. Leaving nice clean smooth engineered banks is even worse. Wetlands with tall vegetation or carr beside brooks or rivers are therefore very suitable for otters. Arable, or even short grassland, is not – unless there is a good buffer strip between it and the river, or the farmland is broken up by patches of suitable vegetation.

Coypu were imported to fur farms from 1929, and, inevitably, escaped (especially when farms shut down in the Second World War). The only serious, long-term pest population was in Broadland, and East Anglian estuaries, where a much enhanced eradication programme started in 1981 and the coypu was presumed extinct before 1990. Unluckily for the coypu (and unlike mink, whose numbers have remained relatively low), they can increase extremely rapidly, and winters cold enough to cause massive deaths are rare. The coypu is a pest partly because it eats crops from farmland, partly because it burrows into banks, damaging them, but mostly because of the vast amounts of reedswamp plants it eats. *Phragmites* and *Cladium* populations, and, worse, those of rare species like *Rumex hydrolapathum*, were widely damaged; the coypu grubbed up rhizomes too, large patches of fen being converted into expanses of black mud and shallow water (George 1992). Coypu, therefore, can change plant communities, alter vegetation type, and alter successional trends.

Reptiles

Grass snakes (*Natrix natrix*) were formerly common in and near, particularly, fens and marshes. Adders (*Vipera berus*) live more just beside these flooded places, and on the drier moors, and going into bogs. (Bogs are not flooded from outside.) Slow worms (*Anguis fragilis*) may be common on banks (e.g. Firth 1984).

Amphibia

All 12 native species of amphibia, frogs, toads and newts may occur in wetlands. Indeed, Beebes (1987) recorded all of them in ponds on one southern heath. There one species, the palmate newt (*Triturus helveticus*), occurred in all pond types, although least in rain-fed ones, while five others bred only in ponds with limestone or concrete fragments: common newt (*T. vulgaris*), great crested newt (*T. cristatus*), common frog (*Rana temporaria*), common toad (*Bufo bufo*), and the natterjack toad (*Bufo calamita*). Eggs may be laid preferentially on some plant species (e.g. *T. cristatus* on *Veronica* spp.; Leicester Polytechnic 1986), so closely linking plants to amphibia. Newts spend more time in water. Toads (*Bufo* spp.) are the most terrestrial, frogs (*Rana* spp.) intermediate.

Amphibia are damaged by impacts such as acidification, drying, and degrading vegetation.

Adding lime is destructive to the vegetation and so, long-term, to the habitat. It is always difficult when different biota require different treatment, as when a mire has acidified with air pollution, and *Bufo bufo* needs lime to survive. If liming is done, it should be downstream, on a small scale, in a place with drainage (and drainage that permits loss of the lime before water reaches the stream) (Haslam 1994).

Microorganisms and fungi

Information is inadequate, some of it being listed in Table 5.5. In acid bogs, fungi are more important than bacteria, which are deterred by the low pH. Most fungi succumb in very wet conditions, and lignin-decomposing basidiomycetes, in particular, are not found

Table 5.5 Wetland microorganisms (from Patten 1990; Verhoeven 1992; Wells 1978)

(a) Groups	*Pseudomonas*
Aerobic free-living bacteria	*Streptomyces*
(rare in wet)	
Anaerobic free-living bacteria	(c) Commoner fungi in peats
Anaerobic nitrogen fixers	Bogs and fens
Arbuscular mycorrhiza (on *Narthecium*	*Cephalosporium*
ossifragum, some grasses, etc.)	*Chrysosporium*
Arbutoid mycorrhiza (*Arctostaphylus*)	*Cladosporum*
Denitrifying bacteria	*Gelasinospora*
Ectomycorrhiza (*Betula pubescens,*	*Geotrichium*
Pinus spp.)	*Mastigomycetina*
Ericoid mycorrhiza (on *Erica* spp.,	*Mortierella*
Vaccinium spp., *Calluna vulgaris*, etc.)	*Mucor*
Glucose degraders	*Paecitomyces*
Heterotrophic blue–green algae	*Penicillium*
Methanogenic bacteria	*Philophora*
Nitrogen-fixing nodule actinomycete	*Trocoderma*
Frankia (*Alnus* spp., *Myrica* spp.)	*Verticillium*
Nitrate-reducing bacteria	
Nitrogen-fixing Cyanobacter in	Fens the species above and (commonly found)
Sphagnum leaves	*Arthrinium*
Nitrogen transformers	*Cylindrocarpon*
Purple photosynthetic bacteria	*Fusarium*
Rhizosphere nitrogen fixers	*Pseudoerotium*
Sphagnum degraders	*Volutella*
Sulphate reducing bacteria	
Sulphur transformers	(d) Fungi in one British fen include:
Symbiotic algae (with e.g. *Azolla*)	33 Ascomycetes, including *Belanopsis*
Symbiotic bacteria	*excelsior, Ciboria viridifusca*
	8 Fungi imperfecti, including *Botrytis*
(b) Commoner bacteria in peats	*cinerea, Menispora ciliata*
Achromobacter	20 Heterobasidiomycetes, including
Bacillus	*Puccinia calcitiapa, Tremella foliacea*
Chromobacterium	62 Homobasidiomycetes, including
Clostridium	*Amanita muscaria, Russula betularium*
Micrococcus	10 Myxomycetes, including *Arcyria*
Micromonospora	*denudata, Trychia botrytis*
Mycobacterium	5 Phycomycetes, including *Arthrinum*
Nocardia	*phaeosporum, Penicillium claviformis*

in water, although a few occur in swamps. In one blanket bog, the microbial production was 59 g/m^2 for fungi, 33 g/m^2 for bacteria (Patten 1990).

Bacteria are very small, and although there are some taxonomic differences in shape, physiological and genetic factors are more characteristic. Bacteria are abundant in sediments, and common in films on surfaces (see Chapter 8). Bacteria are often the dominants in nutrient and metal cycling. They use organic materials. They secrete enzymes, to work beyond their cells. It is now possible for native (not just the faecal-indicating *E. coli*, etc.) bacteria to be estimated and described (Boon, in Gopal *et al.* 2000).

Bacteria abound in more fertile wetlands, with actinomycetes second, and only low numbers of fungi and yeasts. Bacteria can use many different substrates, and vary with soil pH and moisture. (Most bacteria, however, can only decompose soft tissue, carbohydrates and cellulose.) In a Minnesota forest swamp, the aerobic heterotrophs were $(8\text{--}34) \times 10^6/$ cm^2 in submerged sites, 16–103 for levées and 23–241 for cultivated sites. Drying and clearing peat means a rapid loss of soil organic carbon and moisture, and so an increase in biota. On litter, microorganism – and indeed algal – populations vary with the plant species (Patten 1990). In riverside hardwoods in Switzerland, fungi differ with the nutrient status of the wood (Haldeman 1993).

6 The waters of the wetlands

The Netherlands, once a wetland.

(M. Wassen 1990)

A dry and thirsty land, where no water is.

(Ps. 63.1)

The early seventeenth century drainage was to make summer land. The possibility of rendering it winter land was not recognised.

(from Miller and Skertchley 1878)

through fire and through flame, through ford and whirlpool, o'er bog and quagmire

(W. Shakespeare, *King Lear*)

That custom should not be allowed, of cutting straws in low grounds, sloughy underneath, which turn into bog.

(Dean Swift)

Water in the landscape

Without wet, water, now or in the past, there is no wetland. The details of that water, its level, movement, fluctuations and chemical quality determine not just the major type but also the details of the development of a wetland, its exact habitat type, and to what sort of dry land it can be brought by human impact. The general principles and patterns are shown in Basic Figures 1–14, Basic Charts, and Figure 6.1. Regrettably, in view of impact, it is advisable to study both the pre-drainage and the drained landscapes together. Moving from left to right, on the Basic Figures, there is the rain-fed, rain-created blanket bogs, and the former bog of the moor. A small lake (lochan or tarn), run-off fed, has a reedswamp. At the bottom of the hill, gathered run-off creates a flush, wet and, because the water is moving, bathing the soil, giving a higher nutrient status than if this same water was still. The diagram combines several habitats that could not co-exist in such a short and flat plain. The flood plain itself, once regularly flooded, is now not so. The part near the river is still frequently flooded, and receives water from the river in three ways.

First, water floods from the river itself. This contains silt and chemicals collected from upstream, and ensures the land near the river is on mineral soil.

Second, water can seep through underground. As has been shown in the Rhine (e.g. Carbinier and Ortscheit 1987), when the river water is polluted, and the ground below the river bed is porous (e.g. gravel), seepage can enter and pollute the water's wide flood

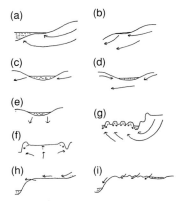

Figure 6.1 Some run-off patterns (based on Novotny and Olem 1994; Wassen and Grootjans, Lloyd and Tellam, Burt, in Hughes and Heathwaite 1995; Wassen, in Wheeler 1995). Arrows show water movement. In (f) and (g) the dotted lines are summer flow when water level is below peat surface.

plain, several kilometres across. The river itself, of course, drains as well as creates the flood plain. After it has been channellised, it drains more.

The poor fen is unsilted but wet, fed primarily by the run-off from the hills above.

Over on the right-hand picture, Basic Figure 5b, the same applies to the flooding and seepage from the river, except this river brings more nutrient-rich water, and nutrient-rich silt. The fen has a spring-line at the base of the chalk hills, and the fen near that is spring-dominated, so is a calcium-dominated fen. Towards the river the calcium domination is relaxed, the peat becomes increasingly silted, and the habitat nutrient-rich. There is here porous rock, an aquifer, below and beside the flood plain. By the hill groundwater discharges. A discharge wetland occurs. Further in, water soaks down into the chalk, refilling it, so it is a recharge wetland. Over to the left, the highland rock below both hills and flood plain is non-porous and impermeable, bogs are neither discharged nor recharged through the impervious rock below. There can, however, be thick soil, say glacial deposits, above impermeable rock. This can contain shallow groundwater, recharging from rain and run-off upslope, and discharging downslope (Figure 6.1).

Raised bogs can form on this rich flood plain as easily as on the nutrient-poor type. Bog is not, and post-glacially has not, formed on soft limestone (which easily dissolves, and neutralises bog acids) in the drier south and east.

There is also the valley bog of New Forest type, fed by water passed through very acid and nutrient-poor sands. This comes both as direct run-off and as groundwater springing or seeping into the bog downslope. The springs do not come from deep groundwater, but from shallow water within the New Forest Sands. The nutrient status rises only by the stream, which carries nutrients. The bed and banks are in mineral soil, so pick up nutrients, even if only little. Stream water moves, so, as already explained, has a more nutrient-rich influence than still water.

Moving next to the drained Basic Figure 7, the bog and moor peat is being eroded. Being dried, cracked, no longer cohesive, the peat is easily washed off in storms. The rate varies with the efficiency of the drainage. The lochan has its water led off. Depending on what is left, there may be more wetland if reedswamp conditions apply over a greater part of the bottom, or less wetland if the lochan dries out. The river is embanked, its water level is now lower, and the river no longer normally floods over the wetland. In porous ground, river water may seep out as before, depending on the relative water levels of river and flood plains.

The flush is drier, but still exists, as there is still run-off. The flood plain is now dry, and, depending on economic and political will, could be used for arable. The peat nutrient status has increased with the drying, but much fertiliser would still be needed.

On the right-hand picture the same applies to the river and flood plain. River flooding has stopped, the plain is dry, and, because it is more nutrient-rich than that on the left, can be more easily converted to arable. Once arable, peat is not only dry but also disturbed, and is oxidised, mineralised by exposure to the air without the protection of water. It is then lost, wasted. Ground level falls – so more drainage is needed. Therefore, more peat is lost, until only the mineral soil below is there. Ground level is so much below the original river level that further works are needed to prevent flooding from the river: further lowering of the river level, further raising of the flood banks.

Grassland is usually wetter than arable, and even if as dry is not as disturbed, so both the nutrient status (excluding fertiliser) and loss are less affected. In pasture, the soil is kept trampled down, so more coherent, so both wetter and less nutrient-rich. Water table can be stabilised by constant spring water, or permanent ditches criss-crossing the land.

Drainage does not change a deep groundwater spring (for decades or centuries). This water is unaffected by what happens up on the surface, and as rain still falls the aquifer is recharged from it. Until, that is, the rain, instead of mostly soaking into the aquifer, runs off because of intensive agriculture or concrete-like surfaces, and mostly goes to river and sea instead of the aquifer. The aquifer level sinks.

That scenario, however, is not the one portrayed in the picture: here water is being abstracted for mains supply – as is happening increasingly over Europe. This ordinarily removes a lot more water than is annually added by rainfall (especially when the recharge water is decreasing). The water level in the aquifer is lowered: lowered most, of course, close to the borehole, and to a decreasing amount further away. Springs above the current level of water in the aquifer, of course, no longer flow. What had been a calcium-dominated fen no longer receives the limestone water, and the calcium dominance relaxes. The fen has also been drained, and from both causes, the nutrient status increases.

Basic Figure 8 summarises the water resources of wetlands like these. The flood and flow, storage and regulation control apply only where flood water can run onto the flood plain in storms. The value is great: but like all such, with wrong planning the reverse can occur. Floods stored in wetlands can be allowed to sweep downstream instead of easing out slowly with no damage to property or people. Erosion control follows flood control.

Water purification is an extremely important function, discussed in Chapter 8. It occurs wherever water passes through soil, around plant shoots or otherwise finds microorganisms.

Water abstraction is for mains water or irrigation (or, sometimes, for industry). Mains water is taken from wherever there is, first, enough to make it worth tapping on a commercial scale, and second, where it is not too polluted (where it is possible and not unduly expensive to clean to drinking water standards). It can therefore be taken from bog run-off (provided the bogs stay wet), from streams and rivers of sufficient size, and from water stored underground.

Clearly there is a conflict of interest between wanting to get rid of water for more farming, construction, and so on and wanting to conserve it for mains supply. This conflict is only being realised slowly. It is too easy to think the aquifer and river water may be lowered, but never exhausted.

However, the previous paragraph presumes that the only point of water, whether present or absent, is for direct human use! What about the ethics of having wetlands? The right of flora and fauna to exist? And the human pleasures of aesthetics, recreation, study? The culture and tradition? This adds a third dimension to the equation. This has at least been recognised for some time: but recognition is not solution, and loss of water is the main cause of wetland loss and destruction. Even conservation interest is still primarily for the individual wetland, for the Site of Special Scientific Interest (SSSI), for the National Trust Reserve. The idea that whole catchments should be managed to give sustainable water to wetlands is in its infancy.

A given site is often governed by hydrology a long way off. Lowering water table for agriculture usually sets off changes in Reserves: increased vertical fluctuations, decreased residence time of groundwater in the root zone, changed chemistry of the soil water, etc. Organic substances in the soil may start releasing nutrients to the root zone (mineralisation, etc.), or absorbing them (wash-out of nitrogen, phosphorus, potassium, fixation of phosphorus). This leads to changes in vegetation (Grootjans 1985).

Soils and drainage

Soils that are wet, but not flooded, have their wetness determined by the relationship between the amount of water coming in and that going out. Peat holds water. This is why bogs can develop, they retain the incoming rain, and can live and increase even when that rain is low. The other extreme is gravel, where water drains out very rapidly. The soil can be wetted, but cannot remain wet unless the general water table is at that level.

Peat and clay are both water-retaining soils. Both have (of course) high overland flow rates, both have high flow rates within the surface soil if this is dry (and, for clay, with holes), and low flow in the saturated zone below. Peat has the two conveniently divided layers of the active acrotelm above (note the three 'a's!), and the compact chemically-quiet catotelm below. Wetness varies in the acrotelm, water level usually rising in winter, falling in summer. The catotelm is permanently saturated, so has much water storage, and low flow. Fibrous peat has higher hydraulic conductivity than amorphous, and completely humified and squashed catotelm rates may be even lower. This is because the acrotelm is thin, and the catotelm has so little water movement, non-storm outflow (baseflow) from bogs is little (Burt in Hughes and Heathwaite 1995).

Soft, muddy soil, and indeed most peats store a great deal of water. When peat is dry, some of this water is lost, and the peat shrinks. In Holme Fen in East Anglia, the seasonal rise and fall is about 50 cm. This is rare peatland where there is a firmly set post with reference levels (Gilman 1994).

In clay, there are pores and cracks, and water movement depends on both. Larger pores mean swift flow when otherwise saturated soil moves water very slowly. These 'holes' in the solid clay are therefore crucial for water movement. When soil is saturated with water, all flow is through these larger holes. The top soil, where drier, is finely structured and permeable, the subsoil is neither (Burt, in Hughes and Heathwaite 1995).

Water soaking in from above, infiltrating, in both peat and clay, moves through these holes, when the soil is not saturated. Rain movement also varies with the type of surface (Figure 6.1), whether rain can collect on the ground before it has a chance to properly soak in, and whether there is much overland run-off. If water is caught into pools, then the later run-off will be mostly from these pools: from the surfaces saturated with water (Burt, in Hughes and Heathwaite 1995).

Drainage in peat is usually by open ditches, but in hill peat may also be by narrow slots (gripping). In clay there is a network of pipes (earthenware pipes, clay tiles, perforated pipes) connecting the ditches. Such under-drainage almost stops overland run-off: water drops into the pipe instead (Burt, in Hughes and Heathwaite 1995).

The vegetation patterns of free-draining and ill-draining wetlands differ (Basic Charts K–O), unless both are flooded, when the reedswamp pattern will be controlled by factors other than the potential water movement. Draining ability is linked to water fluctuations, to some extent. This and the water-holding ability affect drainage.

Free-draining soils have variations in soil texture, and therefore in draining ability, of wetness, of different soil types. Different species have different preferences in water regime, of course, so different tree species occur in different habitats in riverine (flood plain) forests (e.g. Hughes *et al.* 1997).

Vertical fluctuations

Water regime selects vegetation. Vegetation responds with extreme sensitivity (though not necessarily to the present regime).

As water level often does vary greatly in wetlands, the plants and animals have to be adapted to such changes. Water level changes may be natural, due to such factors as weather, and overall seasonal climatic and sea level changes. They are also – and, at present, much more – due to human impact. This takes the form of drainage, either on the wetland or on the river beside (lowering bed level, regulation of flow), and of any impact leading to the silting and filling or erosion and scour of the wetland around. Water level change also comes from peat extraction and from vegetation removal. Both alter the amount of water reaching the soil surface and the soaking-up ability of that soil surface (Gopal, in Patten 1990).

Many organisms, both plant and animal, need different water levels for different stages of their life cycles. Many plants need dry conditions for germination and establishment – and dry at the season of germination, too. Various wetland species tolerate winter but not summer flooding. Many insects need wet conditions for a larval stage, e.g. dragonflies, as also do amphibians. Seedlings and (adult) land plants can be drowned, larvae and wet-loving animals can be dried up. Mosaics undoubtedly favour species diversity. The larger animals, mostly birds and mammals, have the advantage of being able to move away from unsuitable water levels: an advantage that is nullified if there are no suitable ones in the vicinity.

The effect of fluctuations varies greatly with their relation to water level. The most crucial are those around the water level, those that make the difference between a flooded and a non-flooded habitat. Changes in level below that of waterlogging in, say, arable, are irrelevant. Surface-living species do not alter with changes between 1 m and 1.5 m down.

While 5 cm of rainfall, or its equivalent in run-off, raises the level of a flooded wetland by 5 cm, in an unflooded wetland the rise will be much more. The water goes into soil pores, etc., but the soil already occupies space. Therefore, a given input or loss of water has the largest effect on water level around soil level.

Different species have different physiological responses to, and therefore tolerance or preference of, flooding. Such responses, however, may differ in seedlings and adult plants. In addition, the establishment of a given vegetation type depends on there being enough suitable habitats for that establishment both over the surface of a wetland and down the years. Finally, that vegetation succession, change with time in the same place, depends on the geomorphic development of the wetland: ability to develop raised bog, thick sedimentation, and so on (Mitsch and Gosselink 1993). Tables 6.1–6.7 illustrate vegetation and other biota in relation to water level, and the effect is marked. Water quantity, water level, is the primary determinant of wetland, and of any given wetland.

Other habitat factors influence vegetation too. For instance, a middle–high nutrient fen shallow-flooded in winter, waterlogged or moist in summer, may bear:

1 *Phragmites australis* reedbed (with appropriate management, with or without clear lateral flow);
2 *Carex paniculata* fen (with appropriate management, and with a little lateral flow);
3 a mixture of the above (probably after management ceases);
4 *Cladium mariscus* fen (with appropriate management and probably at the more aerobic end of the habitat;
5 *Typha* spp. fen (with appropriate management and probably at the less aerobic end of the same);
6 *Salix cinerea* carr (if management, tussocks or chance summer drying permit establishment);
7 *Frangula alnus* or other shrub carr (as last, and perhaps in more ancient, and aerobic or more calcium-rich areas);
8 *Alnus glutinosa* wood (as 6 above, probably more aerobic);
9 tall-herb fen (with appropriate management or its lack);
10 short-herb fen (with appropriate management);
11 wet grassland (with appropriate management);
12 other allied communities and intermediates.

Here it is specified the water level is the same, and the chemical and other water conditions vary little. Management and other habitat factors differ, so determining the expression of that wetland vegetation. Therefore, there are the overlapping ranges of habitat of so many of these types.

A single fen may bear one, a few, or many plant communities, all within the physiognomic category of 'fen' and separated by one or more of:

water level, average
water fluctuations, vertical
water movement laterally
water source, so water chemistry
soil texture and water-holding ability
soil chemistry
management, past, recent and present
accidental occurrences of conditions suitable for invasion of dominant species plus propagules of those species (Haslam 1994).

Half of these are directly to do with water, two more have probably been determined by the water regime. Management has been determined by water regime (thatching reeds were cultivated in wetter places than osier beds, on average). Management has also determined that water regime (by drainage, flood-wash, abstraction from underground, etc.). Finally, one of the relevant factors for establishment is water regime.

Table 6.1 Typical water levels of vegetation types of the National Vegetation Classification (from Rodwell 1991a,b, 1992, 1995)

Levels given in centimetres above or below ground level. Divisions arbitrary.

(1) Very wet, usually flooded

+25 to +150	*Scirpus lacustris* swamp **S8**
+2 to +100	*Carex rostrata* swamp **S9**
+10 to +40	*Carex vesicaria* swamp **S11**
+10 to +50	*Sphagnum auriculatum* bog pool **M1**
+10 to +50	*Sphagnum cuspidatum/recurvum* bog pool **M2**

(2) Wet

to +40	*Carex elata* swamp **S1**
+40 to −15	*Cladium mariscus* swamp **S2**
+1 to −2	*Phragmites australis* swamp, etc. **S4** (+400 to −400, SMH)
to +20	*Carex riparia* swamp **S6**
to +20	*Carex acutiformis* swamp **S7**
to +50	*Eleocharis palustris* swamp **S19**
+10 to −40	*Carex rostrata–Potentilla palustris* swamp **S27**
+20 to −50	*Salix cinerea–Galium palustre* woodland **W1**
+5 to dry	*Eriophorum angustifolium* bog pool **M3**
+10 to −20	*Carex echinata–Sphagnum recurvum/auriculatum* mire **M6**
+5 to −20	*Carex cursta–Sphagnum russowii* mire **M7**
+40 to −20	*Schoenus nigricans–Juncus subnodulosus* mire **M13**

(3) Very damp

0 (average)	*Phragmites australis–Peucedanum palustre* fen **S24**
0 to −40	*Phragmites–Eupatorium cannabinum* fen **S25**
+20 to dry	*Alnus glutinosa–Urtica dioica* woodland **W6**
Moist to flood	*Cynosurus cristatus–Caltha palustris* grassland **MG8**
Moist to flood	*Holcus lanatus–Deschampsia caespitosa* grassland **MG9**
Moist to flood	*Festuca rubra–Agrostis stolonifera–Potentilla anserina* grassland **MG11**
0 to 30	*Carex rostrata–Sphagnum recurvum* mire **M4**
0 to 25	*Carex rostrata–Sphagnum squarrosum* mire **M5**
Wet	*Carex dioica–Pinguicula vulgaris* mire **M10**
0 (?)	*Carex rostrata–Calliergan cuspidatum/giganteum* mire **M9**
0 to −15	*Calluna vulgaris–Eriophorum vaginatum* mire **M19**
0 to −15	*Eriophorum vaginatum* mire **M20**
0	*Scirpus caespitosus–Eriophorum vaginatum* mire **M17**
0	*Erica tetralix–Sphagnum papillosum* mire **M18**
0	*Narthecium ossifragum–Sphagnum papillosum* valley mire **M21**
Wet to drier	*Juncus subnodulus–Cirsium palustre* fen meadow **M22**

(4) Damp

0 to −40	*Phragmites–Urtica dioica* fen **S26**
−5 to drier	*Salix cinerea–Betula pubescens–Phragmites* woodland **W2**
−5 to drier	*Salix pentandra–Carex rostrata* woodland **W3**
−10 to drier	*Betula pubescens–Molinia caerulea* woodland **W4**
Moist to drier	*Alnus glutinosa–Fraxinus excelsior–Lysimachia vulgaris* woodland **W7**
Moist	*Holcus lanatus–Juncus effusus* rush pasture **MG10**
Moist to waterlogged	*Festuca arundinacea* grassland **MG12**
Moist	*Juncus effusus/acutiflorus–Galium palustre* rush pasture **M22**
Moist to drier	*Molinia caerulea–Cirsium dissectum* fen meadow **M24**
Moist	*Molinia caerulea–Potentilla erecta* mire **M25**
Moist	*Molinia caerulea–Crepis paludosus* mire **M26**
Moist	*Filipendula ulmaria–Angelica sylvestris* mire **M27**
Moist	*Iris pseudacorus–Filipendula ulmaria* mire **M28**

Table 6.2 Water level and ecological requirements of some breeding lowland wet grassland birds (from Newbould and Mountford 1997)

Birds	Water levels[a]		Vegetation structure		Soils	Notes
	Breeding	Feeding	Breeding	Feeding		
Garganey	Damp pasture	Shallow pools			—	
Shoveler	Damp pasture	Shallow pools			—	
Spotted crake	0 to −15 Wet margins	0 to +15 Wet soil/ shallow water	Lush aquatic vegetation	Lush aquatic vegetation	—	Wet margin species
Lapwing	−35	−35	10 to 15 high	10 to 15 high	—	Drier meadow
Redshank	−35 to −20	−35 to −20	up to 50	up to 50	Peat/silt	
Black-tailed godwit	0 to 20	−20 to shallow pools	10 to 15 high	10 to 15 high	Peat	<6 kg force to probe
Snipe	0 to 20	−20 to shallow pools	to 80 high	to 80 high	Peat/silt	
Ruff	0 to 20	−20 to shallow pools				
Curlew	−15 to −30	−15 to −30	Long grass >80 high	Long grass >80 high	—	
Yellow wagtail	−15 to −35	−15 to −35	Tussocky grass to 80 high	Grazed sward to tussock grass	—	

[a] Water levels in centimetres above or below ground level.

Table 6.3 Water level requirements of some reedbed birds (from Newbould and Mountford 1997)

| Bird species | Water levels required (depth of water, cm) | |
	Breeding	Feeding
Bittern	10 to 25	10 to 25
Marsh harrier	10 to 30	10 to 30
Bearded tit	(Not significant)	10 to 30
Cettis warbler	(Not significant)	—
Savis warbler	(Not significant)	10 to 30
Reed warbler	10 to 30	10 to 30
Sedge warbler	—	10 to 30
Water rail	0 to 30	0 to 15
Teal	0 to 30	0 to 20
Shoveler	Terrestrial	0 to 30
Mallard	Terrestrial to 20	0 to 35
Tufted duck	Terrestrial	up to 250

Table 6.4 Dragonfly species vulnerable to small changes in water level (from Newbould and Mountford 1997)

Species	Habitat	Water level (cm)
White-faced dragonfly (*Leucorrhinia dubia*)	Peat bog	10 to 50
Black darter (*Sympetrum scoticum*)	Marsh or peat bog	10 to 50
Small red damselfly (*Ceriagrion tenellum*)	Marsh or peat bog	10 to 50
Scarce ischnura (*Ischnura pumilio*)	Peat bog or soil	2 to 20
Southern coenagrion (*Coenagrion mercuriale*)	Peat bog or soil, stream edges	2 to 10
Northern coenagrion (*Coenagrion hastulatum*)	Marsh or peat bog (Scotland only)	2 to 50

Hydroperiod affects the availability and transformation of nutrients. Waterlogged soils suffer from lack of oxygen and become reducing. Soluble reduced toxins increase. Typical wetland species can bring oxygen to their roots (e.g. *Phragmites*), and typical land plants avoid such habitats. Species tolerance to ferrous ion is roughly associated with their distribution in wetlands (Wheeler, in Hughes and Heathwaite 1995). This lack of oxygen can be avoided by those species growing on tussocks (e.g. *Molinia caerulea* on *Schoenus nigricans*, *Rosa canina* on *Carex paniculata*), or by those rooting shallowly in an aerated top soil (e.g. *Parnassia palustris*). Shallow root systems, however, are a liability for wetland plants in places liable to dry in summer. Plant extension into drier habitats may be limited by competition, management, change of nutrient status on drying, etc., as well as by water shortage. Their extension into habitats wetter than their normal range may be determined by change of substrate (e.g. silt to bouldery), shade (of tall reedswamp species), waves and erosion, change of chemical status on wetting, management, etc., as well as by drowning the roots or shoots.

Each species has a water level range in which it can grow well, in which other factors (chemical, soil texture, etc.) must be really adverse if the species is to be eliminated. On each side of this optimum range is a peripheral belt, one drier, one wetter, in which the species can grow if conditions are good. An example is *Cladium mariscus* in dry fens cut

Table 6.5 Potential value of indicator plant species for assessing fen water level change (after Wheeler and Shaw 1992)

1   Some past records of species occurrences are available for many sites, whereas there are usually no good water level data
2   Species presence can usually be assessed by 'spot measurements' – that is they can be determined on a single occasion – while repeated, and often frequent and long-term, water-level measurements may be needed to adequately quantify water level flux in fens
3   The species record can help to assess the significance of hydrological changes. It may be argued that the conservational significance of hydrological change in fens relates not so much to its *magnitude* as to its *impact* upon the biota
4   The presence of indicator species effectively integrates the overall effects of water-level flux upon the ecosystem. The impact of drying may be less a *direct* than an *indirect* effect, operating upon other component species of the community, or upon other concomitant environmental changes (e.g. nutrient release)
5   The survival of indicator species effectively integrates the long-term effects of water-level flux: many species may be able to survive periods of temporary dehydration (within certain limits); it is notoriously difficult to relate rigorously measured values of a strongly fluctuating variable such as water level to vegetation composition, especially when it is uncertain what time-period needs to be considered

So the problems of selecting species as indicators of water level changes are comparable to those of assessing the susceptibility of fen species to dehydration. The 'ideal' properties of a water-level indicator-species may be compared with the actual properties of potential indicator-species, insofar as they are known.

1   There are several uncertainties surrounding the use of indicator-species to assess hydrological change
2   It is unlikely that a single indicator-species can be found. Rather, the observed responses of a number of species will be needed
3   It may be difficult to disentangle vegetational effects of dehydration from the effects of some other forms of environmental change

every 3–4 years. The cutting is done to get a commercial crop from the *Cladium*, but also prevents the development of the carr, which would shade it out. These woody plants are therefore prevented, by management, from growing in water levels to which they are suited. At the other end, *Phalaris arundinacea*, which normally requires to be summer-dry and to have shoots intermittently above water in winter, can grow totally submerged under water all the year, in limestone streams.

Inadequate work has been done on the ecotypes of most wetland plants. Where they have been looked for, for example in *Phragmites,* they are there: the different strains adapted to, and growing in, different but overlapping environmental conditions. (Even in *Phragmites,* water level adapted ecotypes are not studied.)

Extremes matter. It is the extremes, whether of wetting or drying, that put the greatest stress on an organism and community. It may be these extremes that delineate and create a community, or that mean it does not occur in a habitat that appeared perfectly suitable for it during, say, a decade of study. Or the extremes, by wiping out one community can allow another, more opportunistic and less competitive, to enter. The amplitude of the extremes may be less important as a determinant than frequency and duration. Drawdowns create harsh environments, and reservoirs liable to these are notably short of reedswamp (Gopal, in Patten 1990). Alternatively, the amplitude may be the most important, and one population may be wiped out by an extreme of, say, drought or fire.

Table 6.6 Water levels, soil fertility and plant species (selected from Wheeler, in Hughes and Heathwaite 1995)

	Ellenberg moisture value*	Mean water table (cm)	Minimum water table (cm)	Maximum water table (cm)	Mean soil fertility (mg)
Carex diandra	9	4.3	−14		9.1
Drosera anglica	9	1.9	−6		6.8
Menyanthes trifoliata	9	−0.2	−59		9.6
Carex dioica	9	−0.9	−16		6.1
Galium palustre	9	−2.2	−79		15.5
Equisetum fluviatile	10	−2.3	−50		15.6
Pinguicula vulgaris	8	−2.8	−35		6.8
Veronica beccabunga	10	−3.4			23.7
Sphagnum subnitens		−4.3	−50		9.0
Caltha palustris	8	−45.0	−50		16.7
Species of drier and more disturbed habitats					
Phalaris arundinacea	8	−6.9	−78	+15	20.9
Epilobium hirsutum	8	−13.1	−100	+7	17.7
Urtica dioica	6	−20.2	−100	+11	23.3
Galium aparine		−26.4	−100	+11	24.9

Water table values are expressed relative to the soil surface (−, below; +, above); they are mean values of all water levels measured associated with each species in soligenous fens in Britain. Soil fertility was estimated phytometrically using seedlings of *Epilobium hirsutum*; values refer to mean dry weight of seedlings (mg) and represent mean values of fertility measured associated with each species in soligenous fens in Britain. For a more complete species list, and more habitat data, see Wheeler and Shaw (1992).
* See Table 6.7.

Table 6.7 Definitions of Ellenberg's 'moisture value' (from Ellenberg (1974), in Wheeler, in Hughes and Heathwaite 1995)

Feuchtezahl		*Occurrence of plant species in relation to soil moisture or water level*
1	Starktrockniszeiger	in extremely dry soils, e.g. bare rocks
3	Trockniszeiger	in dry soils
5	Frischezeiger	in fresh soils, i.e. intermediate conditions
7	Feuchtezeiger	in moist soils that do not dry out
9	Nässezeiger	in wet, often not well-aerated soils
10	Wechselwasserzeiger	in frequently inundated soils
11	Wasserpflanze	water plants, with leaves mostly emergent
12	Unterwasserpflanze	submerged plants, usually entirely immersed
X	Indifferentes	with broad amplitude or with different behaviour in contrasting habitats
~	Wechselfeuchtezeiger	in fluctuating moisture conditions
=	Überschwemmungszeiger	in soils that are fairly regularly inundated

Water level fluctuations in the year range from several metres in the flood plains of large rivers (e.g. the Danube) to several centimetres in spring-controlled fens (where, because there is a continual inflow, people have seen to it that there is an outflow, and water is maintained at the level of the outflow). At the near-stable end of the fen range, *Schoenus nigricans* is a typical dominant of calcium-dominated fens. It can live well with more fluctuation, e.g. 30 cm, and judging by older records (e.g. Ashfield 1861, 1862), even more. But the total community of the stable habitat is now found only in that stable habitat. The associated mosses can stand little flooding, and for example *Anagallis tenella*, pimpernel, even less. *Hydrocotyle vulgaris* needs a wet habitat, though not much flooded, and *Phragmites* (if left to competition) would not grow in a much drier place. Therefore, it is the total community, the floral composition, the species balance, the habit of the different species, that integrates the water regime. *Carex elata* likewise grows with, for example, 30 cm annual fluctuations, but does very well with fluctuations of over 1 m (with its associated species, e.g. *Potentilla palustris*). There are plant and animal communities characteristic of all possible wetland water regimes. Some are species-rich, some species-poor, some with generalist (widespread) species, others with specialist species: all adapted to different wetland conditions.

Sideways movement

As well as moving vertically, up and down, water may also move sideways, laterally. Rivers flooding over their plains have water moving laterally as well as rising vertically. This is enhancing for species benefiting from the fertilising silt, or requiring the water. It is enhancing for waders and death to their prey, surfacing to try to avoid drowning (see Chapter 5). In this habitat, the lateral movement is not relevant in itself, only for the silt, etc. it brings laterally.

Spring and seepage waters rise, but pass laterally downstream. Run-off, both overland and underground, moves laterally downstream. These sideways-moving waters bring in their chemical load, from upslope or from springs. They bathe the roots in the waters: so raising their relative nutrient status. They wash out chemical solutes there, whether this is helpful (removing anaerobic toxins, lowering nutrient status, etc.) or the reverse.

Lateral movement also comes from the drainage dykes intersecting so many fens and marshes (except where deep drainage has rendered superfluous all but a few). These are now also found all too often in bogs.

In lakes with fringing reedswamp that are large enough for waves to develop, the soil disturbance and erosion waves may bring, or may hinder and limit, reedswamp. In the larger bog pools, waves that erode the sides cause local increases in nutrients from decomposition.

Vegetation as an indicator of water regime

Plants, as individual species or as communities, are valuable indicators of water regime. The community integrates and assesses habitat, not just water, but all other factors, including the starling roost last year and the horse trampling last month.

Even considering only the water, long-lived and dominant plants may have become established under a different regime. Because such plants control the community, and are long-lived, the community may not change until the dominants die. This applies only partly to open communities like the *Schoenus* one described earlier, where invasion can occur, and the community can be succeeded before the tussocks die (see Figure 10.8). Lack of young plants of the dominant is, necessarily, a feature.

To use vegetation for interpretation, therefore, needs the knowledge of, first, what facets of the community are due to water regime and what to other factors, and second, what time period the community represents, whether this year, a decade ago, a mixture of the two, or whatever. This is the basis of all ecology: to understand plant behaviour well enough to interpret the plant community and habitat of the given site. The principle is:

'the plant is always right' (A. S. Watt). The plant integrates and responds to all facets of the environment.

What about ecotypic variation? Geographic variation (different climates, altitudes, latitudes) causing changes in behaviour? And it should be added here, what is water regime? One measurement? Weekly through a decade? In between? And what should be measured? Ecologists tend to measure water level – the distance the water table is above or below the ground surface. Geographers tend to measure water pressure (with piezometers).

A frequent measure is Ellenberg's moisture values (see Table 6.7, and Wheeler and Shaw's application in East Anglian Wetlands, Table 6.6). The Ellenberg values were developed in central Europe, and take a rather simplistic view of plant response. The Wheeler and Shaw values are based on wide research in Broadland, East Anglia (but not in other regions).

Case studies

Three fens near Mildenhall, Suffolk (Haslam 1960)

The River Lark was developed early (early historic and pre-historic sites; mills, fisheries, meadow in the *Domesday Book*, 1086), and would have had early modifications for transport to the Abbey town of Bury St Edmunds. Channellisation and staunches (sluice and weir at same point) enough to make a legal Navigation were done in the seventeenth century. As there is both open water mud and *Phragmites* peat in the Tuddenham–Icklingham stretch of the flood plain, conditions have varied. In recent centuries, of course, river flooding has been decreasing and then (effectively) stopped, and a deepened river drains rather than waters the fens.

The Cavenham (English Nature) Fen (Figures 6.2, 6.3, 6.8, 6.9) extends from the river (from which it is embanked and ditched) to the sandy heath around. The fen is drained,

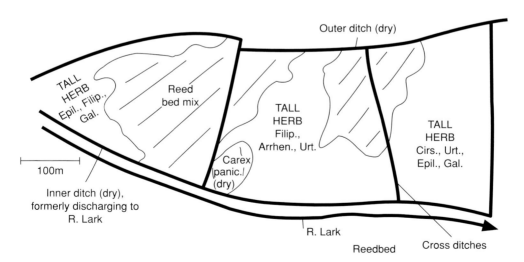

Figure 6.2 Cavenham fen, Suffolk (1958), plan (after Haslam 1960). Abbreviations: Arrhen., *Arrhenatherum elatius*; Car. acut., *Carex acutiformis*; Car. panic., *Carex paniculata*; Cirs., *Cirsium arvense*; Epil., *Epilobium hirsutum*; Filip., *Filipendula ulmaria*; Gal., *Galium aparine*; Urt., *Urtica dioica*.

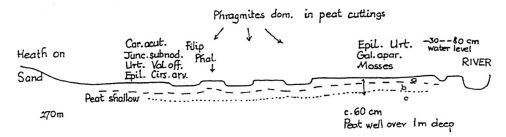

Figure 6.3 Cavenham fen, Suffolk (1958), profile (Haslam 1994). The soil above is dark oxidised, friable, invertebrate-rich. It is above normal water level but normally saturated in winter. The soil below is pale brown fibrous reed (*Phragmites*) peat, always saturated. The water level is 60–90 cm down. There is a freshwater organic clay layer below, in part of the fen. Abbreviations: see Figure 6.2. Val. off., *Valeriana officinalis*; Junc. subnod., *Juncus subnodulosus*.

formerly much used: reed cutting until the 1920s, peat to the early twentieth century, other general crops. The fen has drainage ditches, and is drained. The former intense management prevented carr development in either traditional times or during the first drying. Tall-herb fen, because of its invertebrates and its shading, permits only slow and scattered woody plant invasion (see Chapter 10). In the 1990s, the Fen bears pasture: the drainage and loss threatened in the 1950s has come to pass. The description here is 1950s.

The top 30–40 cm of peat are dark, and oxidised, friable with numerous invertebrates near the surface. Even winter water level seldom reaches the surface. The lowest (i.e. wettest) parts are dominated by *Phragmites*, the rest is tall-herb fen.

The fen is primarily fed by run-off (water levels are low, little comes from the river). Run-off moves only slowly to and across the fen, so that it may take a week for the water from a heavy storm to peak at the bottom of the fen. In one storm, water level rose by an average of 23 cm all over the fen, and two days later the level was falling in the outer part, but still stable near the river, with new water still arriving.

The profile is determined as much by management as by nature, and the water regime is simple.

Icklingham (Poors) Fen (Figures 6.4, 6.5, 6.8, 6.9) is on the other side of the river, and differs in water regime and history. There are two seepage areas along the upland margin, bringing chalk–sand water, aeration and stability. The drier parts have tall-herb fen and dark friable invertebrate-rich upper peat, as in Cavenham fen, although the Icklingham communities are less diverse. Here, however, the level in the centre has been further lowered by peat cutting, and is winter-flooded. It is sometimes also summer-flooded, always summer damp and waterlogged, and communities of *Phragmites australis* and *Carex paniculata* dominate. The former was cut for thatching, fencing, litter, etc., the latter for furniture, litter, etc.

The ditches in this wetter part carry water. Because the fen has seepage inflow, it also has a (working) outflow. The same storm that raised water level in Cavenham fen by 23 cm raised it only by 16 cm here, with a 2–6 cm fall two days later, despite run-off still entering the fen. The through-flow system due to the seepages makes the water level more stable.

In the seepages, water tables are high and stable in winter, with only small rises with heavy rain. Evapotranspiration is little. In summer, however, evapotranspiration losses increase. The inflow is less than this, and water level drops. The less the inflow, the shorter

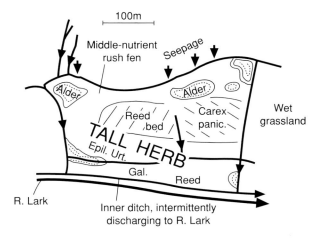

Figure 6.4 Icklingham poor's fen, Suffolk (1958), plan (after Haslam 1960). Arrows show the direction of seepage flow into the fen, of ditch flow within the fen, and of river flow (which is now outside and separated from the fen). Note that the lower-nutrient species are towards the outer edge where seepage water influences habitat (*Carex appropinquata, Filipendula ulmaria, Galium uliginosus, Juncus subnodulosus, Lotus uliginosus*, etc.). Abbreviations: see Figures 6.2 and 6.3.

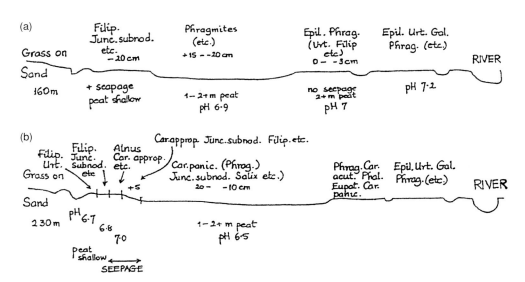

Figure 6.5 Icklingham poor's fen, Suffolk (1958), profile (Haslam 1994). Soil types as in Figure 6.2. Abbreviations: see Figures 6.2 and 6.3.

the wet, stable-level period. Further from the seepage, fluctuations increase, finally becoming nearly as unstable as that of Cavenham fen. Before drainage, the springs presumably ran well all year, and water was high and stable all year.

The river-flooding of the past brought silt and high nutrients to the fen. This source is stopped. The seepage water is fairly calcium-rich, but low in other nutrients, and the

Carex approprinquata and *Juncus subnodulosus–Filipendula ulmaria* communities result. With species such as *Galium uliginosum* and *Valeriana officinalis*, *Filipendula ulmaria*, etc., it is less nutrient-rich than *Epilobium–Urtica–Galium*. The latter is found further from the seepage areas. *Alnus glutinosa* dominates near the seepage areas (more aerobic), *Salix cinerea* away from these (less aerobic). Tree colonisation is easy on *Carex* tussocks, mainly alder on *C. appropinquata* near the springs, mainly sallow on *C. paniculata*, further away. Sallow is even starting to invade *Phragmites* (see Chapter 10). Over the next decade, carr spread fast over the wetter areas. Summer evapotranspiration is about 3 mm/m² per day in (dry) tall-herb fen, 5 mm/m² per day in (wet) *Carex paniculata* and *Phragmites australis*. Evapotranspiration can be less or more than that of open water, 60–121 per cent. Fens lose 40 per cent more water than bogs (Mitsch and Gosselink 1993). Evapotranspiration varies with microclimate, soil moisture, vegetation type and density. Water loss increases with grazing and mowing (dead material decreases loss) (Gilman 1994).

Although *Phragmites australis* and *Carex paniculata* communities both dominate in the wet fen, peat (what is left after cutting) is *Phragmites* only, so the *C. paniculata* is more recent. It dominates where nutrient status is (1) somewhat lower because of more seepage flow and (2) where management encouraged this rather than *Phragmites*. The communities are separated by a ditch: man-made. In the *C. paniculata*, night inflow is less than day evapotranspiration, so summer water level falls.

The water regime and consequently the vegetation are more complex than in Cavenham Fen.

The Eriswell Fens (Figures 6.6–6.9) are near the source of the Eriswell Lode, three miles away, on a tributary to the Little Ouse River. These have source, rather than downstream valley, conditions, with some strong springs. Along the stream, the fen area is mainly wooded, with a series of peat basins, formerly cut for peat.

The strongest springs are near the stream, in basins A and D as shown in Figures 6.6–6.9. Basins B and C, a little further away from the stream, have weaker springs. No doubt

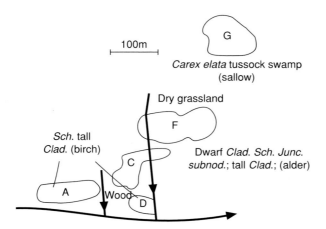

Figure 6.6 Eriswell lode head fens, Suffolk (1958), plan (after Haslam 1960). Basins A and D contain seepages of chalk water, C is less influenced, and G receives no direct chalk or spring water. Abbreviations: Clad., *Cladium mariscus*; Junc. subnod., *Juncus subnodulosus*; Sch., *Schoenus nigricans*.

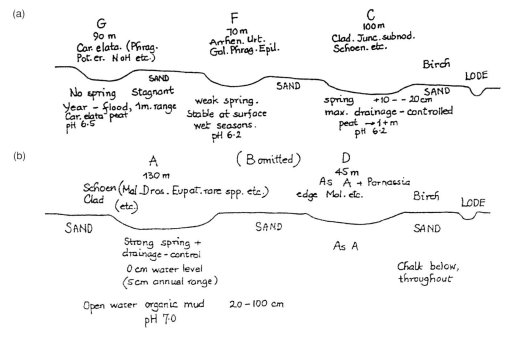

Figure 6.7 Eriswell lode head fens, Suffolk (1958) (Haslam 1994). Upper section through basins G,
F and C, at right angles to lode stream. Lower section through basins A and D, parallel
and beside stream (see Figure 6.5). Abbreviations: see Figures 6.2 and 6.3, and Dros.,
Drosera spp.; Eupat., *Eupatorium cannabinum*; Mol., *Molinia erecta*; Phrag., *Phragmites
australis*; Pot. er., *Potentilla erecta*.

all were much stronger in the past. F is intermediate to G, which is fed by run-off and the
deeper run-off that can also be called shallow groundwater. By the stream, spring water
immobilises phosphate (calcite) and *Schoenus nigricans* communities can dominate, with
nutrient-poor species such as *Drosera* spp., *Sphagnum* and *Parnassia palustris*. *Cladium
mariscus* dominates here, as it does in C also. Calcium is less dominating in C, however.
G is a total contrast, with deep stagnant water and dominant *Carex elata*. Sandy heath
occurs between most hollows. The average water level in all peaty hollows is almost the
same. In the sandy heaths beside, the level is both lower and unstable, astonishingly
unstable so close to the stable levels in the peat.

Vegetation varies with the chemistry and source of the incoming water. Basins A and D
have aerated, particularly high pH spring water with its level stabilised at ground level (by
a complex drainage system). It has *Schoenus* conditions (also suitable for *Cladium*). The
contrasting stagnant G has a fluctuating, non-spring water coming from sand as well as
chalk, and *Carex elata*. In the National Vegetation Classification, these communities, each
so clearly restricted to different habitats here, have, over the whole country, a wider overall
habitat range: there are differentiating environmental factors elsewhere.

In the strongly calcium-dominated basins (A,D) the little woody plant invasion is by
birch (oligotrophic), primarily on tussocks. In the intermediate ones (B,C) it is alder, and
in the stagnant deep hollow G it is *Salix cinerea*, again mostly on the tussocks (here of
Carex elata).

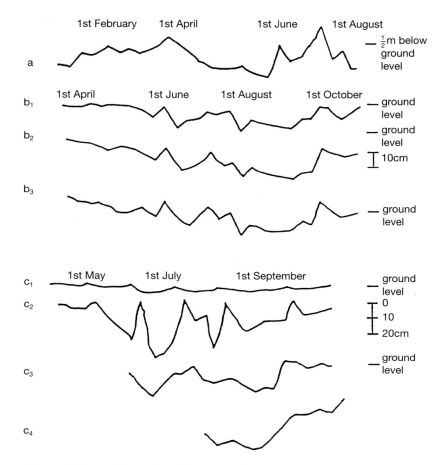

Figure 6.8 Seasonal water level changes in fens (redrawn from Haslam 1960). (a) Cavenham fen, Suffolk: no stabilising groundwater inflow. (b) Icklingham poor's fen: 1, area of maximum seepage, stable high water level maintained until June and after October; 2, less seepage; 3, negligible seepage, but partly stabilised by winter flood. (c) Eriswell lode head fens: 1, basin A, spring-controlled and stable; 2, edge of basin A, on sand, unstable; 3, basin C, weakly spring controlled, not very stable; 4, basin G, not spring controlled, but with (almost) perennial flooding so with longer swings than the sand beside basin A.

The water level in A is remarkably stable, the average annual range being less than 5 cm, with a much stronger through-flow than in Icklingham Fen. Heavy storms raise the level briefly for a few centimetres: very different compared to the effect on the Cavenham and Icklingham fens. Instability increases (spring flow decreases) away from the stream. Further from the main springs the levels become increasingly variable as G has no through-flow. Figure 6.9 shows the evapotranspiration pattern in A: inflow is fully sufficient to replace daily losses.

Soon after this research, the New Lark Cut was constructed through the area, a catchment channel to divert excess flood water from the Fenland.

The water regimes of these three fens show:

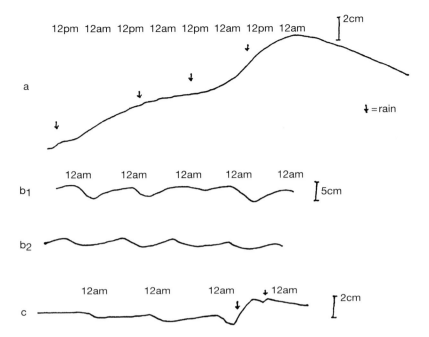

Figure 6.9 Daily summer water level changes in fens (redrawn from Haslam 1960). (a) Cavenham fen: near river, showing slow, high rise after heavy rain. (b) Icklingham poor's fen: 1, tall herb vegetation (*Filipendula ulmaria–Urtica dioica*), May; 2, tussock sedge (*Carex paniculata*) September. (c) Eriswell lode head fens: basin A, showing daily evapotranspiration and the trivial effect of heavy rain in a spring-controlled, drained setting (contrast Cavenham fen).

1   water source is important to water balance, fen chemistry habitat and communities;
2   habitats with regular inflow have (whether man-made or natural) facilities for outflow;
3   different water regimes and water-determined habitats can occur within one fen, these habitats varying in chemistry as well as in water movement;
4   ground level in fens varies with past history (e.g. peat-cutting), and this affects the position of the water table;
5   evapotranspiration losses in summer are substantial, and should be considered in water balance equations. Inflows fully adequate to maintain water levels in winter may be inadequate to do so in summer.

Badley Moor Fen, Norfolk (Gilvear et al. 1990; Gilvear et al., in Hughes and Heathwaite 1995; Figure 6.10)

This is representative of fens on leaky aquifers. Chalk is below, boulder clay around, and the fen is on the edge of a river flood plain in farmland. Chalk-water springs have formed tufa mounds, 4–5 m high, starting 3–4 m down, and these date from around AD 800, when the land was (relatively) densely populated.

The tufa mounds and the flushed slopes beside bear diverse calcareous fen communities. Agricultural run-off flows beside the fen, in the ditches, so has low impact (but see

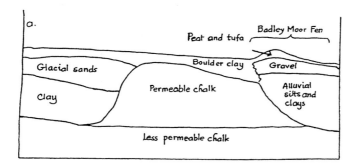

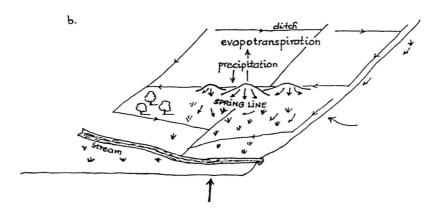

Figure 6.10 Badley Moor fen, Norfolk (modified from Gilvear *et al.* 1990): (a) section; (b) water flow.

Buitengoor Fen, described later). The flood plain is intensively farmed, and watered both by drainage and groundwater.

About 90 per cent of incoming water is groundwater (from April 1988 to September 1989) and 10 per cent is rainwater. The spring areas are therefore very strongly calcium-dominated. The measured flow varied from 366 mm per month per unit area in late summer to 430 mm per month per unit area in April to June. The surface water outflow was 269 to 432 mm per month per unit area, paralleling the inflow. It fell lower with summer evapotranspiration, which varied from 12 to 143 mm, stressing the importance of this in the water balance. This fen occurs because the geology provides springs keeping the surface saturated, and is a calcium-dominated one because of the groundwater chemistry, which also led to tufa accumulation. The water regime is simple, but is due to complex geomorphological patterning. The remaining peat is of no importance in the overall water mechanisms.

Redgrave and Lopham Fens, Norfolk (mostly Harding 1992) (Figure 7.3)

These are at the head of the river Waveney, and are like a larger Eriswell Lode source fen. Early records (e.g. Ashfield 1861, 1867) show that, in the second half of the nineteenth century, there was peat-cutting, the surface was very wet and soft, and in parts could not be passed on foot. By the 1950s (e.g. Haslam 1960, 1965a,b) it had stable surface wetness in the *Schoenus* area upstream, and was – as is usual further from springs – rather more variable downstream. This also was Poor's Land (given in compensation for the enclosure of a common). A major borehole was commissioned, and the river was drained and deep-dredged. A sluice was added, but there was too little input to maintain wetness, and far too little chalk water to maintain calcium dominance (adding rainwater is no substitute, see Buitengoor, p. 143). By 1979 the fen was much too dry. In 1990, there being some doubts among the various organisations, the boreholes were shut down for a trial period. After one week the cone of depression (see Basic Figure 7) had gone, and after four weeks there was once more a dome of water allowing seepage. Obviously all was lost when pumping restarted. With drying and nutrient increase:

Schoenus–Juncus subnodulosus → rush fen meadow
Schoenus–Juncus–Cladium → *Phragmites–Urtica*
Schoenus–Juncus–Carex → *Phragmites* (tall herb fen)
Schoenus–Juncus–Erica tetralix → *Cirsum–Molinia* (shallow peat over acid sand)

Scrub and ruderals also increased.

The 'end point' communities were of course there before, but locally, in dry and disturbed places. There was an equivalent loss in invertebrates, including in Red Data Book species.

The main harm is the abstraction, as that caused the loss of the chemical type of the water, as well as the quantity of water. Drainage added harm. Subsequently, pumping was moved further from the fen.

Catfield Fen, Norfolk

This, like Cavenham Fen, is now mainly fed by run-off (and rain), and has little groundwater. It is partly managed as a commercial reedbed. It has lost much of the original river flooding, which was nutrient-rich. It has decreased its groundwater input, with abstrac-tion. It is kept wet – much wetter than Cavenham Fen – so there is no drying-induced nutrient increase. On the contrary, there has been a loss of nutrient input (including calcium), and quite large mats and bolsters of sphagnum have developed in the centre (furthest from run-off). The Dutch call these '*Sphagnum* reedlands'.

Anglesey Fens (Gilman and Newmon 1980; Meade and Blackstock 1988) (Figure 6.11)

Anglesey has limestone springs giving rise to calcium-dominated and calcium-influenced fens. Rainfall is high in north-west Wales. Springs occur at the bottom of scarps, and open water deposits and peat have infilled basins forming fens. Cors Erddreiniog is well-drained, Cors Gogh only slightly. Past peat-cutting causes variation in ground level, so also in vegetation type. Grazing keeps communities short and open, where cattle can reach and find vegetation palatable. This means shorter *Schoenus* and more *Juncus subnodulosus* than in the Eriswell and Redgrave fens. The usual associates are present, e.g. *Anagallis tenella*, a good moss flora, (fen) *Sphagnum* (*S. subnitens*), *Erica tetralix*, *Carex diandra*, and so on. *Cladium* communities are present too, in wet places. Drier land commonly bears (often species-rich) *Molinia* grassland. *Molinia caerulea* here is associated with some

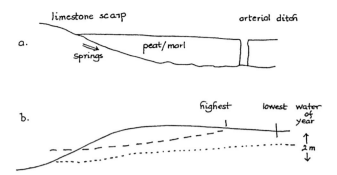

Figure 6.11 Anglesey fens (modified from Gilman and Newson 1980).

calcium influence and partial drainage: less drained than in East Anglia. *Molinia* tolerates considerable water fluctuations, and is favoured by some (not excessive) mineralisation: less than tall-herb fen. While drying peat usually leads to its mineralisation, Wheeler and Shaw (in Wheeler 1995) point out that this does not always happen. It depends on the nature of the tolerance to wetting, the range of water tolerated, the importance of competition in tolerance, the water pattern (timing and duration of flooding), the amplitude of, as well as the mean, fluctuations, whether the existing vegetation is relic, and, finally, what other factors are influencing the water regime. Tall-herb fen is present in drier parts. *Molinia* and tall-herb communities are created by drying and removing calcium domination. Tall herb has peat dry enough to mineralise and become nutrient-rich, and no cattle to squash it down. Both are degraded fen communities.

Ditching is very effective at preventing flooding. However, it intercepts spring water to a depth of about 0.5 m (which, with grazing or mowing, leads to *Molinia*). The low permeability of the deeper peats means that drawdown by the ditches extends over only about 15 m from their edges. For recovery, the limestone spring water would need to be held above ground for long periods.

Wetlands of the River Torridge, Devon (Figure 6.12) *(Lloyd and Tellam, in Heathwaite and Hughes 1995; Murphy et al., in Mitsch 1994; Maltby et al., in Mitsch 1994)*

Devon, like Anglesey, has a high rainfall. These wetlands are without limestone, and only the lower ones, beside the river, have river-flooding. Run-off sinks down in the hills, and seeps out further downslope. This (deep run-off or shallow groundwater, depending on definition) has a residence time of days. In deep aquifers, the time water stays put in the aquifer before it comes back to the surface is likely to be at least decades, and quite possibly (as in the spa spring at Bath) millennia.

The hills are on impermeable mudstone, so in the wet climate much water seeps down valley slopes, and there are many drains and ditches. Most rain in fact comes as overland flow or through ditches. This slope-aquifer is therefore filled (recharged) on the higher ground above the valley, and discharges above the flood plain. These here are perched wetlands. On the flood plain, both rain and run-off soak in, recharging that aquifer. It is a typical (narrow) flood plain, with soils developed from sediments deposited earlier.

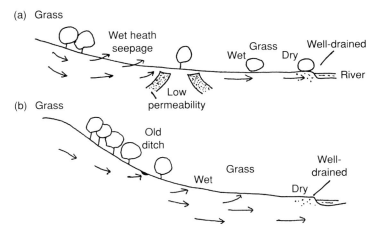

Figure 6.12 Wetlands in Devon (modified from Maltby *et al.*, in Mitsch 1994). (a) Kismeldon meadows, River Torridge; (b) Bradford Mill meadows, River Walden. Water moves on more than one level, seepages are determined by slope, ground texture and human impact.

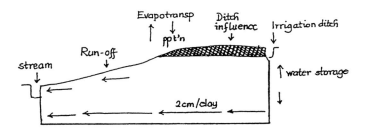

Figure 6.13 Buitengoor fen, Belgium (modified from Boeye 1992).

Buitengoor Fen, Belgium (Boeye 1992) (Figure 6.13)

This is the source of a tributary of the river Nete, in a depression where groundwater reaches the surface, on sandy permeable subsoil. It is a spring fen, a groundwater discharge fen.

Water comes in from seepages, from an irrigation-water channel that crosses one side of the fen (feeding two ditches) and from rainfall. On average there is more rain than evapotranspiration, and much rain runs off the fen. The seepage level keeps depressions and hollows wet, but not higher parts. In dry summers, however, the water table drops, and groundwater does not reach or saturate any of the surface. Run-off is then the only water reaching the depressions. Water level is therefore more variable (depending on uncertain rain, not constant springs). Water chemistry is different, incoming water reflecting the content of the Mol sands through which it runs, rather than that of the deep aquifer below. Because there is then less water, the soil is more aerated (and with a more variable redox). In the long term this would mean the irreversible change to mineralisation.

As the spring water reaches only to the lowest fen levels, just a small further drop would destroy the present hollow vegetation, and sharply change the fen. If the present spring water level was 1 m above the base of the depressions, it could sink, say, 25 cm with alteration only to the detailed mosaics: spring water vegetation would still be there. This again illustrates the importance of relating factors to habitat. Here it is not that a fall of 25 cm would, in itself, be drastic, it is that such a fall would remove all groundwater influence from the fen surface.

To protect the fen, groundwater must be protected. Raising the water level from other sources will not substitute (see Redgrave Fen, p. 141). The fen did not develop under the influence of ditch water or run-off or rain. Only by maintaining the groundwater (preventing damaging abstraction) can the habitat be preserved.

Valleys in the Drenthe plateau, the Netherlands (Grootjans 1985). (Figures 6.14, 6.15)

Only a few near-intact valleys remain. The water sources are:

1   rain;
2   subsurface run-off from farmland, agrochemically polluted, that which stays longest in the land gets closer to groundwater in chemistry;
3   surface run-off in severe storms;
4   groundwater;
5   flooding from the river.

Rain is mineral-low, and is dominated by sodium and sulphate ions. As rainwater goes into the soil, it picks up a small amount of minerals, but conductivity remains low. The polluted inflow has much higher conductivity. As it stays in the ground, it becomes

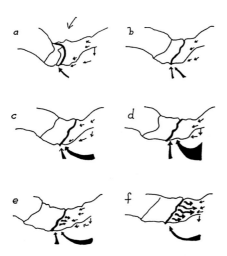

Figure 6.14 Downstream changes (a to f) in a Drenthe Brooke valley, the Netherlands (modified from Grootjans 1985). Arrows show water movement and its quantity. Water sources feeding the flood plain change along the series. Peat depth increases downstream. Upstream vegetation is rain-influenced, midstream is groundwater-influenced, and downstream communities are flood-influenced.

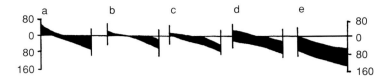

Figure 6.15 Periods of flooding and drying for different plant communities (redrawn from Grootjans 1985). Although each community has its distinct pattern, these overlap almost completely, showing the influence of chemistry, management, inertia and other factors on distribution. Water levels in centimetres. (a) *Caricetum aquatilis* (several subassociations); (b) *Caricetum curto–echinatae typicum*; (c) *Cirsio–Molinietum peucedanetosum*; (d) *Senecio–Brometum caricetosum nigrae*; (e) *Lolio–Cynosuretum lotetosum*.

anaerobic, dissolves much ferrous ion, its sulphate concentration decreases, and its calcium, bicarbonate and conductivity rise. When they have done so, the water is getting close to groundwater. This long-stay deep run-off may remain in the plateau a long time before reaching the main and deep aquifer. That discharges partly into the valleys. The groundwater is calcium-rich, but less so than the East Anglian chalk groundwater. A mosaic of water types results, varying downstream, varying cross-valley. Therefore, the vegetation varies likewise. This is a near-independent hydrological system.

Near the source rainwater is important, percolating through sands with boulder clay drainage. The water is nutrient-poor, with communities of *Erica tetralix*, *Erica–Sphagnum* and *Nardus–Gentiana* (although the last two have nearly disappeared, following drainage). *Carex acutiformis* grows by the stream. In the next stage groundwater is more important and it (not the stream) may flood the valley. Until the 1930s oligo- and mesotrophic communities were well developed. The former *Cirsium–Molinia* has almost gone.

Next, valleyside run-off keeps the slopes oligotrophic, while more mesotrophic water near the brook (from seepage or flooding) keeps the centre mesotrophic. Further down, the seepage water is paramount, even giving a spherical outline to the peat masses (whose centres have dried out, with the drainage). Mesotrophic conditions, with plenty of *Carex* communities, dominate except at the edges.

Next downstream seepage decreases again, the vegetation bands are narrow and there is less wetland. Here for the first time flooding from the stream may be important. In the basal stage such flooding is most important, and groundwater inflow is low. Consequently the peats are rich in silt and clay (and hence nutrients) from the flood water. Communities of large *Carex* spp. are common, and *C. gracilis*, *C. elata*, *Phalaris arundinacea* and *Glyceria maxima* stands show the higher nutrient status. Drainage has removed most seepage water.

Where inflow is mainly rainwater percolation, species diversity is low (*Caltha palustris*, large *Carex* spp., *Phragmites australis*), rare species of emerged aquatics are few, and the more nutrient-rich communities are absent. The calcium-rich seepage areas, in contrast, have high diversity, many rare emergents, and nutrient-rich fen meadows. The polder area in the downstream basin has *Phragmites* and large *Carex* communities, no heath vegetation and even no fen meadow.

Calcium is, again, the most significant chemical, chloride coming next, being high in wetlands associated with flooding and groundwater abstraction. In the Netherlands, with so much interconnection of water channels, this chloride probably comes from the Ijsselmeer, and, regrettably, this water would be replacing deep groundwater from the main aquifer. The hydrological stability of the wetland is being undermined by abstraction. Where deep groundwater flow is undisturbed, diversity and rare species are high, and fen meadows are abundant, giving good species gradients. Removing this means rainwater partly replacing groundwater, more incoming agrochemicals, drying of soil, etc., and completely changing the fen.

If the fen is to be protected, groundwater abstractions affecting it must cease. The obvious alternative is to abstract from the stream instead, so choosing to lose that!

In some parts, correlations were better between plants and the water regime of the 1950s (before the last drainage) when areas now with rainfall and run-off just predominating had groundwater seepage just predominating. Consequently, **in changing habitats, preservation of the existing water table may not conserve or protect the existing vegetation.** That vegetation may be unable to survive more than one generation in its new and unsatisfactory habitat. Study of the sustaining ability of each community is needed.

Otherwise, communities were associated with specific water regimes and man's activities. Twelve plant communities and 17 characteristic species had distributions well correlated with the seven identified water types. In low-calcium seepage areas eutrophic marsh species are more sensitive to drainage and fertiliser than in high-calcium areas. This is a characteristic pattern. Where species are at the extreme of their range for one factor (here calcium), they are more vulnerable to unsatisfactory changes in other factors.

The water pattern at a given site can vary, given the variations of weather, groundwater and run-off, each alternating with season.

By studying water regimes, duration line patterns can be constructed that show the length of time a community spends at any given water level. Those in Figure 6.15 show overlap, but clear differences. Change the water within the pattern, no change in vegetation. Change it outside, and in due time (whenever that may be!) the vegetation and hence animals do change.

When studying the effects of drainage, sites were observed five years before and seven years after drainage: a respectable period.

De Weeribben, the Netherlands (Wirdum 1991)

In the Netherlands, brackish water is often found as groundwater, not strongly salty, but influenced by this. It is often the oldest water present.

A little ancient freshwater has been found. Brackish water now underlies the area in pockets, and brackish influence used to spread occasionally over the wetlands. Freshwater groundwater was upwelling during records taken before 1970. After that, surface water from the ditches (ultimately derived from polluted main waterways) has been the main wetting agent, and has been sinking down into the ground and aquifer. Vegetation changes have accompanied the changes in water sources, of course, this being monitored for the changes after 1970.

This overall pattern has been complicated by changes of many kinds over periods of a decade or so, rendering some even of this pattern suspect as a long-term one. In particular, droughts like that of 1976 have drastic and long-term effects: as do main new pumps. It becomes difficult to determine which are long-term changes due to man, and which are due to extreme weather conditions.

Hydrology is complex. Detailed chemical analyses are needed to place and date each source of water found on a large wetland area. Time also is needed. A single season of measurements would have been misleading.

The Vecht flood and fen plain, the Netherlands (Schot 1991; Verhoeven et al., in Symoens 1988; Wassen 1990) (Figure 6.16)

Here, unusually, there are estimated changes in hydrology down the centuries, and the breaking up of the originally simple water pattern.

Originally, rain recharged the aquifer in the moraine on the higher land beside. This soaked down to become groundwater, which, in turn, flowed to and discharged near the outer edge of the plain, leaving a large stagnant brackish water body further in. With poldering, drainage and abstraction, this simple pattern has been broken into four separate flow patterns, some subdivided. Two have abstraction centres, and groundwater is being lost. Two have surface water sources (a river and a canal), and water infiltrating in. The old brackish water is partly replaced by freshwater inflow. The current seepage areas are on lower levels than before, and infiltration areas on higher ones. There is now a net deficit, annual infiltration exceeding seepage.

In four fens studied, flow patterns differed sharply: between fens, and between wet and dry periods. The crucial portion for vegetation is the root zone. If this is always in one water type, water swinging around below this is of no concern. Water moving above does mix with that below, more above sand than above saturated peat or clay.

Different flow systems can be separated chemically on a very fine scale (e.g. by the amount of the ions coming in rainwater). Vegetation is of course closely related to water by its quantity and movement as well as by its chemistry: by the comparative amounts of river water, spring water and rain soaking down within the wetland.

The Vecht plain is now divided into many small parcels, differing in water status between, and even within, the divisions. All have controls imposed by man, managed regionally, in the Dutch way. Consequently it has a mosaic of habitats.

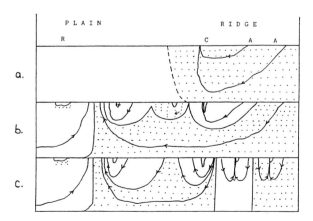

Figure 6.16 Groundwater flow simulations from the fourteenth to the twentieth century. Vecht flood plain, the Netherlands (modified from Schot 1991). (a) Fourteenth century; (b) 1885; (c) 1985. A, main abstraction points; R, River Vecht; C, canal. Dots help to show separate groundwater systems.

The River Vecht itself, through its unfortunate connections to the River Rhine, is increasingly polluted. It is salty, but, unlike the ancient and habitat-proper brackish groundwater below, it also contains high levels of many other substances. These it brings wherever it seeps or floods: which is up from the river. The calcium-dominated water comes from the other end, near the ridge.

One spring area wet all year has a major and complex quaking fen (which has been kept wetter than most British wet fens!) with reedswamp, alderwood and excavation ponds. Nineteenth-century excavation unfortunately destroyed the peat seal to the underside of the fen. Another very wet fen, with a floating mat, receives ditchwater and, in flood, river water. Of these, the groundwater is the least contaminated. Remaining from the peat excavations are the solid peat bars between pits, more solid than any recently infilling peat in the pits. These are near-impermeable, and effectively seal off parcels as independent hydrological units when water is below surface level. These solid peat ridges (often only a few metres wide) are usually a little above the general fen surface: and therefore may not receive surface groundwater. Instead, they receive only rain, and because of the compactness of the peat the rain can collect there, locked in a lens between surface vegetation and peat. Calcium and iron are then leached from the peat. What follows may be *Sphagnum* and low nutrients, or phosphate mineralisation and tall-herb fen. How close are the beginnings of the pathways to such separated ends! Rain of course comes in all. In winter, the rising ditch levels 'push' the rainwater to the centre. This is how nutrient status there is lowered, and – see Catfield fen above – *Sphagnum* can come in.

The springs are scattered along the base of the fluvial sands ridge: there is no continuous seepage line.

8000 BC to AD 1300   At the foot of the moraine, sedge, reed-sedge and mesotrophic brook peat developed (springs). In the middle, *Sphagnum* and, locally, sedge or reed sedge peat (rain) and by the river, nutrient-rich brook peat (river), developed.

1300–1900   Plain embanked, so river floods decrease. Peat extracted, most of the *Sphagnum* raised bog removed (cf. western Fenland). Many lakes were developed. Reeds and sedges were on the sphagnum-removed areas. *Sphagnum*–moss–sedge was in a belt under the moraine ridge. Groundwater discharge decreased, and was replaced by rain-like water.

1890–present   Discharges decrease, as lakes are dried (reclaimed) from the plain, and water abstracted from the ridge. In the past few decades, abstraction and agriculture have lowered the water table yet more, in the plain and in the river.

In the 1940s, species characteristic of this calcium-rich water were abundant: in the abundant seepage areas. More abstraction of this water from the ridge lost much, and puts the rest at risk.

More has been found out about the hydrology of these Dutch fens than in other British wetlands: it by no means follows that the British fens have less to be found out!

Cowles Bog, Minnesota (Wilcox et al. 1986)

The last fen example chosen is a complex of eight water regimes and, appropriately, eight vegetation types. Its peat mound developed when the lake level fell. The water table is higher in the peat than elsewhere (see Eriswell fens). The upwelling water is solute-rich and well-buffered.

Discussion

From these descriptions it can be seen that although few fen examples have been studied, they have been studied with some care. Each has its particular features of interest, and it is a pity all aspects have not been studied together on one fen. Water regime is demonstrated to be all-important: its level, chemistry, source, and movement. The common view that putting any water into a drying wetland will restore it, is false. It needs the water regime under which it developed: and often, in consequence, cessation of abstraction or drainage near it.

In different water regimes, nutrients have different relative importance. Calcium is important in all these examples. Phosphate deficiency (calcite formation) determines the *Schoenus nigricans* community in the Eriswell source-fens. In the Drenthe valley, sodium and sulphate are important upstream (from rain); and chloride downstream (from brackish groundwater) (see also Keddy 2000; Gopal *et al.* 2001).

7 Chemical types and vegetation types

Fish and waterfowl, who feed of turbid and muddy slimy water are accounted the causes of phlegm

(Flayer)

Waterfowl joy most in that air which is likest water.

(Francis Bacon)

Can bulrushes but by the river grow
Can flags there flourish where no waters flow?

(Southey)

Introduction

Soils are made up of solids, both organic and inorganic, and of water that contains solutes. Soils are chemically and biochemically active, not the inert dark mass they appear on first sight. (Deeply buried compacted peats are inert, though.) Chemicals come in, stay in, go out (Figure 7.1). The solids in the soil come from mineral deposition (silt, sand, gravel) or from organic material, humus and peat, accumulated when its input is greater than its breakdown.

Organic life on earth is based on carbon. There is an inexhaustible supply of carbon in the air. Peat and soil organic matter accumulated, so peat wetland soils form immense carbon reserves. They date, in Britain, anything from 10 000 BC (ice retreat) to last year. Organic carbon enters as dead plants, animals and microorganisms. It is vital to many habitat processes.

Into wetlands come substances from (Basic Figures 12 and 13):

1 *The air.* In precipitation, in fog, by direct deposition.
2 *Run-off, below and above ground.* The chemicals in run-off now almost always differ in content and proportions from those in wetland soils. Run-off composition differs with land use and rock type, for example that from blanket bog, oak forest, pinewood, arable, fertilised pasture, road, car park, housing estate and rubbish tip. Most wetlands developed with inflow from traditionally managed land, or even unmanaged land. They then need this for their protection and continuation.
3 *Groundwater, upwelling as springs, seepages, etc.* So far, although groundwater is polluted, it is (ordinarily) less so than any other water source, and abstraction is more of a threat.
4 *Open fresh water (watercourses, lakes and pools)* draining to, or passing through, wetlands. This may be polluted by many and various kinds of effluents and run-off. Sources

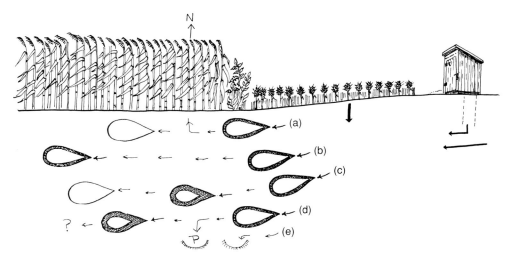

Figure 7.1 Movement of pollutants through a wetland. Entering are: (a) nitrogen and other substances leaving wetland as gases; (b) calcium, etc., which, as in sands and peat bog (Boeye 1990), may, or in silts, etc., may not (Prach 1992a,b) travel and influence long distances; (c) bacteria, mostly deposited and removed; (d) phosphorus, heavy metals, etc. Deposited and removed from the water in the outer band: until this is saturated, when these penetrate further in; (e) silt, with large-molecule pesticides bound to it, and other large molecules, etc. Mostly filtered out before entering wetland.

include Sewage Treatment Works, factories, mines, silage liquor, slurry, chemical farm spills, road spills, and run-off. Flooded reedswamp in lakes receives polluted lakewater directly. (In other wetland types it usually passes through soil so is at least partly cleansed in the edge strip.) If that reedswamp is embanked from the lake to protect it from the lake pollution, it may or may not remain a reedswamp, but it will not be a lake reedswamp. Thought is required before restoration projects are undertaken!

5   *Dumping.* Any type of waste may be dumped in wetlands, legal or illegal, large or small. Fortunately such dumping is infrequent in Britain: although it can include both Council tips and (legal) factory waste, in enclosed or unenclosed sites.

Out from wetlands come:

1   Dissolved solutes leaching out, more when there is more through-flow.
2   Particles, and particles carrying solutes washed out by run-off and drainage, increased by erosion.
3   Gases going to the air, developed in the wetland, e.g. nitrogen, methane and many organics.
4   Excavations, soil and all its contents (e.g. for horticulture or fuel).
5   Wetland products, fowl, thatch, hay, etc.

Entering chemicals may pass straight through or be there for short or long periods while they are inactive or active chemically, to emerge later changed or unchanged in the outflow or as gases (Figure 7.1). Solutes may be stuck to, adsorbed on, what are called 'adsorption sites'. There are not many of these, and if they are all full, no more of the substance can be taken. Changing conditions can, however, either make more adsorption sites or set now-held chemicals loose. Solutes may also be stuck, immobilised, by micro-organisms such as bacteria, and they may be precipitated.

Adsorption, immobilisation and precipitation are active processes. Active processes change conditions around them. Less or more of the substances may be taken out of solution, may come back into solution, or may be totally lost to the wetland. Substances can occur at different times in different categories, for example be active then immobilised, be precipitated then washed out.

As well as the processes developed in the natural untouched wetland, there are all human impacts. Contamination of open waters, rain, run-off and groundwater by domestic, industrial and farm wastes pollutes wetlands. Water loss by abstraction and drainage alters physical and chemical compositions. Other changes come from management, disturbance, and so on.

As impossible as it may seem, there are no full analyses of wetland waters (whether river or wetland, clean or polluted) or soils. Thousands of chemicals are known to occur there, but where and how much are not known. Even the number of chemicals, let alone the name and significance of almost all, is unknown. Even nitrate, perhaps known the best, is far from fully known. The interactions, whether or not synergistic (acting together), are largely unknown. So are the exact nutrient requirements of plant species and communities. That a species needs high calcium may be known, but how does this fit with the number of parts per million (ppm), the form of the calcium, the range of the other nutrients, the rest of the habitat, the season, the stage in the life history? These questions are unanswerable as of now, but the answers are clearly necessary for the understanding of the role of soil in habitat.

Top soils are usually the most active. Dry ones are well aerated, and kept so by invertebrates moving about. Even in flooded soils there is usually a thin layer of oxidised soil, even in apparently stagnant wetlands, and may be only a few millimetres thick (Mitsch and Gosselink 1993). Such oxidised layers also occur deeper down, around roots – that bring oxygen.

Active processes are influenced by temperature (higher temperatures increase chemical activity and so, for example, decreases peat production), water (different in flooded and non-flooded places) and other natural factors, plus the additional impacts of people. The whole is immensely complicated, and is currently understood in only a very superficial way. (Without knowing what substances are present, how can behaviour and effects on the habitat be understood?)

Chemical types in the landscape (Basic Figures 1–4, 10–13)

Introduction

Water takes on the chemical characteristics of the habitat (soil, rock, etc.) through which it passes completely or partly depending on how long the water stays in that habitat, and what chemical processes are occurring. This applies to deep groundwater and to run-off and leachate from the surface. Conifer run-off is acid and toxic to some plants; that from birch wood is certainly acid (its toxicity is less or not established), that from most other broad-leaved wood is neutral to nutrient-rich. Bog run-off is acid and – at least at the start – toxic. The deeper the brown of the water, the lower, often, the pH. Run-off from fens and marshes is more nutrient-rich and (perhaps) non-toxic. Run-off and springs from limestone are calcium-rich. In land with different chemical layers, the water reflects the layers through which it passes, for example acid sands over chalk may have both acid nutrient-poor run-off and calcium-rich springs reaching the wetland beside. Run-off and river water integrate the chemical effect of the upstream catchment, with most emphasis being placed on the land near the wetland. River flooding brings silt near the river, but the flood probably reaches further. The poor fen beyond receives little silt, and maybe not much water. Vegetation pattern within the poor fen habitat varies with water level and impact, small-sedge fen being the most nutrient-low, and fen meadow, on the silty banks near the river, being the most nutrient-rich.

Wetland vegetation type depends directly on wetland water type. There are a myriad of water types: and a corresponding myriad of vegetation types. A wetland may be fed by only one water source, e.g. rain, but even rain varies chemically from one place to another. Run-off, spring and river water differ far more between themselves, so there can be all possible combinations of different proportions (Basic Charts A–K). A continuum of wetland habitats reflect a continuum of chemistry.

Soils

From left to right on the Basic Figures there are first the rain-fed peats, the blanket bogs on the lower plateau, the drier ex-bog, formed in an earlier period of wetness (e.g. the Atlantic period) on the moor above. The wetter bog has the more extreme dystrophic features. The moor shows some mineralisation, and is less nutrient-poor. The tarn is oligotrophic, so *Scirpus lacustris* or, with fewer nutrients, *Carex rostrata* will be the reed-swamp. Where there is more water flow down the slopes and, especially, in the flush at its base, the peat and peaty soil are bathed in flow. With (even low) nutrients passing by, nutrients are higher, and even grass moor can occur on the peat. On steeper slopes, however, there will be less or no peat, but in the postulated grit or gneiss, etc. rock, derived mineral soil will also be nutrient-poor. Pine and birch wood may or may not be on peat.

On the flood plain, the raised bog has the same dystrophic peat as the plateau blanket bog. The poor fen peat or peaty mineral soil of the landwards side has nutrient-low run-off (and rain) only.

Crossing to the (main) lowland picture, a more nutrient-rich flora occurs on the silty bank of the more nutrient-rich (chalk) river. The chalk springs under the slope give a calcium-dominated fen there, becoming calcium-rich further towards the river, with nutrient status rising as the spring influence lessens and that of the river with its flooding and silt increases.

If the rock below is clay, the fen is nutrient-rich throughout.

The New Forest is on lowland acid sands (Figure 4.26). The moving water of the stream (which may also rise in mineral soil) gives nutrient-medium (although not calcium-rich) habitat, with carr and, formerly, reedswamp. The valley bog has water, run-off and springs, altered nutrient-poor sands, and so characteristics of these.

Impact

Basic Figure 14 shows some chemical results of impact; first, air polluted with many and various compounds (see also Figure 7.2). These reach the wetland by direct deposition, in rain and in fog. They also reach it by run-off and by sideways flow in the upper soil. The

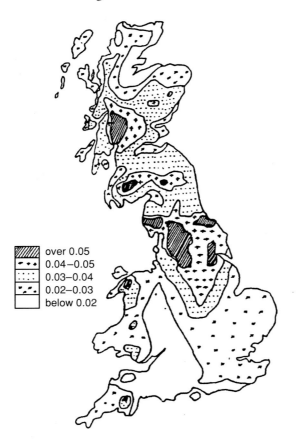

over 0.05
0.04–0.05
0.03–0.04
0.02–0.03
below 0.02

Figure 7.2 Annual wet deposition of acidity. Data in gH^+/m^2 (redrawn from Rimes 1992a).

changes are small and slow where soils have a high buffering capacity. Bogs, dominated by hydrogen ions, are the most vulnerable, with reedswamps in low nutrient lakes also being at risk. Salt and magnesium come mainly from the sea, aluminium, iron, ammonia, other fertilisers and ash mostly from agriculture and industry, many metals and organics from industry, and manganese from soils and rock.

Industrial changes mean changes in deposition. In the south Pennines, *Sphagnum* has been locally lost since the Industrial Revolution, probably mainly because of sulphur oxides. However, increases in farming pollution mean nitrogen is now also damaging. British sulphur dioxide, having increased since the mid-nineteenth century, is now going down, but compounds from vehicles are increasing, so correlations are complex. Bad damage, as in the Pennines, is generally associated with industrial (acid, metal) and farming (nitrogen) practices. Surface water pH in south-west heaths and bogs is, however, correlated with rainfall (M. C. F. Proctor, personal communication). Deposition rises with (1) high rainfall, so Snowdonia and the West Highlands are at risk; (2) high emissions of industry so northern England is at risk; and (3) continental air pollution. High acid rain falling on acid rock areas leads to risk. These occur in high altitudes, and in scattered lowland areas such as the New Forest. Nearly a quarter of the area examined (141 Sites of

Special Scientific Interest) have damaged freshwater habitat, north Wales being the worst, followed by mid-Wales, south-west and north-east Scotland, and north-west England (Rimes 1992a,b; Sketch and Bareham 1993).

On the land, agrochemicals are applied, entering both directly and by run-off. These increase nutrient status, perhaps reaching improper levels for the traditional vegetation, and adding toxic pesticides.

On the land, also, there is drying, from drainage and abstraction. In bog, drainage leads to peat loss (drying, cracking, erosion), and, in part, to some increase in nutrient status from mineralisation. In fen peat (particularly calcium-dominated) nutrient status much increases. Even in marsh, there is some increase. In all, disturbance aids oxidation and increases status.

Conifer planting and harvesting both mean soil disturbance, and erosion. More intensive farming increases these too.

Then there are all the effluents, entering the wetland directly, by surface or subsurface run-off, by seepage from rivers, cess pits and worse. These are excessively varied: road spills of anything from milk to cyanide, Sewage Treatment Works' effluent of sewage, detergents, other cleaners, road run-off and any industrial effluents of the region, untreated road effluent, farm effluents from silage to dairy units. It is far too easy to refer to these as being 'nutrient-rich' or 'eutrophic', meaning rich in nitrogen and phosphorus, perhaps in more. In vast quantities these do have an important effect, of course, but in low quantities organic and mineral additions may well be the more effective at changing vegetation type. In the second place, no pollution contains solely phosphorus and nitrogen, and practically no pollution is solely of nitrogen and phosphorus compounds. Effects due to heavy metals, the non-phosphate part of detergents, the hydrocarbons in road run-off, and the toxaphene permeating the habitat must not be ascribed to friends nitrogen and phosphorus. Nutrient-rich pollution normally has nitrogen and phosphorus as a small amount of the whole. That which has not been measured still exists, and can have a marked influence on habitat.

Increased fertility of the land means increased nutrient status of the waters entering wetlands from it. The affected waters are primarily run-off, so reaching rivers, but may also, regrettably, now be deep water springs. There is only one advantage here. Nutrient-rich wetlands used to be kept nutrient-rich by constant river-flooding and the deposition of new silt. If these are now cut off from the supply of the enriching silt because the river is too polluted and toxic, fertiliser-rich run-off may compensate.

Abstracting groundwater so that spring water is cut off is an important chemical impact by man. It withdraws a chemical influence under which a given wetland was formed. It is a chemical change.

Plants as indicators of nutrient regime

Different species have different nutrient ranges (see Chapters 3 and 4). In the same way, as they can be used as indicators of water regime (Chapter 6), so they can be used as indicators of nutrient regime. There are the same provisos. First, a community is a better indicator than a species (as each species has a different range, and a community occurs in the narrow habitat band where all their ranges overlap). Second, there may be different ecotypes in any species, each occurring in a different chemical range, or there may be geographic variation in response within the same ecotype. Third, not enough is known about responses to habitat, and a community may occur in two apparently different habitats that in fact have the same, but unidentified, determining habitat factors.

Naturally each species has its own range (in a given habitat) for each nutrient, not just for 'nutrient status'. These, again in a given habitat, can be measured and used as predictors. Then the order of species replacement is known: but as these apply to the given site only, with its own unique water regime, etc., data cannot be transferred elsewhere (Bakker and Olff, in Wheeler *et al.* 1995; Keddy 2000).

Plants integrate the whole of the chemical influence on their habitat. They are not – unlike some researchers – influenced only by the regime on one given day, but by the variations in that day and month, perhaps years, or decades. In this, extremes (as noted in Chapter 6) may be as or more important than averages: to remain present, a long-lived plant must tolerate whatever surges or withdrawals of nutrients or pollutants occur during those years. Species occur in a wide range of major nutrients.

Altering calcium, that most important mineral, can have dramatic effects, for example liming a raised bog brought a sedge instead of *Sphagnum* vegetation (Morgan, in Patten 1990). On Dartmoor, Moorland springs and flushes can be very sensitive (Proctor 1989). Typical vegetation is:

Springs on granite:	*Sphagnum auriculatum* dominant, *Montia fontana, Ranunculus omiophyllus.*
Larger flushes:	*Sphagnum auriculatum* prominent in a larger range of sphagna may be (a) bog, *Narthecium ossifragum; S. papillosum;* (b) fen, *Carex echinata, S. recurvum;* (c) runnels, *Hypericum elodes, Potamogeton polygonifolius.*
Limestone run-off:	*Campylium stellatum, Drepanocladus revolvens, Scorpiurus scorpioides* and, for example, *Anagallis tenella.*
Base-rich flush:	*Drosera intermedia, Pinguicula lusitanica.*

In a calcium-dominated Dutch fen, the depth of the drainage ditches affects the future of the fen. Shallow ditches mean groundwater is high, and so these ditches draw off rainwater that sits above groundwater. Wetness and calcium dominance remain. With deep ditches the groundwater is low, and so not moving through the top soil layers. Therefore, rainwater can accumulate on the top, so that bog processes, sphagnum and cottongrass (*Eriophorum angustifolium*) begin (Grootjans and van Diggeln, in Wheeler *et al.* 1995). Alternatively, and in Britain far more commonly, drainage leads to mineralising the dry peat, and tall-herb vegetation.

An example of change with fertiliser comes from Verhoeven *et al.* (in Mitsch 1994). On acid peat ground there was wet heath where there had been no fertiliser. *Molinia caerulea* grassland occurred with past fertiliser. In *Molinia*, with a productivity of 100 g/m², after three years' addition of fertiliser the yield was 18 times higher, 1800 g/m². (Over time, after fertilisation stopped, nitrogen decreased more than phosphorus.) There are two changes: the change in community, and the increased crop. These are different, although an increased crop can mean, and elsewhere has meant, increased competitive vigour and so a change in community.

Case studies

Buitengoor Fen, near Mol, Belgium (Boeye 1992) (and see Chapter 6, Figure 6.13)

This is a good first example, as the fen contains a calcareous species-rich community in a wetland without limestone, where no such fen could exist. The fen has – as is common – a variable surface, and the higher parts bear nutrient-poor *Molinia–Myrica–Pinus*, but the hollows, a *Carex davalliana*-type vegetation (see Chapter 10). This alliance is the more base-rich of the small-sedge, poor fens, the one often considered medium fen. There are three water types:

1   Run-off water from the Mol sands around. This shallow groundwater has no bicarbonate activity, and reflects the nutrient-poor sand habitat. It flows in with rain, and out fairly fast.
2   Deep mildly calcareous groundwater. This moves very slowly, spending longer in the fen than in the example above, but it affects only the very lowest part. What is not known is how much – if at all – the level of this water has dropped since, say, 1800. How much of the fen used to be watered by it?
3   Calcareous and moderately polluted water from a channel constructed in 1850 to bring irrigation water from the River Meuse. (This is not as strongly calcareous as the chalk spring fens of East Anglia.) The pollution comes both from the Meuse and from a settlement of 2000 people. This water is carried in the ditch along the upper edge.

This calcareous ditch water flows sideways, going in between the (top) Mol sands run-off and above the deep unpolluted groundwater. Three water sources feed three different levels in the fen.

The acid water contains much aluminium: such waters, organic-low, unpolluted and acid, have the highest aluminium toxicity. At and below pH 5 there is no bicarbonate (bicarbonate ions do not occur with aluminium ones), iron and aluminium are high, sulphate ions may be even higher (i.e. in excess) and aluminium is active biologically.

The deep groundwater is rich in iron, and in the fen is mostly saturated with iron ($FePO_4 \cdot 2H_2O$). The calcareous ditch water is almost in equilibrium with calcite. Therefore, mixing these in different proportions leads to further chemical permutations.

The nutrient-poor communities, bog, etc., are fed by the acid water (cf. New Forest valley bog in Basic Figure 2). This water probably controlled most of the Buitengoor wetland in the past. The calcareous fen receives ditch water (filtered through soil), alkaline, base-rich, high-nutrient. Species of calcium-rich habitat came in along the channels and the irrigation ditches they feed. There are excavated sand pits, some containing deep groundwater, others the channel water – that therefore has opportunities to sink to the aquifer.

The deep groundwater forms the source of a small tributary, and this tributary has alder carr around. Nutrient status, as usual, is increased by the moving water. This bears poor fen, where wet enough, the water being nutrient-poor.

Finally, along the nutrient-rich irrigation ditches and in recently felled alderwood there are nitrate-favoured species, and a nitrate-favoured community, of, for example, *Urtica dioica*, *Eupatorium cannabinum*, *Typha latifolia* and *Rubus* sp. Dry fen soils increase in nutrient status, and the ditch contains sewage pollution.

Peat growth in a fen raises that peat so it is no longer in contact with groundwater, and the conditions for bog can develop. In this fen, alkaline water entered through human activity, and converted those levels to calcium-rich fen.

The calcium, and alkalinity, bicarbonate, potassium and chloride from the ditch have reached the main fen, and made it calcareous. Nitrate and phosphate, however, so far remain close to the ditch. Most phosphate disappears within about 50 m, nitrate within 200 m (unusually far, see Chapter 8). No doubt many other pollutants are lost similarly.

The ditch water therefore supports two communities, calcareous fen and nitrogen-favoured tall-herb vegetation. The time scale here is important. After 140 years of ditch water, calcareous communities were established. The time needed to achieve this is not known, nor whether further change is occurring. Again after 140 years of ditch water, nutrient-rich tall-herb fen is established along the ditches, etc. The approximately 200-m band of nitrogen-rich habitat, in stable conditions, is not likely to increase (as nitrogen is lost to the air). The roughly 50-mm phosphate-rich band, however, would do so, and these conditions gradually spread. The difference in mobility and effect of the different pollutants is noteworthy.

Habitat chemistry has been substantially altered, as an unintended side-effect of making irrigation channels. There is also the unintended but common result of abstraction and drainage of the groundwater: lowering of the level of the groundwater. Spring water now drops in dry summers, so water supply in the spring area at the lowest part of the fen is rain and surface run-off. Coming from the Mol sands, this run-off water is acid, nutrient-poor: and with aluminium as a major cation makes a sharp difference to the groundwater.

A few dry weeks once in 10 years are unlikely to have much long-term effect, but if groundwater is lowered further and leads to a few dry months every year, habitat and hence community would change. The fen centre would be more acid, perhaps bearing bog or wet heath, or perhaps even dry land.

The effect of drainage in wet meadows, Drenthe, the Netherlands (Grootjans 1985)

A new drainage ditch lowered the water table (except near the original stream). In the now-dry peat, this meant more oxygen and higher pH so more nitrogen mineralisation. The quality of the organic matter changes. This influences, for example, whether mineral-ised nitrogen can be taken up by plants (net mineralisation) or whether it is immobilised by microorganisms (no net mineralisation). Dry peat loses structure (humification, etc.) so it loses the ability to keep rainwater, so water sinks down, rather than staying in the upper peat and keeping it wet. This increases microbial breakdown and so the breakdown products increase (e.g. amino acids). If soil structure is lost, the peat is likely to have rapid nitrogen mobilisation even when groundwater is high.

Cattle lessen these changes by keeping soil compact and flattened down.

Near the new ditch, nitrate-favoured grassy vegetation increased, with *Urtica dioica*, *Cirsium arvense* and *Anthriscus sylvestris*. This stabilised after two years.

In other *Cirsium–Molinia* meadows on fen peat the floristics showed less increase in nitrogen (less *Carex pulicaris*, *C. panicea*, etc. but more *Holcus lanatus*, *Succisa pratensis*, *Festuca ovina*, etc.). Nitrate was produced, as in the first example, but for some unidenti-fied reason it was not taken up by the plants. This is a valuable example of the fact that unknown factors can be crucial!

Even so, however, the drained peat was irreversibly harmed. The meadow was later flooded. Marsh plants, instead of increasing and flourishing, mostly died. (Results from decades of flooding, when new soil and soil processes would be in place, are not available.)

Vecht flood and fen plain, the Netherlands (Schot 1991; Wassen 1990) (Figure 7.3)

There are polders on the plain. (Dutch polders are embanked and drained wetlands.) The moraine ridge to the side has towns, and groundwater abstraction. Vegetation is clearly related to water type and water supply.

Groundwater comes from the moraine ridge. It is not old deep aquifer water, but more local and short-lived. Calcium increases with age and phosphate decreases. (Phosphorus is

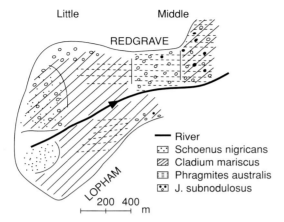

Figure 7.3 Redgrave and Lopham fens, East Anglia (1958) (Haslam 1960). *Schoenus nigricans* is in the most calcium-dominated, spring water area. *Cladium mariscus* and *Juncus subnodulosus* increase further away. The fen is much dried this century (see text and Chapter 6). Reeds (*Phragmites australis*) are sparse, restricted and yellowed in the spring area, increasing only where nutrients increase, behind.

precipitated out by calcium and iron.) This gives the calcareous (not full limestone) low-phosphate water under which spring and seepage areas bear calcareous fen communities. As abstraction increases, the supply is decreasing, and the quality is threatened by pollution.

Interestingly, suitable water can also be developed by rainwater moving in fen peat: it becomes calcium- and bicarbonate-rich, other-nutrient poor. To be useful and replace groundwater, there must be land upstream in the catchment, from which water can reach the relevant wetland, and this peat must not be polluted (even with agrochemicals).

The *Carex davalliana* alliance (with, for example, *C. diandra, C. lasiocarpa, Epipactis lacustris, Equisetum fluviatile, Valeriana dioica*, etc.) occurs only with this calcium-rich, other nutrient-poor seepage water. This is reduced to remnants, because of abstraction and drainage, drying and raising the nutrient status of the peat.

The River Vecht water is polluted from the Rhine, and from rainwater pumped out of polders (with agrochemicals). It carries groundwater. *Phragmites–Cicuta* vegetation occurs in peat now fed by this water. The water is slightly brackish.

Then there are the rain-fed areas. These are kept wet partly because the peat is deep enough to prevent rapid water loss downwards, and partly because there are wet areas (fed by other sources) around, so there is little sideways flow. There are, therefore, areas in the fen wetland now fed mostly by rain. The *Carex rostrata–Stratiotes aloides* community is one of these.

Sphagnum reedlands are also mostly rain-fed, but with high nitrogen, probably from high Dutch aerial deposition, from nitrogen fixation by root nodules of alder, and from nitrogen mineralisation in the peat in drought. Here the *Phragmites* roots and rhizomes reach far down, doubtless with good nutrient supply. The sphagnum, in contrast, is solely on the surface, and grows because of the nutrient-poor conditions created by rain.

There are other waters, and other communities, e.g. *Caltha–Agropyron–Rumex*. This is drained, so relatively dry, and mowed monthly, giving a high-productivity meadow with mineralised peat, etc. Matured regional groundwater is the water source.

Caltha-small sedge. This is fed by rain and the polluted river water, but – obviously, from the flora – with less river water than the *Phragmites–Cicuta*. Both are permanent recharge communities.

Carex curta/nigra is the third along the continuum of rain/river water habitats, here with the least river water, and again a recharge fen.

Thelypteris-reedlands are with groundwater, and are divided into those with calcium-rich nutrient-medium water and those with nutrient-rich brackish groundwater and also lake surface water. Calcium here is saturating.

Lolium–Potentilla anserina is in a deep polder near the river, with permanent seepage. Because of drainage, though, this high-calcium water seldom touches the rootzone, which mineralises to be of middle to high nutrient status.

Large-sedge–*Filipendula*, *Phragmites–Filipendula* are both with brackish river water; the latter, with a stronger river water influence, is mainly in permanently recharge areas.

Carex davalliana alliance is in a calcium-rich and wet habitat.

Carex curto-nigra community is often a lower-nutrient stage to the preceding. If instead there had been regular annual mowing, this might have been *Juncus–Molinia*.

Agriculturally polluted rain raises nitrogen input, and acid rain may leach iron and calcium, so releasing phosphate. Both raise nutrient (non-mineral) status.

Rainwater seeping past a rootzone takes up carbon dioxide, dissolves calcium carbonate, and becomes richer in calcium, magnesium and bicarbonate, and pH.

With different original water sources, changes in the composition of those waters, changes in the sources themselves, patterning and repatterning the surface, it is not surprising that the picture is confused, and careful analyses are needed to separate and untangle these. [0^{18} is a valuable tracer, essential for distinguishing main groundwaters. Tritium analysis can separate waters differing in age by a few decades, and assess mixing. Chloride can be used to trace brackish water (including sewage effluents). Nutrient analyses help to trace recharge areas.]

The dominant water source for a fen is deducible from the chemistry and dynamics. Polder construction leads to many groundwater flow systems that become nested (Figure 6.16). The chemical changes between polders are therefore sharp. Dutch work is carried out to a very high level of chemistry and hydrology, studying the proportions of a basket of ions, movement of different water types, and the relationship between these and vegetation.

Acidification of Dutch moorland pools (van Dam 1988)

The results of aerial deposition were studied over 70 years. The pollution varies in content and concentration with: place, wind direction, domestic fuel changes, year (weather), industrial changes, and farming practices in the area. There was no change in the sulphur component after 1933–7, but inorganic nitrogen increased with livestock units (more slurry on land, more ammonium in the pools, so probably acidifying).

The habitat and vegetation change from year to year with the amount of water in each pool. Acid rain lowers pool pH, but by how much depends on weather and water level as well as chemical input.

If 20 per cent of the pool is dry, the pH dropped about 0.5 over 70 years, which (except around pH 5) is hardly significant. If, however, 70 per cent of the pool is dry, the pH dropped by 2, which (except above pH 8) is enough to alter habitat and vegetation. In dry years, when pH is lowest, *Juncus bulbosus*, *Sphagnum* spp. and the diatom *Euanita exigua* increase, and *Lobelia dortmanna*, acid-loving, neutral and humic-water diatoms decrease.

As often in ecology, the principle, once established, is simple but its application is not.

Three fens near Mildenhall, Suffolk (see Chapter 6, Figures 6.2–6.7)

In the fen basins at the source of the Eriswell Lode, the pattern, also seen elsewhere, is for calcium dominance with chalk springs, with *Schoenus* dominance in the vegetation. Further away, with less such spring influence, there is more of a *Schoenus–Juncus subnodulosus–Cladium mariscus* mixture. Calcium dominance is less, water stability is less, carr invasion is alder, not birch. Finally, with no calcium dominance, in hollows fed by groundwater in the sand (not calcium poor, but not calcium dominant either) *Carex elata* dominates, with sallow as the carr invader. Tall *Cladium* can occur, and dominate in any of the three habitats. These are discharge fens.

In the Icklingham fen, run-off tributaries and seepages from sands over chalk add a lower nutrient influence. They enter along the landwards side of formerly river-fed wetlands. Spring influence used to be localised, but is now spread further afield by the absence of the inflooding river water. *Carex appropinquata* dominates along the spring line, with alder as the invading carr species. Further in, there is less (sandy) spring water. There are more nutrients, more fluctuating water level, and the water is more stagnant, *Carex paniculata* dominates, with sallow invasion. In the nutrient-rich half *Phragmites* dominates (*Phragmites* forms the peat below). *C. paniculata* is – in peat terms – transient, perhaps present because river nutrients decreased after drainage and embanking? This fen has a discharge line near the outside. Otherwise it is a recharge one, formerly recharged from the river and incoming land water and kept wet, now from the land only, and much less wet than before.

The Cavenham fen is and was a recharge fen, like the greater part of Icklingham fen.

Other Breckland fens (Figures 7.4)

These examples show chalk-water dominance by the springs or seepage areas on the outer edge (*Schoenus nigricans*), with this dominance and influence relaxed further away, where reed (*Phragmites australis*) becomes abundant (in the calcium-influenced) or dominant.

Luznice River flood plains, Czech Republic (Figure 7.6)

These flood plains have had a chequered history of management and abandonment, wetting and drying. Nutrient status rises with increased fertiliser and run-off, with drying, and with natural factors. Vegetation interprets nutrient and other patterns.

Broadland (Parmenter 1995)

Catfield fen (also see Gilvear *et al*. 1990). This fen receives (1) acid, low-nutrient run-off from Norwich Crag and sandy brick earth, (2) calcareous river water, (3) brackish tidal water, (4) surface flow from nearby fens, (5) groundwater discharge that is declining, through abstraction. The fen is divided into various beds by dykes, and a few banks. Below, estuarine clay brings in slight salt also. The fen was marsh into the early nineteenth century.

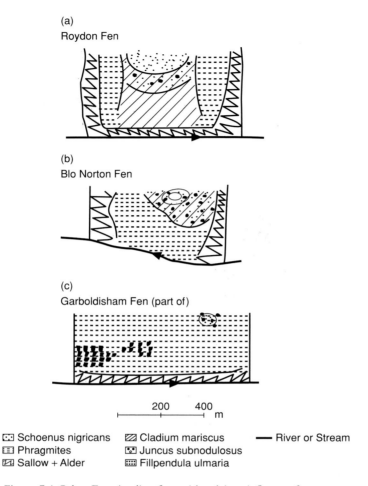

(a)
Roydon Fen

(b)
Blo Norton Fen

(c)
Garboldisham Fen (part of)

200   400
⊢———⊢———⊢ m

⌐⌐ Schoenus nigricans   ⊠ Cladium mariscus   ▬ River or Stream
⊞ Phragmites   ⊠ Juncus subnodulosus
⊠ Sallow + Alder   ⊞ Fillpendula ulmaria

Figure 7.4 Other East Anglian fens with calcium influence from seepage near the land (1958)
(Haslam 1960). (a) Roydon fen, with calcium-dominated seepage areas along the
outer (land) edge, bearing *Schoenus nigricans*. The *Cladium mariscus*, towards the
stream, is calcium-influenced, and *Phragmites australis* occurs further from the springs.
(b) Blo Norton fen, with a smaller spring area and more reed. (c) Garboldisham fen,
with only a small spring-influenced area. Before drainage, there was probably more,
in all these fens.

Hickling area. There is a spring line coming from the Norwich Crag at the outside. The
vegetation in this nutrient-poor water is birch–*Sphagnum*. The salty water table, to Hickling,
increased after the drying of the 1950s. In Hickling Broad marshes the salt increases from
west to east (towards the sea), pH from east to west, and these patterns influence the
vegetation. Added to the chemical pattern is the influence of the dyke, management and
water level. Acidity increases further from the dykes (where rainwater can accumulate). Scrub
has invaded, with neglect. The carr has overrun even the more acid fen nuclei. The start of
bog conditions here has therefore been halted by carr invasion: not by peat mineralisation,
or flooding. Birch–*Dryopteris cristatus* patches do occur, but are hardly incipient bog (like,
for example, Wicken Fen, rather than the raised bog birch wood in Holme fen).

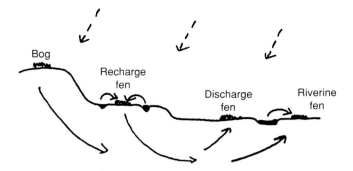

Figure 7.5 Pattern of Dutch wetlands (modified from Koerselman and Verhoeven, in Wheeler 1995). Bogs are rain-fed. Recharge fens are rain- and river-fed. Discharge fens are rain- (perhaps river-) and groundwater-fed. Agricultural areas lie between the 'wilder' wetlands.

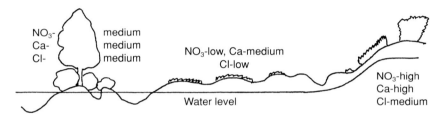

Figure 7.6 Nutrient patterns in the Luznice River flood plain, Czech Republic (redrawn from Prach 1993).

Crestwick marshes. These are mainly fed by chalk seepages. The flushes have short rush-meadows and are very species-rich. They are calcium-rich rather than calcium-dominated, with, for example, *Anagallis tenella, Briza media, Carex echinata, C. pulicaris and Valerian dioica.* There are swards of *Holcus lanata, Carex riparia*, etc., all with rich floras. Nearby abstraction is probably already decreasing fen discharge, and the calcium communities are at risk. In what was, in 1990, valley meadow of good structure, base-rich flushes have already gone, and species have been lost.

Martham Broad marshes. Salt leads to *Oenanthe lachenalii, Samolus valerandi, Juncus maritimus*, etc., a brackish assemblage. Salt overrides management: vegetation is like the differently managed Breydon marshes.

Woodbastwick. This is partly separated from the river, and so mainly fed by chalk water from the outside, from seepages, and rain. Rain can accumulate in central parts where it cannot easily soak down, and where dykes and chalk-water prevent it flowing sideways much. There is a wide range of communities, well patterned even within dyked-off sections, giving variations with past management, past water regime (river flood), present water level and past peat-cutting. The communities include alder carr, *Carex paniculata* alder carr, fen meadows, litter fen, sedge, reed, rush-fen: all species-rich. Most management is for reedbeds. There are small *Sphagnum* areas – as expected with this slow-drying and chalk-water loss. Will this fen, over the next half century, stay as fen, develop in part to bog, or become more mineralised?

Oulton marshes, Whitecast marshes. Interestingly, Oulton marshes, well downstream on the River Waveney, are mainly fed by spring water, and so are made wettest by these. The area of the species-rich fen meadow has lessened dramatically, but is still of much interest. Horse grazing has led to some of the most interesting communities. The nutrient-medium fen includes: *Thelypteris thelypteroides, Lathyrus palustris, Ranunculus lingua, Eriophorum latifolium.* The Whitecast marshes nearby receive some salt from the river, but are freshwater. These were much influenced by silt and nutrient-rich river water. *Glyceria maxima* dominates by the river, *Phragmites australis*, on the less silted land behind. As well as the contrast in nutrient status to Oulton, Whitecast has parts still treacherous to pass. This is now rare even in Broadland, almost unheard of in, for example, Fenland, Pevensey levels. Until drainage and abstraction it was common (see Chapter 1).

Beccles marshes. Here the River Waveney is tidal, with water fluctuations, and entry of both salt and river water. There is very wet alder carr, tall-herb dry fen (*Urtica dioica, Filipendula ulmaria*), and tall-sedge swamps. Calcium is not high. Open fen is mown.

Broadland fens therefore differ to the Breck and Dutch ones here presented. The Breck fens are simple and small – river water, chalk springs, sand (and chalk) run-off, some nutrient-low, some medium – and drying. Broadland water was decreased by drainage and embanking in the past few centuries, and in the late twentieth century became sufficiently polluted to be able to chemically alter the fen.

The Dutch wetlands were less cut off, and the river pollution is penetrating further. The British wetlands are drying out more (there is no 'right' solution!).

Broadland fens may have acid or chalk springs and run-off, or have neither; or have agrochemically polluted run-off.

The Dutch ones described here have no full limestone, nor acid water, so are without that range: but have more *Sphagnum* lands within the plain, and more patterning with the various water types and their mixing. Settlement pollution is threatening them too.

Abstraction lowers spring water, and is damaging in the Vecht plain, Broadland and elsewhere (the Breck fens were damaged before this became relevant).

Both are threatened by abstraction, drainage, loss of clean river water, presence of dirty river water, and loss of traditional management.

Chemical impact

How much pollution can a habitat tolerate before it is changed? This is a concept troubling governments and ecologists alike. The widely accepted definition is that of a critical load. A critical load (of a pollutant) is the input at which the ecosystem is damaged, and below which it is not damaged.

The concept is fine. Like all such, the application is not. What is damage? May other species survive while crayfish die? If a 20-m band round a ditch is polluted with nitrogen, and the rest of a large fen not, is that damaged? If a metal is precipitated and remains there inert for a century, and then becomes toxic with changing conditions, is that damage? If so, at what point: when the metals entered or when they became active?

Then, with an agreed definition – or even, at this stage, without one – what methods are to be used? As yet they are only being developed (see e.g. Smith *et al.* 1992).

Each community differs in its degree of tolerance to any pollutant, as it does to each habitat factor influencing the community. A community in the centre of its tolerance range for nutrients can of course tolerate quite a change in those nutrients. More, because there is no nutrient stress, it has more tolerance for being stressed by pollutants, water regime, grazing, etc. Physical and chemical factors act and interact to maintain or to destroy. A community stressed by, say, burning will be more susceptible to pollution. A change of pH 1 may alter a stressed community but not an unstressed one.

Habitats looking the same are often non-uniform. Species patterning may often reflect chemical micro-patterning (e.g. Hayati and Proctor 1991; van Wirdum 1991) allowing species to grow well in habitats appearing unsuitable, as when deep roots are bathed in alkaline water and fen plants can grow in a surface bog, or as when small changes in chemistry in that bog lead to different species of *Sphagnum*.

The most widespread significant chemical impact on British wetlands is through drying, drainage and abstraction. These may:

1 mineralise and oxidise peat and peaty soils, increasing nutrient status;
2 remove the groundwater under which a wetland developed, and without which it cannot retain its type;
3 gather and mix water types formerly with separate influences, altering chemical regime.

Table 7.1 extends these conclusions.

Table 7.1 Water quality comments (after Haslam 1994)

1	Removal of the quality and quantity of water under which a community developed will, in time, lead to the loss of that community
2	Water composition depends on water management and weather as well as on water resources
3	Man's impact on groundwater systems fragments these, so increasing chemical and hydrological separations
4	For the protection of a drying wetland, more water is needed of the original quality, not just water of any type
5	Communities are very sensitive to chemical composition, and vary with this. Mosaics and other patterns of vegetation develop with corresponding patterns of water quality
6	Drying (from drainage or abstraction) leads to irreversible changes in peat chemistry
7	Many fen wetlands developed with calcareous groundwater, which is being abstracted from the fen supply for domestic use
8	Polluted groundwater is increasing, and is likely to become unsuitable for sustaining some wetlands
9	Air pollution, now only locally severe, may increase, and lead to a wider range of influences (e.g. nutrient-rich, acid, heavy metal)
10	Mild or moderate pollution may take decades or even centuries to develop its full influence on an affected wetland, although severe pollution starts to act quickly
11	In the (in Britain, probably sparse) habitats where it is possible, rainwater filtered through peaty soil elsewhere in the catchment may be similar in composition to clean groundwater, and be substituted for it
12	Wetland communities, particularly those with large perennial dominants, may maintain themselves for decades in changed water regimes in which they can neither reproduce themselves nor become re-established after disruption
13	Community viability must be known before changes intended for conservation in chemistry (including pollution) or hydrology are approved

8   The power to purify

As reek o' th' rotten fens

(W. Shakespeare, *Coriolanus*)

Principles and definitions

The wetland plant controls, and so benefits from, the chemical and biological environment around its roots (its rhizosphere). It does this by numerous and complex chemical interactions (Neori *et al.* 2000).

Wetlands are chemical factories where many chemical transformations take place. Many inactive substances entering remain inactive, many active substances become immobilised, precipitated, broken down into simpler substances (that may or may not be less active). If, however, a wetland that has received pollution is drained or much disturbed, its stores of nutrients, heavy metals, organic substances may become available, and damage the wetland and wash out to damage rivers and beyond.

Chemical pollution is best defined as alterations by man to the chemical habitat that give rise to changes (varying according to the intensity and type of the chemical substance(s)) in the nature or structure of the plant or animal community (Haslam 1991; Descy 1976).

Pollution is shown by the effect on communities of substances added or removed by man. Pollution – like other impacts – begins changes moving habitats further from what they would be without people. So pollution is a process, a process acting on living things, be those things deer, mosses or microbes. It is an absolute (in theory: though in practice it may well be relative).

This is, however, far from the only definition. In the USA, structural damage (channelling, removing bank trees, etc.) is also pollution: a useful definition, combining all effects of humans. The *Oxford English Dictionary* gives two definitions, of which the earlier is 'to render ceremonially or morally impure' and the later 'to make physically impure, foul or filthy'.

By definition, therefore, there is no such thing as natural pollution. If it has been put there by birds, it is not pollution. As with all definitions, there is a grey area between. When a flock of geese or starlings deposit faecal material enough to alter the vegetation (and therefore fauna), then what? Today this may be due to or exacerbated by loss of habitat, so concentrating birds, but when waterfowl were a thousand times or so more abundant, and habitat likewise, flocks would still have come and caused changes (see, for example, the description by the twelfth-century author, Gerald of Wales).

Wetlands have the power to stabilise chemically. Water passing through is cleaned, is supplied to the river, clean. This power has always existed, always been used. It is only recently that its economic value and chemical processes have been recognised, and they are now made use of deliberately (rather than by accident or custom) for their value to people. This is in two ways. Buffer strips are strips of unpoisoned land (whether or not wetland) beside streams, to clean run-off coming from the land above. Such land has traditionally been present in flood plains: wet grassland, rush pasture, fen meadow, etc. In the past century, much has been lost: to intensive farming, roads, settlements, and other construction. Putting buffer strips back – in a few places – does a little to reverse this trend. It is to be hoped that projects organised by DEFRA (e.g. Environmentally Sensitive Areas), the Countryside Agency (e.g. Water Fringe Scheme), and others will come to develop these greatly. Wetlands are sensitive, finely tuned, adapted to their chemical status, and extremely responsive to man-made changes in that chemical regime.

The second use is as constructed wetlands for cleaning effluents. These differ in cleaning far greater concentrations of pollutants, and being specially made, not just using stream land already present. While in one sense this has always been done – leading the sewer to the pond or marsh – its modern development dates from the pioneer work of Seidel (e.g. 1956–7) on, mainly, *Scirpus lacustris*, showing the impact of plants on then-untreated sewage effluent in rivers. Constructed wetlands have now spread across the globe. For the same cleaning power, much more space but less cost is needed than with conventional sewage treatment works. Their use is therefore specialised but widespread, and still spreading.

There is overwhelming evidence for the efficiency of swamp-marsh constructed wetlands for purification. This suggests that when these do not work properly, something is wrong, and the wetland should be altered. Research on those that do work is available! The biological properties are those of the natural wetlands.

The term 'enrichment' is sometimes used to describe man-made increases in nutrients (such as fertiliser) or in organic wastes. 'Enrichment' infers improvement. Most people think that being rich is better than being poor, so that an 'enriched' site is a 'value-added' site. These places are polluted, though: they have been moved further from their proper chemical state. And a site 'enriched' by organic waste is 'enriched' not just by nutrients but also by numerous organic substances, likely to include detergents, pesticides, polycyclic aromatic hydrocarbons and many other poisons. Sewage and farm wastes pollute. So does pure milk.

'Eutrophication' and 'acidification' originally described natural processes, increasing, respectively, food supply and acidity. Natural eutrophication occurs when, over time, a wetland starts to receive nutrient-richer water after a natural change in hydrology or catchment fertility. Man-made 'eutrophication', as used in this book, is used for an increase of inorganic nutrients leading to changes in habitat and community. Other definitions include increases in (1) nitrogen and phosphorus; (2) nitrogen, phosphorus and potassium; (3) organic status; and (4) any pollution including nutrients or animal food, e.g. road run-off.

Acidification strictly means changes in acidity (pH), not in food supply, although the word is usually used as the opposite of eutrophication. Natural acidification occurs when rain-fed bog develops over fen or marsh fed by surface water. Man-made acidification is now mostly from air pollution acting on solute-poor habitats.

As there are no full analyses of any wetland soil or water, there must be many, many polluting substances that are not known. Therefore, lists of hazardous or significant pollutants are incomplete – and, because of too few analyses, the stated harmful concentrations are but preliminary. Within the next decade information should much improve.

Unknown, also, are interaction between pollutants including synergistic ones. Little of the enormous research has been done on wetlands.

How fens, marshes and reedswamps (natural and constructed wetlands) act chemically

These filter, transform, and vary in absorption and storage (Reddy and Smith 1987).

For rapid breakdown, appropriate microbes, good substrates for them, and good other habitat conditions are needed. Conditions for breakdown are different for different substances. A variety of plant species and of organic materials give 'hot spots' for the speedy decomposition of a multitude of substances. These include, for example, aerobic/anaerobic and acid/calcareous conditions. Transformations are usually most when there are mosaics, as in an anaerobic waterlogged soil criss-crossed by roots of different species, all with oxygenated rhizospheres, each with different substances exuded from their roots. Alternatively, rising and falling water brings anaerobic/aerobic conditions. In this demanding habitat, many interesting, and often unique, defensive and aggressive chemical and biological–chemical processes have developed in and around the rhizosphere. These influence plant–pest and plant–plant competition (Neori 2000).

Oxygen in anaerobic soils comes in mainly through the root tips, before they are thickened further back. Oxygen diffuses from the root and reaches the root from the air and shoot by a variety of physical mechanisms (Kvet and Westlake, in Westlake *et al.* 1998). Roots with through-flow have longer lengths giving off oxygen, so are the more effective (e.g. *Phragmites, Typha* spp.). Wind powers this convection, and increases gas flow in the growing season (Armstrong *et al.* 1992; Brix in Moshiri 1993). This oxygen controls many biochemical and biological activities, e.g. methane release from peat, perhaps (Neori 2000).

Many chemicals are bio-active, altering the behaviour and activity of organisms. In wetlands these occur in the soil, the plant and the plant exudate (rhizosphere) (Neori *et al.* 2000). The chemical structure of humic material influences breakdown. Herbicides, for instance, may be degraded via microorganisms, via the non-organism route or be transported down the soil layers, and which of these happens depends on the humic material and how it is linked to mineral soil present. Glyphosate, when adsorbed in dissolved humic substances, may be easily transported, so is likely to reach fresh waters (Piccolo *et al.* 1996).

Different root exudates, so different microbial populations, are found in different species. Good species for degrading toluene, for instance, are *Iris pseudacorus, Eleocharis palustris* and *Juncus inflexus*, while *Scirpus lacustris* is particularly good at removing bacteria and phenols. The rhizosphere organisms influence nutrient availability and plant growth regulators. They stimulate growth, produce allelochemicals, and alter the root environment to increase growth (Gunnison and Barko 1989; Reddy and Smith 1987; Haslam 1987).

Exudates include, among many more (Shimp *et al.* 1993; Neori *et al.* 2000):

Amino acids	Derivatives of fatty acids	Elemental sulphur
Enzymes, e.g. ferric iron chelators	Ethylene	Flavonones
Glutanases	Growth factors, other hormones (gibberellin, etc.)	Hydrogen cyanide

Invertases	Long-chain fatty acids	Nucleotides
Organic acids, e.g.	Peptides, metal-binding	Peroxidases
binders of heavy metals	Phytoalexines	Proteases
Phosphatases	Steroids	Sugars
Rotenones		

They vary with:

Plant species	Microbes	Light
Age of root	Water	Nutrients
Competition	Temperature	

Microbes are much denser in rhizospheres: because of the good substrate and habitat. Mycorrhiza may help in degradation, e.g. *Salix* spp., *Populus* spp. There are nitrogen-fixing microbes in root nodules of *Alnus glutinosa* and *Myrica gale*.

Soil microbes are very widely distributed. Generally, given proper conditions, the species explode. Very toxic substances (such as effluent from chemical factories) can be degraded, as can glyphosate (herbicide) and many aromatic compounds, by *Rhizobium* (Shimp *et al.* 1993).

Alternative microbes may be available to do the same process in different conditions, for example *Azotobacter* fixes nitrogen well at over pH 6 and with abundant organic matter; *Clostridium* fixes it at low pH.

In wetlands flooded enough for a good algal flora to develop, periphyton – thick on plant parts in moderate pollution – add an extra purifying component. The algae remove and store many nutrients, heavy metals, etc. While these return to the system on death, it is often in more acceptable forms (e.g. Lakatos and Bio 1991; Hammer 1989).

Roots take up nutrients, even surplus ones. They take up metals, organic compounds, etc. As noted in Chapter 7, many metal and organic pollutants stay mainly in the roots, although nutrients, necessarily, move up and are used in the growth of the shoots. Some uptake in land species is shown in Table 8.1 (wetland data are not yet available). The least mobile molecules hardly enter roots (low water solubility and low fat solubility, like polycyclic aromatic hydrocarbons).

Purification

A very wide range of pollutants has been tested at the in-and-out level of Table 8.2, and the best results from wetlands are better than those from ordinary sewage treatment works. Obviously effectiveness varies with wetland type, plant species and ecotype, design and construction of the wetland and its maintenance, and the climate (Table 8.3 shows microorganism loss). (See volumes edited by Athie and Cerrio, 1987; Cooper and Findlater 1990; Mitsch 1994; Moshiri 1993; Reddy and Smith 1987; Rubec and Overend 1987; Vymazal *et al.* 1998; Vymazal 2001.)

There are many non-quantified data, for instance that atrazine is relatively stable, and that cyanazine degrades faster in nitrate-reducing conditions (Gee *et al.* 1992).

When an effluent changes, as when a chemical factory switches to a different product, it takes about three weeks for the microbes to change and be as efficient purifiers of the new as of the old effluent.

The time the water stays in the system to be worked on, its residence time, is important. So, of course, is whether the system is intended to deal with solids as well as liquids (e.g.

Table 8.1 Some organic compounds taken up by the roots of land plants (from Shimp *et al.* 1993)

Substance	Species	Substance	Species
Sulphonamides	Broad bean	Polycyclic aromatics	Many species
Anthracene Dichlobenil	Bush bean	2, 3, 7, 8–Tetrachloride benzo(p)dioxin	Many species
Organochlorine insecticides		Bromacene Dichlorobenzonitrite Phenol	Soybean, barley
Benzenes Substituted benzenes	Soybean	Nitroquanicline	Soybean, fescue, brome grass
PCBs	Many species, e.g. Loosestrife	Nitrobenzene	Soybean, barley, poplar, Russian olive, green ash
Atrazine	Barley		
DDT Chlorobenzene (*o*)-Methyl carbamal Phenylurea			
Asulam Bromacil	Maize, bush bean	Dieldrin Chlordane Heptachlor	Eight field crops

Table 8.2 Examples of chemical removal in constructed wetlands (see e.g. Athie and Cerri 1987; Hammer 1989; Reddy and Smith 1987; Rubec and Overend 1987)

	% removed
Nitrogen	25–97
Biological Oxygen Demand (B.O.D.)	55–90
Suspended solids	54–98
Iron	82–99
Toluene	99
Chloroform	32
Benzene	99
Tetrachlorethylene	25
Phosphorus	20–99
Chemical Oxygen Demand (C.O.D.)	85–95
Manganese	9–98
Pathogens[a]	86–99
p-Xylene	99

[a] Faecal streptococci remain the longest, so they, not *Escherichia coli* should be measured.

raw sewage), with just liquid effluents or as the third-stage (tertiary) treatment, to improve an adequate effluent to a polished, good one.

Phragmites is the commonest species in constructed wetlands in temperate Europe. It is ecologically and physically tough and resilient, with good root and rhizosphere structure and chemistry. *Typha latifolia* is commoner in North America – in natural as in constructed wetland. In warmer countries, floating species are favoured, e.g. *Eichornia crassipes, Lemna* spp.

Table 8.3 Factors that affect the survival of enteric bacteria and viruses in (dryland) soil (from Novotny and Olem 1994)

Factor		Comment
pH	Bacteria	Shorter survival in acid soils (pH 3–5) than in neutral and calcareous soils
	Viruses	Insufficient data
Predation by soil microfauna	Bacteria	Increased survival in sterile soil
	Viruses	Insufficient data
Temperature	Bacteria and viruses	Longer survival at lower temperatures
Sunlight	Bacteria and viruses	Shorter survival at the soil surface
Organic matter	Bacteria and viruses	Longer survival or regrowth of some bacteria when sufficient amounts of organic matter are present

Design of constructed wetlands varies from the simple pond with flow control to the complex, with specified and varied substrates, and flow directed, downwards or sideways in the soil. Unfortunately, there is very little information on the associated flora and fauna, or even whether the habitat is or should be mown annually, so removing a small but significant repository of nutrients. (Mowing of course removes less of those substances staying mainly in the roots, such as heavy metals and various organic pollutants.) Although processes are slower in the cold, biological degradation occurs at less than 5°C (Jensen *et al.*, in Mitsch 1994), and in really cold climes as in Canada, effluents are stored in winter (Lakshman, in Mitsch 1994).

Such constructed wetlands are used for: domestic sewage, urban run-off, livestock unit effluent, agricultural run-off, factory effluent (milk, chemical, vegetable, potato, meat processing, sugar, jam, paper, etc.), mine drainage (acid, neutral, sediment-rich), farmyard run-off, dairy run-off, manure, silage run-off, landfill run-off and leachate and similar.

Good removal occurs for nitrogen because it leaves for the atmosphere, and for toxic complex products that are broken down to simple ones (e.g. water, nitrogen, carbon dioxide, methane), substances either going to the air or becoming normal constituents of natural wetlands and not altering properties. Particles of all kinds are filtered out well, but their accumulation may alter the system and its functioning, especially the through-flow. Then there are the substances immobilised in some way, such as heavy metals, which may require soil treatment or removal at intervals to prevent the soil becoming over-saturated and so no longer removing the substances from the incoming flow. Phosphorus has been much trouble (see Chapter 7), and led to constructed wetlands being refused for the treatment of effluents containing this. Recently, however, designs are appearing that do seem to be satisfactory. Dosing with alum (or alum plus lime) increases phosphorus removal, with the bed expected to last for 40–80 years (Davies and Cottingham, in Moshiri 1993).

It is suggested that clay is the best substrate: good for adsorption, ion exchange and chemical reaction to inert insoluble forms. There is little removal in a gravel or sandy bed.

Another criterion determining design is whether the incoming flow is steady, as with sewage, or very variable, as with storm water run-off, which may indeed be grossly toxic, but comes only after rain, and the system must accommodate such fluctuations. Such need careful design (e.g. Mitsch 1993).

Mine effluents are, of course, metal-rich. Long-term metal retention depends on transforming oxidised metals to reduced mineral phases. Chemical treatment of the effluent may be needed before it passes to the wetland (Brodie, Lon *et al.*, in Cooper and Findlater

1990). From the nature of the country, *Sphagnum* is often the most available local wetland type, although tall monocots such as *Typha* spp. may be planted. Obviously, cleaning poisons that should never be there in the first place is a second-best, and careful and stringent measures should be taken to decrease this source of pollution to wetlands.

Complex patterns may be needed to remove the trace metals from acid mine drainage, for example wet meadow, sedges and grasses, and *Typha latifolia* with straw at the bottom for increased sulphate reduction (Eger *et al.*, in Moshiri 1993).

In North America, particularly, there is field landscape planning for water purification (see Table 8.4). Natural existing wetlands should not, of course, be put at risk of pollution. New, or renewed, wetlands may be placed in strategic places to catch poisonous water (poison that in theory should not be there, but in practice is: farm spills, rural garage and village run-off, etc.) as well as those waters unfortunately still intended to be polluted, such as run-off from busy roads (most unpleasantly poisonous). These should be placed to be as near-natural as possible (e.g. adding a small dam in an existing hollow), of natural shape, needing minimum maintenance (e.g. Mitsch, in Olson 1992). They need not be single habitats. They can be patterns of, say, run-off passing down a wet meadow slope, into a pond, then to a marsh: a simple 'natural' pattern, common everywhere in the past! Such replaced wetlands, however, unlike the originals, are receiving pollution. Pollution sinks down as well as moving with water flow downstream. When placing these, care must be taken not to put them where aquifers could be polluted.

Constructed wetlands are not just microbial systems. Microbial systems are well known, used in conventional sewage treatment works, and if only these are wanted, there is no need to taken up the extra space needed for constructed wetlands.

The plants are a necessary part of the constructed wetland proper. *Phragmites,* for instance, is particularly useful in collecting and converting sediment and providing pathways for filtering down from the top. The root system grows and extends each year, and dead as well as living roots keep the filter permeable. The plant litter brings in cellulose, etc., so brings more humus. The oxygen supply to the soil comes from root oxidation, and pores formed by the combination of root presence and growth, and from the water moving up to the shoots for transpiration. There is a large surface for microorganisms. *Phragmites* mediates aerobic, anoxic and anaerobic zones, giving best conditions for mineralisation, nitrification and denitrification. The plants are indeed valuable (Hofmann 1991)!

If wetland-treated water is not fully clean, it may yet be good for hydroponic crops, e.g. sugar beet, broad bean, sunflower, safflower, cotton, sachem, maize, etc. (Butler *et al.*, in Moshiri 1993).

Buffer strips

Beside the river, or indeed beside the wetland, uncontaminated soil, with the vegetation on it, aids the cleaning of run-off and so decreases the contamination of river or wetland (Figure 7.1). It protects the river (or wetland), to some extent, from the pollution from the land around. The main economic use is to protect water needed for domestic supply. Buffer strips filter both overland and subsurface flow, and transform many input chemicals to more acceptable and less polluting forms. Sediment removal can reach up to 90 per cent, as can total nitrogen removal. Phosphorus removal is less, but in the short term can reach to 75 per cent of total phosphorus. Sometimes, however, no phosphorus is removed in the growing season. Dry grassland removes less than taller mixed vegetation. Pesticides are removed, too, but the amounts are not known (Novotny and Olem 1994). Where the

Table 8.4 Selecting best management practices by pollutant: rules of thumb (after Novotny and Olam 1994)

Pollutant	Methods of control	Vegetative
Sediment	Control erosion on land and stream bank Use best management practices that capture sediment Dispose of sediment properly	Cover crops and rotations Buffer strips
Nutrient and miscellaneous effluents and run-offs Animal waste	Minimise sources Take into crop all that is applied to the land or contain and recycle/re-use	Crop rotations and management Cover crop, buffer strip, change crop or grass species to one that is more nutrient demanding
Pathogens (bacteria, viruses, etc.)	Minimise source Minimise movement so bacteria die Treat water	Buffer strips Constructed wetland/microbial filter
Metals	Control soil sources Control added sources Treat water	Crop/plant selection Crop selection Constructed wetland/microbial filter
Salts/salinity	Limit availability Control loss	Crop selection, saline wetland buffer, land-use conversion
Pesticides and other toxins	Minimise sources Minimise movement and discharge Treat discharge water	Plant variety/crop selection Buffer strip, wetland enhancement Constructed wetland
Physical habitat alteration	Minimise disturbance within 30 m of water Control erosion on land Maintain or restore natural riparian area vegetation and hydrology	Buffer strips Wetland enhancement

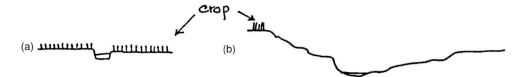

Figure 8.1 Stream banks as buffer strips (Haslam 1994): (a) of little use; (b) effective.

Figure 8.2 Buffer strips of tall herb and carr vegetation.

buffer strip is itself a reedswamp, etc., the behaviour has been described above. That of simpler, usually drier systems, is described here.

Buffer strips were – and often still are – thought of as good farmland wasted. The protective value is being recognised. If it is assumed water should be cleaned to the highest current standards, the cost of doing so by sewage treatment works and by buffer strip can be compared. Using buffer strips 10 m wide beside the stream, the cost of the buffer strip is recouped within two or three years, in Sweden (Torstenson, in Boon *et al.* 1992).

Banks with soil and vegetation purify (Figure 8.1). Beside the bank there may be a hedge, or a narrow tall-herb strip. These, too, purify. There may be riverside vegetation of various kinds (Figure 8.2). The strip may be set aside for this purpose, or may have been left as inconvenient to plough, etc.

If the water level is high in the buffer strip, much of the subsurface run-off moving downslope through it to the river will pass through the rootzone. In drier places, most run-off moves below the roots. Even so, the vegetation on the surface helps the carbon supply, and makes the soil better for degradation.

In farmed lowlands, run-off from grassland will contain at least nitrate, and that from arable the full range of fertilisers and biocides appropriate to the crops. In town centres, run-off is usually sent to the sewage treatment works, polluting the river but not the riverside. Outside the town centre, run-off from roads, drives, car parks, house tops and industrial sites brings a poisonous cocktail of pollutants, some of which may come through buffer strips.

There are two separate, though linked, processes. In the growing season, nitrate is removed by plants, using it in the structure of the plant. In the winter, nitrogen leaves the land as gases, water level in riverside damp grassland rises, the soil becomes anaerobic, and there is much denitrification. To enable much nitrogen processing in this unstable habitat, both the summer and the winter types are needed.

Mowing (and taking away the mown vegetation) removes the nitrogen – and other nutrients – in that vegetation, but prevents the organic carbon of those plants going back to the soil. The decision varies with the circumstances. Riverside grass in narrow buffer strips should be unmown. In wider strips it can, at least, be grazed (manure returning). Wetlands subject to eutrophication may be mown annually to keep down total nutrients.

Nitrate loss is poor if there is too little organic carbon (little or no vegetation), or if there is too little of either or both of summer-drying and winter waterlogging (water levels too regulated or too low).

Most (uncontaminated) riverside land in Britain is grass. Poplar wood has better resilience, for cleaning. When nitrate-rich water flows in, in conditions where a (chosen) grassland retains 84 per cent, poplar wood keeps 94 per cent. Grass retains the most water after peak flows, over 90 per cent. The least is when the winter water level rises, only 20 per cent–30 per cent. Poplar is better: the more diverse, and the larger the plants, the better. Willow has more recently been shown to be very effective, at least for nitrates, some pesticides and perhaps phosphate. Coppicing (removal of pollutants stored in plants) improves removal. Willow woods are a more traditional riverside community than poplars so, if effectiveness is proved equal, they are to be preferred. Alder, the other common riverside tree, has root nodules that fix nitrogen, up to $22.5 \text{ g/m}^2/\text{year}$, so alder carr can be a source of nitrogen, not a sink. In one example, the denitrifying power of the buffer strip was 30 kg/ha/year. The incoming nitrogen, however, was 40:10 from the alder, 30 from run-off. This meant an annual surplus of nitrogen of 10 kg/ha/year to be incorporated into wood or sent outside the strip (i.e. to the stream) (Boon *et al.* 1992). Without fertiliser, or without the alder, the strip would have disposed of the nitrogen. With both, it could not. Alder carr is not, therefore, the best choice for nitrogen disposal!

Nitrogen is lost as gas. Most other substances are not, but may be rendered harmless or washed out. Calcium may be taken out as soon as nitrate, or may, as in Figure 7.1 penetrate a large area (Boeye 1992; see Chapters 6 and 7). Phosphate retention is variable and not predictable (it depends on water contact with soil, soil composition, discharge duration, iron and aluminium content, etc.). Soluble phosphates in the run-off (from fertiliser, for example) is easily taken up by plants. In seasonally waterlogged rough grassland (Devon) soluble phosphate was removed: except where it came through a ditch. (Ditches must be obstructed, for maximum cleaning.)

There are Nitrate Sensitive Areas (NSAs) where farmers can be compensated for using less nitrate fertiliser. The scheme, although excellent in theory, still needs fine-tuning. It, for instance, does not separate lands contributing nitrate and those reducing it. There needs to be provision for protecting all water resources, ditch, brook, river, lake and aquifer.

Most work has been done on nitrate, because of its commercial importance and its easy research. However, land where the varied processes of nitrification and denitrification are encouraged is likely also to be land where other pollutants are broken down or immobilised.

Drainage in wetland makes nitrogen available that was immobilised. This means drained and semi-drained lands can have more denitrification, provided they are still waterlogged in winter. The maximum potential denitrification in wetlands has been estimated as up to 300 kg/ha/year, which is five to ten times that in other habitats.

When run-off comes downslope to the buffer strip, it does not flow straight down to the river. It wiggles, and can in fact have a flow path 150 m long in a buffer strip just 15 m wide. This means that the width of the strip does not matter much for denitrification: it is the waterlogging and the organic carbon that are needed.

Site is important. Waterlogging is needed in winter, from river, run-off or spring lines. Restoring the old winter-wet water levels can be worthwhile. Good areas are those with good denitrifying habitat: wet and organic carbon (e.g. old ox-bows). Dryland sandy areas may leave nitrate almost unchanged. The wet, denitrifying environment is the one to concentrate on (based on the knowledge available in 2000).

In summer, the width is important: the wider the strip, and the larger the bulk of vegetation, the more nitrogen is taken up. The time water spends in the buffer strip affects the outcome: the longer, the better the cleaning. This residence time is determined by the soil hydrology, hydrology and drainage. These are controlled by the farmer or the internal drainage board.

How wide should the buffer strip be? Widths are often given on what pleases the land managers. A 5 m strip is useful, 20 m often adequate (for grass or wood) and 200 m has always removed nitrate (in suitable habitats). These are usually recommended downstream. However, it is the upper tributaries, running directly through crops, that are at least as much at risk. Pollutants are high, dilution is low. Downstream, dilution is more, and there is often riverside grassland: giving less pollution than arable.

Bulkier, more diverse vegetation (giving quantity of vegetation and variation in rhizospheres) and variable (suitable) soil types presumably clean most. Wetland peats are potentially the most effective, but because of the various possible limiting factors (including low dissolved organic carbon, type of humic substance, percolating rain stressing the microbiota, local acidity, etc.) a mineral soil may be as good as, or even better than, peat. The sink capacity is important: humic substances can be crucial. As these vary, so does the sink capacity.

Buffer strips protect the river: provided, that is, the run-off comes through them, not through underdrains! Underdrains should be stopped at the buffer strips, their water passing through the strip or through (open) vegetated ditches. Small effluents can also be cleaned well, with narrow seepage zones between them and the stream. Buffer strips make rivers less sensitive to land use.

Riversides may have dry soils, because of drainage, and these are of course not as good for denitrification. They may have gravel or other free-draining soils. Here the denitrification depends on the structure, whether there are, for example, soil aggregates that can retain water and so become anaerobic.

Buffer strips can also be designed for overland flow, but in Britain this is less common (it needs much tall, thick vegetation). (Material in this section is mainly from Boon *et al.* 1992; Edwards 2000; Gaffney and Rose, in Heathwaite and Hughes 1995; Haycock and Burt 1993; Haycock and Pinay 1993; Haycock *et al.* 1993; Pinay and Decamps 1988;

Prach and Rauch 1992; Reddy and Smith 1987; Russell and Maltby 1994; Weller *et al.*, in Mitsch 1994.)

Conclusions

In general terms, denitrification accelerates as the soil becomes waterlogged in autumn, and slows as the saturation level falls in spring. The increase of nitrate by nitrification is most in the other half-year.

In general terms, run-off brings agrochemicals most when, first, there is enough rain for run-off (more in winter than in summer) and, second, when the agrochemicals are most easily washed off (that is, soon after application). (It should be remembered that the farmer is not the intentional villain. Farmers do not want the waste of losing agrochemicals to run-off any more than conservationists or hydrologists want the consequences of that waste.) The same applies to road run-off: after rain, and most when the poisons are most. This happens after a long period without rain. Effluents are likely to be independent of rain, and may run part-time or full-time.

These simple principles may become more complex in their application, and this also must be stressed (Table 8.4). In the Somerset moors, the fields are sources of agrochemical pollution to the ditch and ultimately to the river systems. When the water flow is from field to ditch there is therefore pollution risk. This happens in winter and spring, when ditch water is kept low, and there is plenty of rain and so run-off. In summer and autumn, flow, if any, is generally from ditch to field, so the ditch water is unlikely to become polluted. Here impact and also climate dictate the water levels and their flow, as is common in large, drained wetlands. This must therefore be taken into account.

9 *Phragmites*: a study in plant behaviour and human use

With his ear to the reedstems, [the Mole] caught, at intervals, something of what the wind
wet wispering so constantly among them.
... the wind playing in the reeds and rushes and osiers

(Kenneth Grahame, *The Wind in the Willows*)

Even now, there are only few that go down into the reedbeds for pleasure or profit (profit
being either income or the advancement of science). Those of us who do, find the
pleasure as well as the profit. 'Solitary' is the proper word to describe being all day
surrounded by reeds, seeing only reeds and sky, hearing only reeds and birds. A unique
experience.

And *Phragmites* is a unique plant. It is found as a native in all five continents. In
Europe, it is the chief peat-forming reedswamp species, this peat testifying to its past
extent, which was even greater than its present spread. (In other continents other species
are of equal or greater importance.) *Phragmites* has substantial, though minor, economic
uses. The main one is, and has been, thatching. Smoke-blackened thatches go back to
Anglo-Saxon and mediaeval times (Letts 1993). These are thatches from pre-chimney
days, when the hearth fire blackened the roof. No doubt ever since early Britons lived near
reedswamps there have been reed thatches. Old uses, now largely discarded, included
fencing, inside walls, drink (rhizomes: as coffee), medicine, stuffing bolsters (fruiting
heads), animal bedding, packing, insulation. A renewed use is to prevent river bank
erosion, either by the living plant (where erosion is minor enough for protected reed
plants to survive) or as, for example, thick matting made of reed bundles, kept in place in
the banks by stakes. The new use, sprung up across Europe, is as the main plant of
constructed wetlands for water purification (see Chapter 8). For this, *Phragmites* is not
just effective, many species are that, but it is tough, tolerant and easy to grow (in North
America *Typha* largely replaces it, for the same reasons).

Phragmites is remarkable in one unfortunate way: its name. It ran across an odd rule
of botany, that the earliest mention in the literature of a plant by two Latin names
(binomial) is definitive, the plant's proper name. From the early nineteenth century up to
the 1970s this plant was *P. communis*, perfectly proper. Then trouble! The Australian reed
had been called *P. australis*, to distinguish it from the European one. When looked at
more closely (without genetic testing) it seemed the same. (I agree, I tried it too.) When
P. australis had been named in 1799, however, the European reed was still known as
P. communis europaeus, three names, so invalid. So instead of being 'common', all reed is
now 'Australian'.

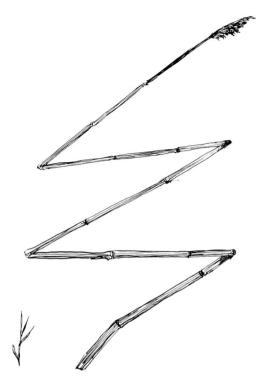

Figure 9.1 The long and the short of it. Two fully grown British shoots (the Giant *Phragmites* of the Danube delta reaches over 8 m).

Reed is extraordinarily variable. It varies in size; Figure 9.1 shows small and large British reed (not as large as the more than 6 m of the Danube delta, though). It varies with genes (with clone), with all aspects of habitat from temperature to history (Table 9.1). So many features vary, whether leaf tips are thin or hard enough to scratch skin, the shape and number of the fruiting heads, the colour the plant is on infrared colour photographs (!), and much more.

Reedbeds and reedswamps (dominated primarily by reed) (Table 6.1 and Figures 4.18 and 4.19) grow in a range of shallow waters and – with management to prevent invasion by land plants – intermittently dry places. Where it dominates, it casts a shading canopy, and can build up a litter mat (poisonous to some other species, perhaps also to itself). But *Phragmites* is found over a much wider range too, in ploughed fields, dry land wherever roots can reach water (several metres down), deep water (wherever enough shoot can be above water for good photosynthesis), bog (with a little minerals), brackish rivers and estuaries, polluted waters, and sparsely in carr (where it is weak and pale), tall herb (where it is strong but usually not tall), short sedge/rush, grazed meadow (short). It even comes up through (poor-grade) tarmac and sedge tussocks.

In any European wetland, the question is more likely to be 'Why is *Phragmites* absent?' than 'Why is it present?' It is absent, for instance, from flood plain gravel with fluctuating water, from areas with heavy grazing, from banks of swifter rivers, from (full) bog, and

Table 9.1 Variation in *Phragmites* (Haslam 1995)

Character	Type of variation					
	Clonal	Spatial[a]	Transplant	Management of reedbed[b]	Annual[c]	Climatic[d]
Stem, summer						
height	+	+	+	+	+	+
width	+	+	+	+	+	+
density	+	~	+	+	+	+
colour	+	+	+	+	+	+
Leaf						
size	+	+	+	+	+	+
texture	+	+	+	+	+	
shape	+	+	+	+	+	
number	+	+	+	~	+	
autumn colour	+	+	~	~	+	
Inflorescence						
density	+	+	+	+	+	
shape	+	+	+	+	+	
size	+	+	+	+	+	
seed germination	+	+(?)	~	~	+	
Timing						
emergence	+	+	~	+	+	
growth rate	+	+	~	~	+	+
flowering	+	–(?)	~	~	(little)	+
fruit dispersal	+	–(?)	~	~	+	+
leaf fall pattern	+	+	~	~	+	+
longevity	+	+	+	~	(+)	+

Legehalme					
production	+	+	+	n/a	+
Rhizome					
density	+	+	n/a	?	+
cross-section shape	+	+	+	?	?
depth in soil	+	+	n/a	?	?
branching of verticals	+	+	n/a	+	+(?)
stimulation of buds on upper verticals	+	+	+	+	+
Infrared colour					
(in late summer)	+	−	?	−(?)	−(?)
Rhizosphere/root, etc.					
factors in competition with other species	+	+	n/a	?	?
breakdown of pollutants, etc.	+	?	?	?	?
tolerance of salt	+	−(?)	−(?)	−	−
Reed (dead)					
straightness	+	+	+	+	+
strength	+	+	+	+(?)	+
wall thickness	+	−(?)	+	?	+
wax, sheen	+	?	+	?	+
colour	+	?	+	?	+
appearance	+	+	+	+	+
nutrient content	+	+	?	?	+

n/a, not applicable.

[a] Variation across the population, from any cause (e.g. competitive species, water depth, unseen clonal variation).

[b] Management for reed production only.

[c] Variation from year to year from any cause (e.g. summer temperature, grazing intensity, flood).

[d] Britain (latitude 50–58°) and Malta (36°) (that contain all other variables also).

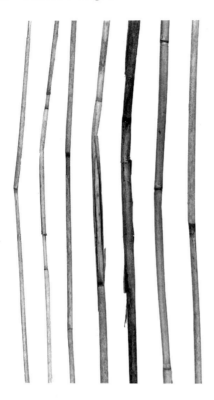

Figure 9.2 Differences in reeds from different reedbeds.

from communities of over-competitive species: although few in this last category have no *Phragmites*. Most have it sparsely.

What makes *Phragmites* able to tolerate South America and Finland, near-bog and rich silt, near-dry and flooded, carr and water lilies? First, the inherent nature of the species, although this is not yet understood: nor why it is so much less tolerant in North America, where it is only locally abundant, and more typical of estuaries, and in most wetlands is quite absent. Europe is the centre of variation (e.g. Rudescu *et al.* 1965; Kühl and Neuhaus 1993; Continental Floras) and so, presumably, of origin. It has numerous types (Table 9.1, Figures 4.19 and 9.2) in terms of the kind of stems, leaves, etc., and quite possibly far more for chemical and other metabolic tolerances and preferences, too. Ample variation facilitates the wide habitat range. *Phragmites* clones can live a long time, which helps abundance and stability. Reedbeds are estimated as up to 1000 years old (Rudescu *et al.* 1965). This gives great resilience, acres of reedbed persist vegetatively. If one part is disrupted, it can be recolonised, by rhizomes. There is much variation, a longer life than most trees, a great habitat tolerance; and it is an abundant plant of high biomass.

The plant unit

Underground, a *Phragmites* plant from a dry, sparse stand is depicted in Figure 9.3. A new length of horizontal rhizome grows each year (from July, when the new food starts going

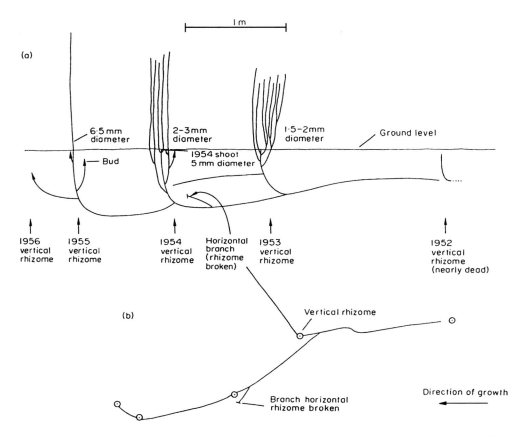

Figure 9.3 Vertical (a) and horizontal (b) plan of a reed plant in a dry, sparse, short stand. The horizontal rhizomes determine the pattern of growth. As vertical rhizomes age, their aerial shoots arise closer to ground surface, and are narrower and more numerous (Haslam 1969a).

down from leaf to rhizome); in about November, the tip turns up, and next spring bears the widest and tallest shoot of the plant, the king reed. The next year this same vertical rhizome bears more and shorter shoots – each year the buds grow from higher up the rhizome: so are narrower, with shorter shoots. The individual plant unit lives approximately four years (can be between three and seven years), dying at the back, and growing on and on through the stand, and as far out as it can go in present circumstances.

In better conditions, rhizomes are much denser, more branched, with horizontals and verticals less clearly separated (Figure 9.4) and far more middle-sized shoots. The depicted plant does not come from a well-grown stand! It is restricted by a dry habitat and by competition in a tall-herb stand. It is restricted down to the minimum for survival. In this state, *Phragmites* moves around: and if it meets better conditions, it can branch and grow and dominate.

These three types of stem, horizontal and vertical rhizomes, and the aerial shoots have different functions. The horizontals maintain, renew and so control the structure of the stand. Their leaves are small, their roots long, and they stay wide. Verticals, near the soil surface, bear short much-branched roots that in dense stands form a root felt. They

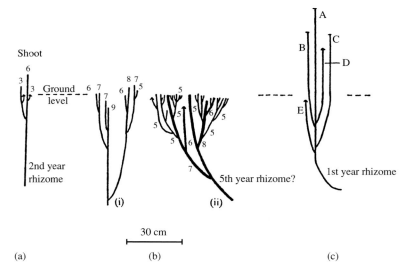

Figure 9.4 Pattern of vertical rhizomes (numbers are widths in mm). (a) A sparse, dry stand, showing high-level branching, buds (often) similar in width on each rhizome, but differing in width with the age of the rhizome (winter). (b) A dense stand, showing part of the complexity of branching and variation in level of bud origin (winter). (c) Advancing into a wet habitat, showing development of several shoots during one summer on a first-year rhizome (September). A, first shoot, 6 mm wide at base, 150 cm high, mature and flowing. B, second shoot, 5 mm wide, 75 cm high, mature. C, third shoot, 5 mm wide, 70 cm high, growing slowly. D, fourth shoot, 3 mm wide, 25 cm high, growing rapidly. E, fifth shoot, 6 mm wide, a bud for next year's growth (Haslam 1970b).

narrow as they grow up, so cannot maintain or renew themselves (except by bearing horizontals). Aerial shoots narrow even more. This means that their potential height is determined by the width at emergence. (There is a limit to narrowing!) This potential height may not be reached: conditions may not be good enough. It is very rare that this is exceeded. (One instance is by excessive internode growth on reeds within reasonably lit bushes.) Aerial shoots have large bladed leaves, flowers and fruits. There is a fourth stem type, *legehalme*, intermediate between aerial stem and rhizome. They are found in some few clones growing from underground. Clones with this genetic ability exercise it only in some habitats, for instance on just one side of a clone where the other sides are bounded by dry soil or thick vegetation. *Legehalme* also occur when growing aerial shoots have fallen onto water and continue growth. They grow about 1–10 m long and can bear lots of aerial shoots. They may also bear rhizomes, so thus spreading the plant very quickly indeed, where habitat permits. It is possible all clones can bear *legehalme*, but there is a genetic variation in the conditions in which they can grow, so most clones have restricted conditions that are rarely met.

The seasonal cycle (Figures 9.1 to 9.11)

The first buds – mostly wide ones at the ends of long rhizomes – are formed, near the soil surface, in autumn. Development is very slow in the cold winter, few buds develop, and those already there make nodes with short internodes (Figure 9.7). Main emergence takes about a month, usually starting in March or April in Britain (it can be late February to early May). The time varies with the clone, the temperature (later in colder weather), and other factors. Then, on average, around half the total buds develop in spring (more if frosts, litter removal, etc. encourage buds, less with a litter mat, stable temperatures, etc.). Generally, wider buds – taller shoots – come up first, each week's intake being progressively narrower in the bud and shorter in the final height. Wider buds come up first: and look different to shoots from narrow ones (Figure 9.5).

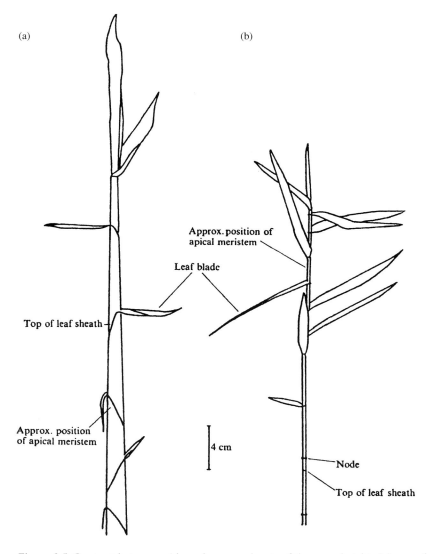

Figure 9.5 Contrast between wide and narrow shoots of the same height: (a) grew from a wide bud, and has the potential for growing tall; (b) grew from a narrow bud and is already almost full-grown.

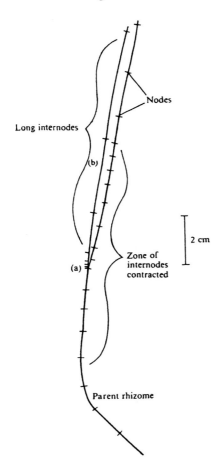

Figure 9.6 Growth of buds at ground level. The parent bud was formed and grew near the surface in the autumn. It grew very slowly during winter dormancy, with many short internodes. In late spring, a side bud (a) grew quickly and, without dormancy, formed an aerial shoot (b) without the stage of short internodes (Haslam 1969b).

Emergence takes about a month (although it can be prolonged if there are many replacement shoots after, for example, many frost deaths). Once started, it cannot be stopped (even if each shoot is killed as it comes up).

Young shoots are fed by rhizomes. Only when they have much green leaf can they make their own food, and their own minerals have to wait until their new roots grow, after about a month. Figure 9.7 shows the way a shoot grows. There is an overall S-shaped curve, both in the height of the top growing internode and to the top of the upper leaf. Shoot length depends on (1) the number of nodes, which is mostly determined by bud width at emergence, but which can be slightly altered by environment, and (2) the length of each internode. These are longer in the lower three-quarters (see Figure 9.7), and more easily altered by environment.

After the main emergence, when just a few small buds are coming up, some newly-formed ones, nearly as wide as king reeds, can appear. These 'summer shoots' can be absent, or can be up to a third of the whole crop. They do not do well. Narrow late shoots grow to the proper (short) height for their width (Figure 9.5), and within that, do well. Summer shoots, however, stop growing long before their potential height, often die early or do not harden in autumn, and are 'waste' for thatching. It seems they came too late to get the rhizome food support such large buds need, and succumb to internal competition.

During the time food is, basically, moving up from rhizome to shoot, young shoots killed (by frost, reedbug, grazing, etc.) are replaced. With much temperature fluctuation (burn, severe frost) two or more replace one, otherwise it is roughly one for one. This continues well into June. After July, however, food moves down: and damaged shoots are rarely replaced. This makes sense. June shoots that have enough summer left to make food for the rhizomes, and to flower, do so. July ones do not: killing autumn frosts arrive first.

If fertiliser is added, replacement shoots – whether for June or for July – are definitely denser. This explains why *Phragmites* tolerates grazing better in fen and marsh than in bog: bog is short of the nutrients needed for replacement shoots. Figures 9.8 and 9.9 illustrates populations developing. In (a), a lot of shoots are killed by frosts in May, and almost all the rest survive to maturity. In (b), only a few shoots are frost-killed, a few more are killed by reedbug (eating the fragile tip), and then a few are killed by late (internal) competition: the total dead is much less. 'Late competition' causes the death of summer and other short shoots. Although shaded, shade does not kill them: there is just as much shade in, for example, tall-herb vegetation, where they remain in good condition.

Figure 9.8 shows a managed population growing. When young, all shoots are close in height, although there is some variation. By the end, the shoots are spread out. There are a few tall shoots, a wide peak of medium ones, and a tail-back of short ones. The peak is high; the stand looks uniform. This is characteristic of managed, or ex-managed, reedbeds. When unmanaged (in thatching terms 'wild') reed is brought into production it becomes more uniform (perhaps shorter), denser and straighter. This is partly because harvesting means open soil, more temperature fluctuations, so more buds emerge (so making stands denser). It is difficult to see why the reeds become straighter, however, and even more difficult to understand the timing effects. Cutting and perhaps burning for, roughly, (one to) three years creates a managed reedbed. Fair enough, after three years most buds are arising on rhizomes grown since management started. But, after management stops, the same managed reed growth patterns can persist for, anyway, 60 years. How is this? Unsolved: but it shows that

1   reed is influenced, certainly by one, so presumably many, factors operating many decades before;
2   there is a reason other than genetic/clonal why reed from different reedbeds differs (see Figure 9.2).

There are four different types of population in reedbeds, all with intermediates (Figure 9.10). First, there is the population with most reeds tall (plentiful wide rhizomes: best habitat). The peak is to the right on the graph. The stands can be called 'optimal'. These may be 'managed' or 'wild', but the 'wild' ones have the wider spread of height. The second type is the population with the peak in the middle (rhizomes dense and branched, but fewer wide ones: adequate habitat). This can be termed 'suboptimal'. In these the reeds may be dominant or frequent within other vegetation, but something is amiss: drying, competition, excessive frost deaths, etc.

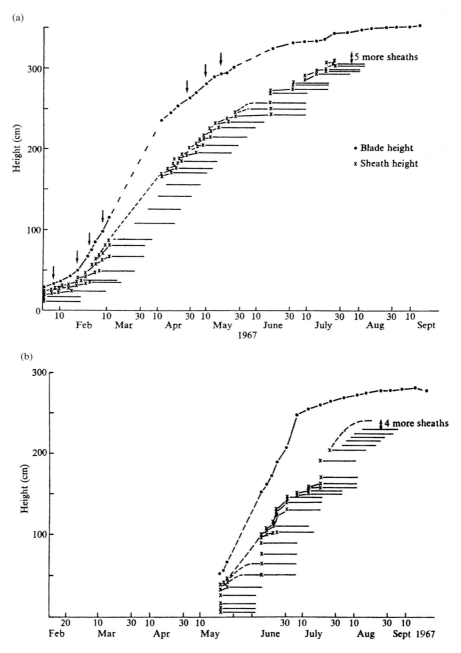

Figure 9.7 Shoot growth: (a) large, 15 mm wide, early-emerging shoot, Malta. The arrows show where new blades come to the top of the shoot in two short periods; (b) large, 7 mm wide, Braughing, Hertfordshire (Haslam 1969c).

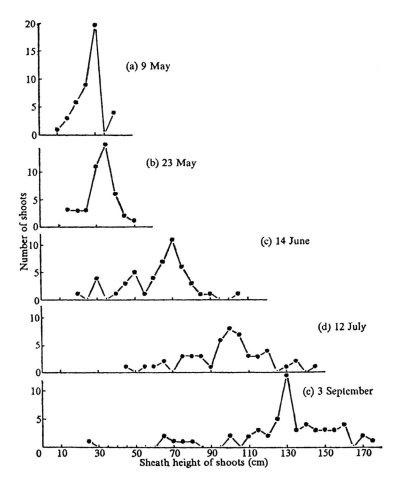

Figure 9.8 Population development in a fairly uniform (ex-managed) bed, Catfield, Suffolk (Haslam 1970b).

The third type is short reed, depauperate because of grazing (repeated shoot death), and with the peak to the left of the curve. Finally, there is the restricted population, as in Figure 9.3. There is a wide range of height, but no peak. The rhizomes are little-branched, shoots from the back (hinder) end of the rhizomes are small, usually by competition, but sometimes by drying, salt, too few nutrients, stony substrate (the two last may go together), etc.

From the population type, deductions can be made about rhizomes and habitats. Deductions can also be made about changes over time. Figure 9.11 shows two stands, drying over nine years. Both have lost height, with correspondingly narrower buds. Both have also changed in population type. The peak in both has moved from the right (optimal stands) to centre (suboptimal).

Within stable communities there are considerable annual variations in height and density (with temperature and water regimes) and lesser ones in bud width and shoot internode number. A whole subject in its own right.

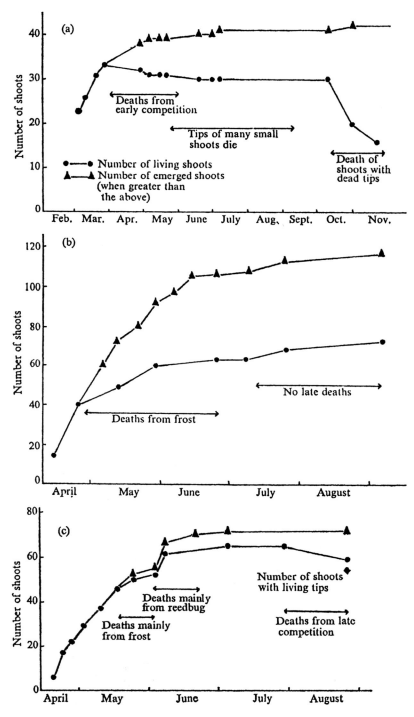

Figure 9.9 Shoot death before maturity (Haslam 1969c): (a) optimal stand, Salina, Malta; (b) considerable frost, Cavenham, Suffolk; (c) little frost, Icklingham, Suffolk.

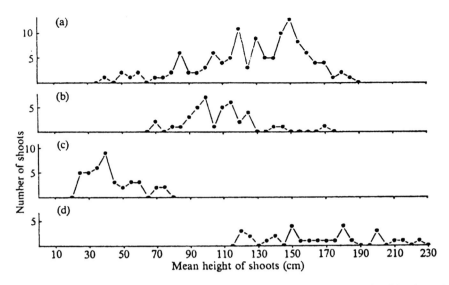

Figure 9.10 Different population types (Haslam 1970a): (a) optimal stand, Icklingham, Suffolk; (b) suboptimal, Cavenham, Suffolk; (c) depauperate, Clashnessie, Sutherland; (d) restricted, Icklingham, Suffolk.

Inflorescences ('feather' in thatching terms) emerge within a fortnight of late July. They are more frequent on the taller 10 per cent of shoots. A second set, usually smaller (e.g. 1–3 cm) may come a few weeks later, and perhaps do not grow fully. Flowering is in late August and early September (varying with clone, habitat and weather), fruits ripen by November and are dispersed by spring (varying with clone, and, less, with weather).

Shoots start to die in late September and October (later again if all growth has been delayed by, for example, spring cold or unusual drying). They harden in November, and ripen (mature for thatch) before early January. By this stage the reeds are 'final', no longer influenced by the plant or general environment except insofar as breakdown starts (Table 9.4). Dead reeds in winter have a vital function in the (natural) flooded habitats of the reedswamp. Although rhizomes are dormant in winter, they still live, so need oxygen, and that oxygen comes down the dead reeds into the rhizome. Typically, dead reeds fall in their third spring, later if particularly strong, or earlier, in their first spring, if very weak.

No reed has been recorded falling before the first spring. Therefore, it has done its job – oxygenating the rhizomes until next spring's young shoots can do it. Reed weak enough to fall in its first spring, 'crumbly reed', is often referred to by thatchers as deteriorating. However, the *Phragmites* is as vigorous and as competitive as that with stronger and longer-lasting dead reed. It is the thatchers who do not like it, the *Phragmites* is fine.

Each phase of the plant's growth cycle is essential to the whole: rhizome growth, bud development, shoot growth, flowering, ripening. Except, roughly speaking, for the last. Each can be affected by environmental factors – temperature, water, nutrients, grazing and other management, competition. All act on all phases – and with different effects and combinations. This is a complex pattern, but one that can be unravelled with some understanding of the plant, and study of the habitat. Once known, a quick look at *Phragmites*, and the habitat is interpreted. A versatile plant!

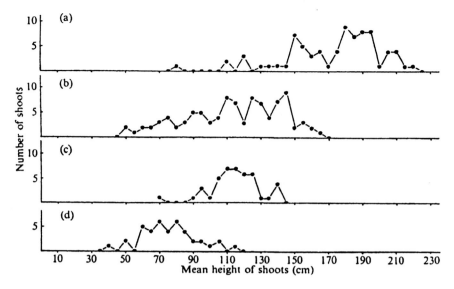

Figure 9.11 Deteriorating stands (Haslam 1970a): (a) 1959; (b) 1967, Icklingham, Suffolk; (c) 1958; (d) 1967, Cavenham, Suffolk.

Figure 9.12 Seedling growth (diagrammatic). In unsuitable conditions, the two-leaf stage can persist for at least two years.

The seedling and young plant (Figure 9.12)

Phragmites does not only spread by rhizome. It bears seeds, fruits, in profusion. These do not continue a given reedbed, but they spread the plant to different sites and regions, especially when environment changes.

It is strange that seedlings are found not to grow in reedswamps, the main natural habitat of the adult plant! Seeds must arrive on soil, not water, rock, or road, etc. They must find adequate nutrient status, not be on bog or, interestingly, calcium-dominated peat. They must be well-lit, and the water level must be high enough for the young root to be thoroughly wet (usually not more than 1 cm below the surface), without allowing the young shoot to be drowned (usually not more than 1 cm above the surface). The temperature must also be suitable (in Britain, as high as possible!). Rather drier, darker, cooler and chemically worse conditions may not kill, but they slow seedling growth so that death from drowning, frost or competition (shading) become much more likely. The shorter the period the seedling is at risk from these, the greater the chance of adult life.

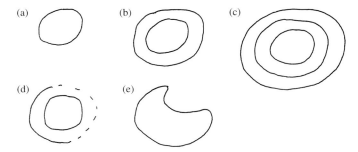

Figure 9.13 Advancing clones in plan (diagrammatic), Rosyth, Fife. (a) Circular, last year's size; (b) this year's growth; (c) either this year's late summer growth, or next year's (ordinary) growth; (d) part of this year's growth as (b), part advancing into ground occupied by a competitive species (*Agrostis stolonifera*, with a little salt), and shoots sparse, not dominant; (e) non-circular clone resulting from differential growth (different habitats) on different sides of clone.

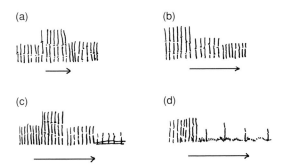

Figure 9.14 Advancing clones in profile (diagrammatic), Rosyth, Fife. (a) A single ring this year. Last year's shoots are the tallest. There are other patterns. Each year's shoots may be the same height, or this year's shoots may be shorter than the rest, etc.; (b) and (c) show a second and shorter ring of shoots, grown in late summer. In (b), this second ring developed from rhizome growth. In (c), the second ring developed from legehalme growth which is usually even shorter and sparser. In (d), this year's and last year's shoots both grew into ground occupied by an effective competitor (here *Agrostis stolonifera*). Shoots are, in consequence, very sparse.

The seedlings grow (Figure 9.12) by each shoot bearing a new rhizome, each deeper and longer than the one before, and ending in a taller shoot than before. A plant is probably safe once rhizomes and roots reach a depth of about 30 cm. It is protected from frost-death, both because, in regard to freezing, its state is almost that of the adult frost-resistant horizontal rhizomes, and because freezing is seldom this far down in the soil (in Britain). The shoots, reaching a height of over 30 cm, are safe from shallow flooding and from shading by litter or short plants. The whole plant occupies some space, so is starting to assert itself in competition. It has enough underground food stem to survive being grazed. A plant may reach this 10-shoot stage in three months, or not have reached it in three years. (In calcium-dominated peat, without added phosphate, it can remain in the Figure 9.12 (a) stage for over 18 months.)

It has been rightly said that such plants need 'a window of opportunity' to germinate and become established. Given this window, all conditions suitable, colonisation by seed can be rapid. Disused gravel pits with plenty of shallow shore-line are particularly good, or artificially reclaimed land such as Dutch polders.

Once full size, and in the absence of effective competition, plants can grow and cover large areas: become reedswamp, in fact. One or many seedlings can contribute, and depending on how much they differ and mingle, the different clones may or may not be visible in the reedbed.

The advancing plant

To make new reedswamps or reedbeds, new plants must spread: from small plants to many, many hectares. This has been watched as new small clones grow, and as old ones expand into newly made reed habitat (Figures 9.13 and 9.14). Reed can spread as a dense advancing wave, with advancing vigour (a physiological change) or as just restricted rhizomes growing along.

Even in poor conditions (e.g. just-tolerable salt, dominant *Agrostis stolonifera* with mild salt), there is more 'impetus' in advancement than occurs in an old population. Shoots are clumped: long rhizomes with short laterals. Advancing into good conditions (e.g. bare wet soil) there is a well-marked advancing margin. Here there are not only out-thrusting long rhizomes but also very rapid branching within these, enough to be equivalent to the shoots of more than one year's ordinary growth. Young rhizomes here branch profusely. In some clones, even 10 laterals may develop on a king rhizome before emergence starts. Clonal differences are much greater than in old stands. These waves advance 1–2 m/year, with up to 200 shoots/m^2. This is irresistible by any vegetation of equal or lesser size: it succumbs. But any vegetation able to stop bud production with root toxins (like brackish *Agrostis stolonifera*) merely prevents the wave developing. Simple. In some clones there is, in summer, a second advancing zone, 1 m or so wide. The shoots here come up after the main emergence, and are unlike anything that happens in mature stands. In a few clones, this second wave is of *legehalme*, with up to 10 m or so advance.

Shoots in the first-year zone are rather shorter, those in the third-year zone are as tall as they come in the stand. Those in the second-year zone may be either.

Energy-food is being poured from the older plant into the newer and advancing part. The older plant is doing well, by no means harmed by this. (Advancing vigour is seen in human populations too, e.g. the Vikings, the Americans invading their West.) Just the spectacle of pouring energy is well worth observing, in such a large plant! Once more, *Phragmites* variation is shown only in some habitat conditions.

Chemistry and competition

A very interesting pattern is shown in calcium-dominated fen. Here some nutrients are too low for proper *Phragmites* growth. The plants are restricted, with short shoots (few over 1 m high), and very sparse. Internodes are unduly short, leaf blades are tough and, significantly, shoots yellow very early in autumn. This habit has nothing to do with competition, from root toxins or shade. It occurs on ground bearing only moss, or indeed bare.

If fertiliser is added in spring, the first year there is no change: shoots had already been programmed with low nutrients, and no change is made. In the second year, rhizome pattern is unchanged, shoots are still as sparse, but they are taller, greener and with more

Table 9.2 Influence of salt on *Phragmites* (Haslam 1995)

In different habitats, at different seasons and in different concentrations, salt increases can alter:

1   Hardness (increases)
2   Brittleness (increases)
3   Fibrousness (decreases)
4   Internode length, and so shoot height (decreases)
5   Bud width, and so potential shoot height (decreases)
6   Bud density, and so reed density (decreases)
7   Emergence date, and so probability of maturation. (If reeds are frost-killed in autumn while still immature, they are also weak) (delays)
8   Leaf size, thickness, toughness and hardness (become smaller, thicker, tougher, harder)
9   *Legehalme* production (more clones with this potentiality grow *legehalme* if salt is influencing)
10 Health and death (too much salt kills)
11 Infrared colour of clones typical of salt (from red, normal, via pink to green, in brackish places)

normal leaf blades. The third year shows more rhizome branching, and again has normal shoots: within the confines of the fertilised plot, that is. Only in the fourth year are green normal shoots found outside the plot, nutrients having been, at last, passed from the fertilised peat. *Phragmites* developed badly solely due to inadequate nutrients.

A different extreme form is shown in a damp *Juncus subnodulosus* community. Here *Phragmites*, though less restricted than in the last, is short, sparse, coming up two weeks later than in the reedbed alongside and dying back in autumn two weeks earlier – a month short on the growing season. Reed buds in all habitats narrow around ground level but here they narrow much more, therefore potential height is less. Although shoots are short, they are as tall or taller than the rush, so they are not subject to competition by shading. There is no restriction by nutrients, either. What is happening here is root competition, rush root exudates inhibit *Phragmites* growth. Cut the rhizome mat, damage it, and the inhibition is relaxed – *Phragmites* grows taller and has the same growing season as in the reedbed (see Chapter 10).

Another *Phragmites* form, caused by habitat chemistry, is that due to salt (Table 9.2). A little chronic salt, such as is found in the back of saltmarshes, on the banks of tidal rivers and similar, favours *Phragmites*. In rivers, salt seems to give reed more tolerance to flowing water and bank erosion. Increased salt concentrations alter features such as emergence pattern, leaf type and reed brittleness. Yet more salt kills. Seawater floods are generally fatal to shoots – if the salt was there long enough. A brief flood can show complete recovery next year. A long one leaves considerable salt, and recovery depends on when this is washed out.

Dry fen and marsh communities, tall herb, etc., often have sparse *Phragmites*. These places are dry, too dry for *Phragmites* to dominate easily. Rhizomes do not branch much in dry places. Also, the land plants can grow tall. So there will be competition by shading. Shaded small shoots usually live, unlike shaded small reeds within reedbeds. So these shoots must be supplied with food by the rhizome – or at least not have it removed. There is also root competition between plant species. This appears to be the main reason for keeping *Phragmites* sparse. Where land plants can grow well, *Phragmites* is sparse. Other species take up the space, produce toxins, and shade.

Phragmites may occur in damp woodlands, where heavy shade prevents the ordinary tall fen, etc. plants (but allows shade plants such as fen ferns). It is pale and rather limp, tall, not dominant: but an important member of the shade flora. However, both here and in dry fens it is much easier for *Phragmites* to remain while the site is colonised by trees and dries than for it to invade such a community from which it was absent when wet or when open.

In Malta, there is no tall-herb or sward vegetation to compete with *Phragmites*. In the Mediterranean climate most short plants are green in winter. Reedswamp plants are green in summer, so no root or shade competition occurs. And *Phragmites* can have very frequent shoots, in dominant patches, 4–10 m above water level. This depth is about as far as the plant roots can reach. Certainly there may be different genetic varieties there, but that is not the whole story. On, for example, railway banks in Britain and elsewhere, well above water table, there are patches of *Phragmites*. As also in rather odd places, such as hedgerow bushes above former damp meadows, and heathy places above and outside bogs. Without competition, *Phragmites* can occur in the dry; with competition, it is absent or very sparse.

In nutrient-low places, *Phragmites* has much less trouble with either shade or root competition. On the contrary, it has much less resilience, less ability to tolerate grazing or other disturbance.

Phragmites, therefore, is often lessened but not excluded by chemistry: by the chemistry of the soil (nutrients and salt) and by the chemistry of root exudates.

The water

Reedswamps lie between deep open water, on the one hand, and dry fen, carr or more managed (e.g. arable) habitats on the other. On the drier side they are limited by competition: and can be prolonged by management specifically for reedbed (see Chapters 2 and 3). On the wet side, they are limited by deep water.

Most reedswamps and managed reedbeds lie between water level 2 m above to more than 0.5 m below ground, but can occur down to more than 4 m where shoots can grow tall enough to have sufficient green leaf above water to photosynthesise food to support the plant. Roots usually have contact with saturated soil or water, but *Phragmites* can grow where it is too dry for this. However, such drier stands are typically in the wetter west, where rainwater penetration is greater. Fluctuations in water level may be as much as 3 m (Rudescu *et al.* 1965) or as little as 5 cm. It is generally considered that some water movement (lateral) is beneficial. Most flourishing British stands have their water between 1.2 m above ground level to being at ground level. In fact, in watercourses, where the bed is at least 1 m deep, little management is needed to restrict *Phragmites* to banks. Reeds are sparser and shorter in nutrient-low stony lakes. Here the plant may grow in water up to a depth of about 75 cm (summer level) only. The deepest water that *Phragmites* can colonise thus varies. There is an absolute depth (because there must be ample green shoots in the air). But within this, the actual depth found is determined by nutrient status, substrate texture, erosion and perhaps temperature.

Phragmites is unusual in having both:

1   a very wide range of possible water regimes;
2   a need for a regular water regime.

If the annual pattern is disturbed, particularly that around +5 cm and −10 cm of soil surface, and particularly but not only in spring, the stand is traumatised. Emergence will be late (by a week to more than 2 months, depending on the severity). Shoots will be sparser and shorter (in extremes, reduced from dominance to sparse (perhaps 20 shoots/m^2) short (not over 1 m high) shoots. Autumn maturation will start late, probably meaning frost-killing before ripening. This is the normal response with any sort of trauma. After brief, mild disturbance the growth is all right the next year. After severe disturbance, normal growth is not resumed for four years. By four years on, all or most shoots are from rhizomes that grew and developed after the trauma.

An unusual autumn drought brings early yellowing and death to affected habitats. In a normal year a dry stand next to a normally flooded one yellows first, in a drought autumn, it is the other way round.

Lowering water level moves populations from optimal to suboptimal to depauperate in type. Water level influences bud width, and so (potentially) shoot height. Buds tend to be wider for the same rhizome level in wetter places, e.g. 5 mm in wetter, 3 mm in drier places. In unflooded places, horizontal rhizomes are typically cylindrical, and rather tough. In flooded soil, particularly if nutrient-rich, they tend to be flattened and less tough.

Flood insulates from frost, so gives earlier emergence and less frost damage. However, frosting increases bud density. Flood means rhizome and roots must be aerated through aerial stems. Cutting below water level in summer decreases bud formation, and therefore also reduces stand density.

Wetter dominant reeds are usually sparser and taller (other factors, including clone type, influence this too). Adaptation to new conditions is faster in wet stands (it can be half the time). There is, therefore, better robustness, resilience and ability to react to circumstances in wet stands. This may help to explain the sensitivity of *Phragmites* to competition in drier places.

Changes in water regime, like other habitat changes, can alter *Phragmites* form, for example a drop of 75 cm in water level leads not just to fewer tall shoots, as would be expected, but also to the loss of prominent ribs on the leaf. Such a characteristic as prominent leaf ribs is therefore seen only in a satisfactory habitat. Otherwise, such reed can be confused with an always flat-leaved clone.

The thatching reed

Various reedbeds are managed for reed good for thatchers (Table 9.3). This keeps reed dominant in drying beds, where, without management, other vegetation would take over. It also produces the height, density and uniformity required. (Height can be decreased by exposure to frost, or by cutting young shoots. It can be increased by spring flooding and harvesting alternate years, 'double wale'). This sort of management is well known (e.g. McDougall, in Haslam 1972).

Cut reed is brought to dry land, and stacked there until collected (Figure 9.15). Then it is stored at the dealers, and (after at least six months) used on the roof. All of these processes, from growing to thatching, must be right if the thatch also is to be right, and to last for its hoped-for 80 years.

Bundles are typically 60 cm in circumference measured 30 cm above the base, the butt. They are placed butt outwards on the roof, so the strongest part of the reed is that exposed to sun and storm, and the (usually) weaker part forms a snug habitat for fungi and bacteria (Table 9.4).

The stronger a reed is when put on the roof, the better it will last (other things being equal). 'If it crumbles in your hand, it will crumble on the roof.' Much less is known

Table 9.3 Requirements for good thatching reed (Haslam 1995)

1	Length (height) must be suitable for its use. Short reed is up to 4′6″, about 140 cm; medium is 4′6″ to 6′, about 140–180 cm; and long reed is greater than this. Medium reed is the most wanted
2	Length should be fairly uniform. No reedbed has all reeds of the same height, but a bundle should have an obvious wide centre (a near-parallel lower part) and then taper to the tip
3	Straight or fairly straight reeds. 'Wild' reed, bending in all directions, is difficult to work with
4	Reeds should be strong and durable, when cut on the bed
5	Reed storage should be good by all concerned: in reedbed; by dealer (if any); by thatcher. (Well-stacked, airy, dry and out of the sun)
6	Reeds should be properly chosen for the site of the house and its roof type. (This includes stronger reed for use in damp places)
7	Competent thatchers are necessary
8	Roof should be protected from animal damage

In addition, for a reedbed to be cost-efficient:

9	Reeds must be dense, and other tall species sparse
10	Reedbed must be easy to harvest (with suitable access, water depth, substrate hardness, etc.), and the cut reed be easily movable to the road

Figure 9.15 Storage stacks of reed (far), sedge (*Cladium*) (middle), and half-used sedge, front.

Table 9.4 Reed breakdown (Haslam 1995; partly from Bosman 1985; Kirby and Rayner 1988)

Resistance to breakdown is mainly from lignin.

1	Stronger reed takes longer to decay. Fibrousness is important. Brittle reed can more easily be broken by thatchers, birds, etc. Invertebrates eat less of strong reed
2	The outer wax layer resists microbial infection, particularly when the layer is complete
3	The silica ring protects, particularly when the outer layers have peeled off
4	Microbial decay (the most important) increases with damp and warmth (20–25°C), and leads to rot and probably brittleness. Most such decay is 2–20 cm from the surface. The outer 2 cm weathers. The tip remains strong. As weathering proceeds, the base is lost and decayed reed is weathered faster. Decay is increased by invertebrate grazing, and is most on north-facing slopes
5	Weathering from ultra-violet and sunlight breaks down reed, faster in brittle reed. (Brittleness is given temporarily by warmth, permanently by sunlight)
6	Wind, driving rain and birds, etc. damage thatch, more if the reed is soft or brittle

about the causes of strength than of those influencing crop density, another important reedbed factor.

For centuries both strong and weak reed are deduced to have occurred and to have been used for different purposes: strong for thatch, weak in walls, for bedding, etc. Reed strength for thatching is assessed by: *hardness*, tested by squashing the reeds; *brittleness*, tested by snapping the reeds; and *fibrousness*, tested by twisting the reeds. These are hand-tested as Poor, Fair, Good, or in between. Ordinary engineering tests are not applicable. Those tests developed on the continent (e.g. Binz-Reist 1989), specifically for reed, may correlate.

Reed strength varies. It varies between the reeds in a sample (unless all are Very Poor!) it varies between the samples on a bed, between beds in a region and between years in a bed. Different beds vary differently (Tables 9.5 and 9.6). In the decade of study, there has been no overall change in strength, but while some beds stay fairly stable, others show quite remarkable swings. A demonstration example of the necessity of long-term research!

Reeds from different reedbeds look different (Figure 9.2). There is bed difference, clone difference and year difference. Differences are greater in years with stronger reed. 'General Poor Norfolk' is a useful term to describe most East Anglian reed in bad years!

Reed width affects performance on the roof. Very narrow reed prevents the air circulation necessary to slow microbial breakdown. Very wide reeds look coarse, and may let water through. When there is only one reed type, thinner walls (less xylem) mean weaker reed (e.g. Sukopp and Markstein 1989). But looking at a wider range of reed types, this does not apply to either reed or wall thickness. Reeds only 1 mm wide can be very strong, and those of 10 mm can be weak and easily broken by hand. There can be annual as well as clonal variation, as so often with *Phragmites*.

Strength has nothing to do with nutrient content (although the silica skeleton structure is correlated). This may come as a surprise to those who have heard that 'all those nasty nitrates' were weakening Norfolk's reedbeds.

When is the strength of reed determined? This is difficult: poor reed can result from influencing any phase of growth from rhizome to autumn drying, unfortunately. When is strength observable? Young shoots are fairly strong, then in autumn some remain strong, others become weak.

The maintaining of the stand

How can a reedbed or reedswamp, with dominant *Phragmites*, maintain itself? In lakes, there are few potential competitors. Submerged or floating plants do not interfere, and are reduced themselves in the shade of *Phragmites*. Reed size, together with the space taken up, makes invasion (by other reedswamp species) unusual.

With intermittent drying, reeds are dense enough to cast heavy shade. In summer little or no light reaches the ground. Invasion of other species by seed is therefore hindered. Invasion by rhizome is thus also hindered, unless the invader has a large food supply to feed its shaded shoots and so withstand such shade. The litter mat is thin, as decomposition occurs in the flooded seasons, but is there, and is some hindrance to seedling establishment. Once a gap in the canopy appears, species, woody and otherwise, tolerating wet places and low light can colonise such a gap and, once there, spread. Invasion is less resisted than in lakes. This is how the successional transition to carr works (see Chapter 10).

Table 9.5 Year-to-year variations in reed strength, East Anglia (selected from and added to Haslam 1995)

Site	Strength				
	Poor	Poor–Fair	Fair	Fair–Good	Good
Cley Mill	(1999/00)	1998/9, 1989/90, 1991/2	1986/7, 1987/8, 1997/8, 1998/9		1995/6, (2000/1), 1999/00
Dunwich	1992/3	2000/1	1987/8, 1994/5, 1999/00	1998/9, 2001/2	
Hickling	1985/6, 1991/2, 1994/5, 1995/6, 2000/1	1988/9, 1987/8	1990/1, 1992/3	1995/6, 1996/7	
How Hill	1986/7, 1987/8, 1988/9, 1989/90, 1991/2, 1993/4, 1994/5, 1995/6, 1996/7, 1997/8, 1998/9, 1999/00, 2001/2				
Ranworth	1992/3, 1993/4, 1995/6, 1999/00, 2001/2	1989/90	1996/7		

Strumpshaw A	1994/5 1997/8	1993/4	1992/3 1996/7	1986/7 1998/9 1999/00 2001/2	1989/90 1990/1	1985/6 1988/9	1991/2	1995/6
Strumpshaw B	1986/7 1994/5	1992/3 1993/4					1995/6	
Strumpshaw C	1986/7 1987/8 1988/9 1990/1 1992/3 1993/4 1994/5		1989/90 1991/2	1995/6				
Strumpshaw L	1997/8	1998/9	1999/00 2001/2			1996/7		
Strumpshaw F	1996/7 1997/8	1998/9 1999/00			2001/2			
Walberswick	1994/5	1989/90 1991/2	1992/3 1993/4 2000/1	1990/1 1999/00	1987/8 1988/9 1998/9	2001/2	1995/6 1996/7 1997/8	
Wicken	1989/90 1994/5 1996/7	1987/8 1991/2 2000/1	1986/7 1993/4 1997/8	1995/6 1999/00	1985/6 1992/3 1998/9 2001/2	1988/9	1990/1	

Sixty samples of 15 to 30 reeds are collected per reedbed. Each is graded on a five-point scale. The value given is the average strength for the bed. '1989/9' means reed growing in 1988, harvested in 1989.

Each of these parishes or marshes is or was harvested for reed. The reed areas are large and complex. More than half of the places chosen for this study are not currently commercial beds. Most were regularly harvested up to the 1920s.

Table 9.6 The three elements of strength in East Anglian reedbeds (after Haslam 1995)

In 10 reedbeds over 10 years	No. of records
Hardness as best (or co-best) feature	32
Fibrousness as best (or co-best) feature	26
Lack of brittleness as best (or co-best) feature	5
Records with no 'best' feature	17

In reedbeds that are almost unflooded, a thick litter mat generally forms. When it is complete, colonisation from outside is virtually prevented. However, an over-thick mat weakens *Phragmites* itself, leading to the breaking of the mat. More commonly, the litter mat is opened by accident: deer, pheasant, trampling, wind, and many more (as well as burning, cutting and livestock grazing). These gaps, even if only tiny, allow colonisation. Colonisers can be associated species of reedbeds, no threat to the *Phragmites*, e.g. *Galium palustre*. They may also be potential competitors and successors, e.g. *Epilobium hirsutum, Filipendula ulmaris, Salix cinerea*. For these to invade, there must be soil above water level during establishment, some light, viable seed and young plants tolerating shade.

Human interference to perpetuate drier reedbeds is for thatching and, recently, for conservation. A harvested reedbed has no thick litter mat, and is open to light, frost, and competition. It must therefore continue to be managed to favour reed, and, where necessary, managed to remove other species.

Conclusion

More is known about *Phragmites* than about most other wetland species and only a trivial amount is described here. This does not mean other plants have less to be learnt about. *Phragmites*, at least, is more like people in its sensitive and complex response to environment than is generally realised! It is demonstrably and predictably influenced by a vast range of factors, acting both now, and years, even decades into the past.

That quiet reedbed, soughing in the wind, has more to it than first meets the eye.

(The information in this chapter not otherwise cited comes from the publications of the author listed in the Bibliography.)

10 The silent battlefield: vegetation changes

Among the moorish fens
Sights the sad genius of the coming storm

(Thamer)

The willow-wren was twittering his thin little song,
hidden himself in the dark selvedge of the river bank

(Kenneth Grahame, The Wind in the Willows)

Vegetation develops

Wetlands, even small ones (down to half an acre, if screened by tall vegetation) feel wild. With this goes the idea of stability, it is like this, it comes like this, it always has been like this. That is, of course, untrue. Earlier chapters have shown the vast changes in wetland area, wetland habitat and wetland vegetation, at first by mainly natural causes, now mainly due to human activities. Within the overall movements, this chapter considers how plants advance and retreat from one another, compete with and succeed one another. It considers only slow change (not the results of creating unvegetated habitats like gravel pits and excavated bogs, or even the rewetting of grassland or arable land).

Successful competition cannot occur against the overall habitat change. *Nitella*, a submerged alga, will not invade, let alone dominate, a drying sallow carr, for instance. Alder might well do so.

To say a species cannot enter – what does that mean? It means that even though seeds of the species may be there, they cannot develop to established young plants. Seedlings appear in rich-fen plots, in rich fen, when these are cleared and available for establishment. Species include ones never seen in the community over years. It is only in a cleared and disturbed habitat that they develop. Calcium and indeed drying show the effect of disturbance in an even more extreme way. Disturbance relaxes the calcium dominance. Nutrients are mineralised. Species requiring higher nutrients appear, from seeds already in the peat, e.g. *Urtica dioica*, also *Poa trivialis, Epilobium hirsutum, Cirsium* sp. Over four years of careful study, *Urtica dioica* was never seen (at or after the opened cotyledon stage), nor were the others, at seedling stage in the undisturbed fen.

Many species therefore have seeds present in habitats that they cannot colonise. Seeds can travel further than rhizomes, but rhizomes carry large food reserves, so the plant can be supported from outside the community while it is becoming established, for example until it has grown tall enough to receive good light. Various species (such as *Phragmites australis*, Chapter 9) can live in a far wider range of habitat as adults than they tolerate for germination and establishment.

The quiet 'stable' communities observed in any one year are in fact always under attack, always actively resisting invasion: and succumbing to it after conditions change.

This is one reason for the difficulty in classifying and categorising communities. Is the observed vegetation stable over decades, or changing over months? (The same pattern may, of course, be stable in one habitat and unstable in another.) Is a community that which is in this place today? A decade ago? A decade hence? Are all? But then there are too many!

Sallow (*Salix cinerea*) carr invasion

Sallow carr is common in fens and marshes, and tolerates stagnant conditions (but not calcium dominance). It invades herbaceous vegetation, and – if allowed to dry slowly over a long enough period – is replaced by alder. It can become dominant in anything from a few decades to over a century, depending on conditions.

Spring seed production is profuse, and so is the early summer germination. Seedlings are dense wherever they can flourish. This is in good light, preferably not with the numerous invertebrates of (drier) tall-herb communities, and in damp but not flooded places.

Good light within vegetation can occur from:

1   animal damage, whether livestock grazing, deer walking, pest infestation, or pheasants pushing through, etc.;
2   severe frosts creating open patches in June (and then not continuing, so seedlings live!);
3   death or decline of existing plants;
4   growing on the top of a tussock;
5   other human activities, e.g. bonfires of waste, trampling.

Damp but unflooded conditions come from:

1   water normally a little below ground in early summer;
2   unusual drought in early summer (more likely to be effective for sallow, as the prior vegetation is likely to be weakened);
3   unusual flood raising water level to 'damp'. Not a likely habitat for established sallow;
4   being raised above ground level, e.g. on a tussock, on a clod of soil, raised by deer, people, etc.

Given that the sallow seeds are densely and widely distributed, much still depends on accident creating the 'window of opportunity' allowing not just germination, but establishment to a viable size, to occur. The accident must take place at the right time (for germination), have the right duration (for establishment) and not be accompanied by adverse factors (such as livestock eating the young plants). Established sallow can tolerate year-long flooding, and considerable shading. Seedlings cannot.

Sallow invasion therefore takes time. Windows of opportunity may come this year, or not for 25 years and then in only a few places. Invasion is likely to be quicker where there is a suitable tussock species, such as *Carex paniculata*, *C. elata* or *C. appropinquata* (Figure 10.1). Even closed carr can develop with all the shrubs on tussocks. Unless the habitat dries, however, there may be a difficulty in continuing such a community. Tussocks are weakened and eventually die in shade: new sallow seedlings would find colonisation very difficult.

Invasion is likely to be slower in unflooded (or little-flooded) tall-herb communities, because of the very dense invertebrates, leading to the eating of seeds.

(a)   (b)

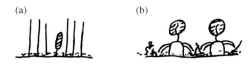

Figure 10.1 Woody invasion: (a) on the ground, shaded, perhaps flooded – difficult; (b) on tussocks, well-lit, dry (though firm when young) – generally easier.

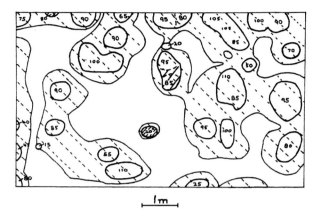

|— 1m —|

Figure 10.2 Carex paniculata community, plan, Icklingham, Suffolk. Living tussocks and litter, dashed shading, dead tussocks, full lines. Numbers are the tussock height in centimetres (Haslam 1960). Main species, as number of shoots (on tussock, on ground) are *Angelica sylvestris* (15, 1); *Phragmites australis* (47, 135); *Stachys palustris* (9, 14); *Epilobium hirsutum* (4, 24); *Carex acutiformis* (0, 104); *Juncus subnodulosus* (0, 34).

Carex paniculata and *Phragmites australis* (Figures 4.17 and 10.2–10.6)

This interaction was studied in Icklingham (Poors) fen (see Chapters 6 and 7). The wet part was on *Phragmites* peat, so there is no long history of *Carex paniculata*. It was divided by a ditch, with *C. paniculata* dominant on one side, *Phragmites* on the other. Peat had been cut, the vegetation being determined by that and the value of the two species to the villagers. (*C. paniculata* trunks, for furniture, leaves for general household and farm use.)

C. paniculata could probably dominate here only because of the lower-nutrient seepage water. Because of this, both species could perform well. The water regime was usually +15 to −15 cm over the years in the *Phragmites* part, rather wetter in the *C. paniculata* area.

C. paniculata relies on seeds; one tussock grows from one seed, and so it needs constant recruitment if the population is to survive. The tussocks reach 1 m high, and more. *Phragmites* relies on rhizomes, which can travel through any fen or marsh soil. It is far more widespread, and far more widely dominating, than *C. paniculata*. It is also taller, so potentially can shade *C. paniculata*. *Phragmites* has a shading, dense canopy. It also has a litter mat, which prevents invasion by seed. (The litter mat is thin, as the habitat is wet.) Surely, then, in a central part of the *Phragmites* range, yet in the *C. paniculata* habitat, if both are together, *Phragmites* will suppress *C. paniculata*? Not necessarily.

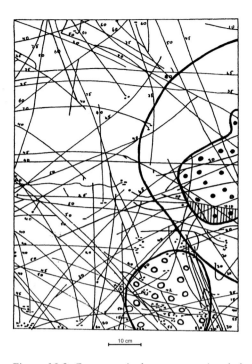

Figure 10.3 *Carex paniculata* community, below-ground plan, Icklingham, Suffolk (Haslam 1960). Closed circles, living tussock; open circles, dead tussock; thick line, root zone of tussock; thin lines, *Phragmites* rhizomes (depths in centimetres); dashes and dots, shoots of different years (dense in dead tussock).

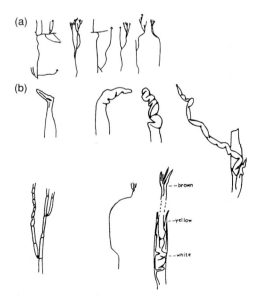

Figure 10.4 *Phragmites* within *C. paniculata* tussock. (a) Rhizome pattern (arrows, growing buds), which may develop new levels of horizontal rhizomes; (b) tips of *Phragmites* buds failing to grow through firm living tussock (Haslam 1960).

Figure 10.5 C. paniculata succeeding Phragmites.

Figure 10.6 Sallow (*Salix cinerea*) colonisation and succession of *C. paniculata*.

C. paniculata germination is mostly in late spring–early summer, in conditions like those for sallow (above), except *C. paniculata* can manage with less light. Doing fairly well, the plants develop into a tuft of 3–5 shoots, 15–30 cm high, after 5 months, i.e. before their first winter. At this stage the seedlings can of course stand considerable flooding. (As long as the upper parts of leaves are above water, aeration takes place.) As the tuft grows bigger, the short rhizomes (each of which have terminal shoots) are no longer horizontal, flat above the ground. They turn obliquely, then almost straight upwards. This eventually creates the tussock. The tussock bears shoots only on its crown. The shoots mostly develop in spring, living 13–14 months, bearing fibrous sheaths outside – good for tussock making – foliage leaves and often a flowering stem inside. The crown gets wider, year by year, and the extra width for the trunk is made up of the new roots, growing down from the crown (to about 1 m below ground). The dead leaves decay slowly, so there is a thick ring of tough dead leaves around the crown – unless the stand is grazed. If grazed, the community has lost one of its important weapons to resist invasion.

Within the dominant *Phragmites*, *C. paniculata*, once established in a window of opportunity, continues under the *Phragmites* canopy as spindly, weak young tussocks, looking as though they will be dead in a year's time. What a difference 15 years can make! After that time small tussocks had widened as well as grown upwards, the tough leaves were thrusting sideways: and pushing the *Phragmites* leaves away. *C. paniculata* was replacing *Phragmites*.

In dominant *Carex paniculata* – one stage on from this last – *Phragmites* dominated only between the tussocks (in half the community, *Juncus subnodulosus* in the rest). It was, therefore, much subordinate to the *C. paniculata*. No *C. paniculata* establishment was seen here over 4 years: no window of opportunity happened to occur. Vigorous living tussocks resist the direct invasion of *Phragmites* rhizomes, but when the tussocks weaken with age, *Phragmites* rhizomes come in and do very nicely (Figure 10.5). Dead tussocks may in fact be densely covered with reeds. The resistance to invasion certainly comes from toughness, but maybe also comes from a chemical released by the living tussock. Other species growing on tussocks also increase as the tussocks weaken. These include rose, ferns, pennywort and hemp agrimony. In this pattern, *Phragmites* is about to replace a *C. paniculata* community. In the two, therefore, *C. paniculata* can succeed *Phragmites*. *Phragmites* can succeed *C. paniculata*. All depends on the window of opportunity for the establishment of *C. paniculata*. With it, this dominates; without it, *Phragmites* does so.

On the tussocks, sallow colonises too (Figure 10.6), with a few alder (in less stagnant habitats, alder is the principal woody coloniser). When enough bushes are there, this is carr. A heavy bush can make a tussock fall over. Shade and age eventually weaken tussocks. Sallows are much larger plants, and sallow is left dominant. Sallow cannot survive in the second generation unless it – and not alder – can also re-colonise. If tussocks are not available for colonisation, this means new seedlings growing on the ground. And in such wet places this is difficult, windows of opportunity are rare. Sallow carr can invade *C. paniculata* more easily than it can invade *Phragmites*, as it can colonise the dry tussocks, which are open enough for slow but easy colonisation. Colonisation of the ground in reedbeds depends on a rare window of opportunity.

Galium aparine (goosegrass) in tall-herb fen vegetation (Figure 10.7)

Goosegrass is the only annual dominant, so is of much interest. These fens usually have water level ranging from around 5–30 cm to 30–80 cm below ground. Peat has oxidised

(a)

Figure 10.7 Tall-herb vegetation with (a) little, (b) dominant *Galium aparine*.

(b)

Figure 10.7 (Cont'd).

and mineralised, turned black and crumbly, and is teeming with invertebrates. Goosegrass is usually important in the species-poor types (see Chapters 3 and 4).

Three other co-dominants here also establish well by seed – *Epilobium hirsutum*, *Filipendula ulmaria* and *Urtica dioica* – but all, of course, are perennial, so less dependent on seed.

Goosegrass does not do well if water reaches the surface in a wet year. Seedling growth is poor, the root system is superficial, and many die. Not a year for it to be co-dominant.

In late winter, these tall-herb fens are semi-open, with bare peat and mosses, good for germination. In one fen, seedling (not grown plant) numbers (in 0.25 m^2) were:

1957	6/2	28/2	4/4	17/5	19/6	2/8	13/9	5/11	13/12
	50	200	125	100	30	1	2	7	6
1958	12/2		25/3	6/5	6/6			21/11	
	74		123	85	80			0	
1959		3/4	28/4		2/7				
		50	70		25				

Numbers, therefore, vary from year to year. Main germination starts in winter or late winter whenever the weather turns warm. Early growth means early shading ability, which means a competitive advantage (and see Chapter 9). The rhizomatous species are more consistent, sprouting at nearer the same time each year. When goosegrass does well, it has large numbers of early shading shoots before these others do so, and so is 'in place' for good later growth also. When it does badly, seedlings have been few or late or many killed by e.g. frost. (Seedlings are more frost tolerant in February than in June.)

Goosegrass grows tall, becomes too heavy for its stem, and sprawls over neighbouring plants. In its good years it grows taller and heavier. It can, surprisingly enough, sprawl over, and bring to the ground quite large plants of *Epilobium*, *Urtica* and others. Goosegrass can be left the sole dominant, albeit in a patterned fen with other combinations as well.

It is extraordinary that an annual can dominate over these large, tall, bushy plants. This dominance is determined not by the behaviour of the other species but by the effect of the weather on goosegrass germination (early warmth means potential dominance) and on growth (absence of frosts and floods means potential dominance). A cold spring holds up other species more, allowing goosegrass to get ahead. Therefore, a mild late winter and cold but nearly frost-free spring are the most likely to lead to dominant or co-dominant goosegrass.

The *Schoenus nigricans* community (Figures 10.8–10.13)

In the much wetter conditions of the nineteenth century, *Schoenus* was very abundant in calcium-dominant eastern fens; *Cladium*, now the main dominant, was only frequent (e.g. Ashfield 1861, 1862; Bunbury 1889). In fact at Eriswell *Cladium* had formerly been abundant but, when these papers were written, was restricted to the margins of water bodies. The wetness is shown by the description of, for example, pools that cannot be walked through, varying up to several feet of water, peat cuttings. The depth and permanence of the then pools is also shown by records of aquatics (*Chara, Utricularia vulgaris,* etc.). *Cladium* generally does best in a water level of about −15 to +40 cm, *Schoenus* goes deeper, and both do well in fairly stable waters. *Schoenus* better tolerates both burning and cutting. Burning and cutting both tend to lead to many separated tufts instead of the original good-sized (e.g. 0.5 m tall and wide) tussocks.

By the 1950s, however, the calcium-dominated *Schoenus* areas had stable water levels at about ground level, spring-controlled (see Chapters 3 and 6) with few pools (peat cuttings), those pools reaching 30 cm deep near the river only.

Seedlings are most common, in autumn, where the water table is close to the surface, and shading is minimal. Seedlings are killed with prolonged flooding. The first shoot is small, with green leaves, and a short horizontal rhizome ending in another shoot. Each shoot is larger than the last, and sheathing leaves develop outside the green one. Leaves are 9–12 cm long after 3–4 shoots have grown. When there is a tuft 3–8 cm wide, the (short) rhizomes turn obliquely up. If it is too dry, or they are burnt or cut, however, they stay growing flat along the ground. Rhizomes turning up lead to tussocks. These, unlike those of *Carex paniculata*, bear shoots all over them. The tussock is mostly made of shoot bases, with rhizomes and roots that help to compact the tussock. New shoots mostly come up in early spring. Flowering is late May to early June, but the fruits remain in the compact flowering head until the following June, when the stems start dying as the new ones grow, and by July the heads have fallen and the fruits are set free.

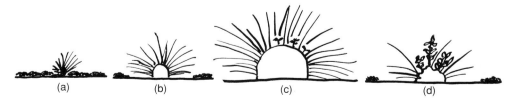

Figure 10.8 Schoenus nigricans community (not peat building), with tussocks in the (a) pioneer, (b) building, (c) mature and (d) decaying stages. Other species colonise older tussocks. Mosses, etc. on open ground.

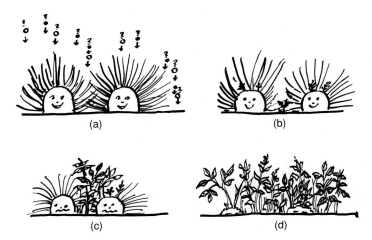

Figure 10.9 Habitat control of *Schoenus* community. (a) Community in good condition, seeds of many species arriving; (b) calcium-dominated habitat prevents species of other communities developing; (c) drying relaxes calcium dominance, outside species (here tall herbs) enter; (d) tall herbs succeed *Schoenus.*

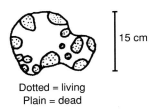

15 cm

Dotted = living
Plain = dead

Figure 10.10 Regrowth of *Schoenus* tussock after fire, which will result in several new tussocks.

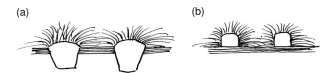

Figure 10.11 Growth of *Schoenus* tussocks in (a) peat-building and (b) non-peat-building communities.

Because shoots grow from the base, *Schoenus*, unlike *Carex paniculata*, can propagate vegetatively. Where rhizomes are growing flat along the ground, normal ring growth occurs (Figure 10.9), one tussock becoming several, by tussocks growing in, usually, two to three directions, to make 2–3 new tussocks around the edge of the original one, which eventually dies, leaving the new ones as independent tussocks. This happens only when the ground level is not regularly flooded. In the opposite habitat, very wet with peat accumulation, peat accumulates between existing tussocks. Peat therefore smothers the lower part

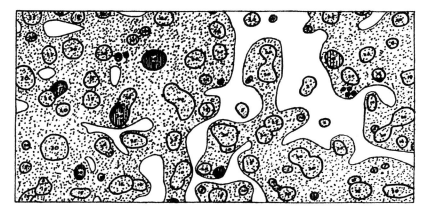

Figure 10.12 *Schoenus nigricans* community, plan, Eriswell lode head fen, Suffolk. *Schoenus*-covered parts dotted, dead tussocks shaded, numbers, height of tussocks in centimetres (Haslam 1960). Main species, as number of shoots (on tussock, on ground), are *Molinia caerulea* (151, 15); *Potentilla erecta* (21, 2); *Phragmites communis* (17, 89); *Equisetum palustre* (11, 15); Abundant mosses (0, 120).

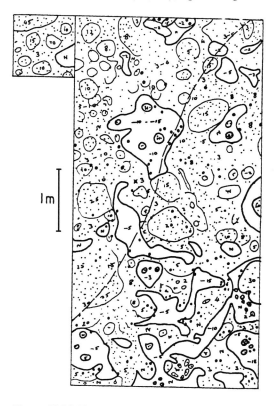

Figure 10.13 Fire-controlled *Schoenus nigricans* community, plan, Redgrave fen, Suffolk. *Schoenus* mostly in tufts and short indefinite tussocks. *Molinia caerulea* frequent, on tussock and drier ground, mosses frequent on ground. Large dots, *Cladium mariscus* shoots. Above the dashed line, fen was burnt the previous year (Haslam 1960).

of the tussock, which then cannot grow shoots. So the tussock, getting wider upwards, is carrot-shaped (triangular).

There is Wattian pattern (Watt 1947) (Figure 10.8). Between tussocks or, rather, between the shade cast by tussocks, in the open, a range of species grow (including rare ones such as *Anagallis tenella, Drosera* spp., *Epipactis palustris, Gymnadenia conopsia, Parnassia palustris,* and *Ranunculus lingua*). There are a range of Bryophytes, including fen sphagnum. Mosses are common in the open and light shade, decreasing in heavier shade. Sphagnum is, as usual, in small hummocks. Open ground is the first stage.

Among the species colonising the open mossy places is *Schoenus*, the *Pioneer* stage of the *Schoenus* community (stage 2). Colonisation can be by seedlings, or by small bits growing vegetatively. As the new tussocks grow and their shading power increases it is in the *building* stage (stage 3). Gradually most other species are shaded out. First, annuals cannot develop, then many perennials are gradually weakened by the shade, and die. These species, however, colonising the new places opening up elsewhere in the community, are still present. They are in the same pattern. It is just that the open spaces have moved as *Schoenus* tussocks grow and die. When the tussock is *mature* (stage 4), leaves cast dense shade. In this are found only species supported by food from outside the shade, or growing above the 0.5–1 m shading leaves (e.g. *Phragmites*, which is here sparse and restricted) (see Chapter 9). The very firm, younger tussocks age and become less firm. *Molinia caerulea* (discussed later) has probably colonised, and perhaps *Potentilla erecta* and others also. In the *degenerate* (stage 5) stage, shading is still heavy, but the tussock is even less firm. *Molinia* may become dominant, other species frequent. When the tussock finally dies, the litter and, later, the tussock gradually break down. As the habitat opens to sunlight, mosses increase, the species of sunlight re-colonise, and the cycle starts again. *Schoenus* controls (dominates) the community, through its life cycle, because it is able to shade others out.

Birch is the commonest woody species to colonise the tussocks, potentially leading to birch carr (with the reservations described earlier for sallow on *C. paniculata* tussocks for second generation continuance). If the place is a little drier with rather less calcium dominance, alder is the main coloniser. Alder then colonises the ground rather than the tussocks, the soil being dry enough: at least intermittently. Ash and pine occur sparsely. Tree colonisation is through a window of opportunity, of course. A year of local burning, cutting/grazing, drying, etc. can lead to a generation of young trees. A different window of opportunity in the next-door habitat can lead to another uniform-aged tree stand, of different age.

Where fire is intermittent (as happened at Redgrave and Lopham fens), *Schoenus* cannot form tussocks, as the higher level growth is constantly killed. Instead *Schoenus* occurs in tufts, with shoot bases hardly above ground level. Generally, the greatest heat of a fen fire is a little above the ground (though some fires do scorch the upper soil). The same burn temperatures damage the older parts in the middle of the tussock more than the outside, so new tufts grow at the edges of old ones, and tufts spread. Burning also destroys the dead leaves (litter). This means associated species are not controlled by the shading by the dominant. Burning keeps the whole plant shorter, usually under 0.5 m, with only a few flowering heads and low crop yield.

Therefore, in a regularly burnt community, there is no tussock habitat well above water level. If old peat cuttings are present, however, these substitute. They provide higher and drier as well as lower and wetter places, and the species that, in the unburnt community, grow on tussocks (e.g. *Molinia caerulea, Potentilla erecta, Equisetum palustris*) grow instead on the higher peat. Mosses grow where the summer water level is down to −5 cm, algae grow in wetter levels.

This *Schoenus* community, whether of the unburnt or burnt type, is self-sustaining. So are the wetter, peat-forming communities of, for example, Oxfordshire (Dawkins 1939). Tussocks reproduce by seed (above water level) or vegetatively. In the drier communities described, birch can invade, unless stopped by fire – or, in the past, no doubt by peat-cutting or direct scything.

However, if *Cladium mariscus* is around, and uncontrolled, that is a different matter (discussed later). The succession to birch carr is expected as habitats dry. That to *Cladium* is less expected. In addition, the impact of drying plus disturbance destroys the whole community, as is all too plainly seen in fens such as Redgrave. These relax calcium dominance. They prevent species of wetter habitats re-colonising. The higher nutrients and dry conditions allow *Urtica dioica* seeds and those of other tall-herb communities (that are already in the soil) to colonise: and to destroy.

Schoenus nigricans and *Molinia caerulea* (Figure 10.12, 10.13)

Molinia, like *Schoenus*, is also tussock-forming. Its tussocks are somewhat shorter, though equally broad. It grows widely in nutrient-poor conditions, both acid and calcium-dominated ones. *Molinia* grassland is common in drier moorlands, fen meadows, etc., and in calcium-influenced drier fen. When an example of the last dried more, *Molinia* had been replaced by *Calamagrostis canescens* (a species of higher nutrient status than *Molinia*).

Molinia, although growing well with high rainfall, is seldom where the permanent water table is likely to reach its rhizomes. This is a much drier habitat range than that of *Schoenus*, therefore. *Molinia* also grows with more nitrogen mineralisation than does *Schoenus*.

Molinia develops from tufts to tussocks, like *Schoenus*. Its rhizome, however, branches more, and it has frequent longer rhizomes, with longer internodes than *Schoenus*. These mean that *Molinia* can spread quite fast vegetatively and even nearly form a sward.

Molinia can occur on *Schoenus* tussocks, maybe on only a few, maybe on most. Once established, it can, in its long-internode form, spread over the tussock both sideways and downwards quicker than the *Schoenus* itself can grow. *Molinia* may overgrow and dominate the tussock top, forming several centimetres of tussock above the *Schoenus*. Young *Schoenus*, with the toughness of the young tussock (and any chemical toxin?), bears little *Molinia*. Ailing ones may – but need not – bear much. (It is unclear whether the *Molinia* is responsible for the ailing.) The vigour of the young tussocks prevents *Molinia* dominating. So does wet ground between tussocks. Burning also lowers *Molinia*, as it prevents *Schoenus* tussocks growing much above water: too wet for *Molinia*. In only a little drier habitat, however, *Molinia* can grow on the ground. Once *Molinia* has invaded at ground level, its tall tussocks on the ground and also its ability to smother many *Schoenus* tussocks would lead to its dominance. Unless, of course, birch or *Cladium*, etc. arrived first! Neither in fact happened in Redgrave Fen, as drying was too quick, peat mineralised and nutrient-rich communities took over. Their taller shoots were able to shade *Molinia* and (depauperate) *Schoenus*.

Schoenus nigricans and *Cladium mariscus* (Figures 10.14–10.18)

Cladium has increased at the expense of *Schoenus* in the twentieth century, as shown by distribution records and by the presence of dead *Schoenus* tussocks under Cladium. This

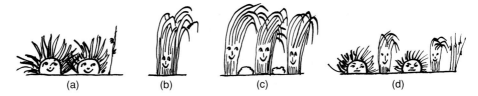

Figure 10.14 Schoenus nigricans and *Cladium mariscus* (a), (b) dominant communities of the two, separately; (c) *Cladium* large enough to form a shading sward and kill *Schoenus*; (d) *Cladium* relatively shorter, unable to form a shading sward, and co-dominant with *Schoenus* and *Juncus subnodulosus*.

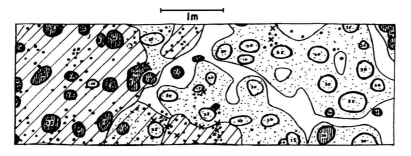

Figure 10.15 Wave advance of tall *Cladium* into *Schoenus*, plan, Eriswell lode head fens, Suffolk (see Figure 10.14c). *Cladium*–shaded area shown with oblique lines, *Schoenus* shaded area dotted. Dead *Schoenus* tussocks shown dark (Haslam 1960).

contrasts with the *Phragmites–Carex paniculata* example described earlier. There, the tussock species potentially dominates. Here it is the rhizomatous. *Cladium* has a wider and drier habitat range than *Schoenus* (in the east of Britain). *Phragmites* has a wider range than *Carex paniculata*: but then *Phragmites* has an exceptionally great habitat range, extending into both *Cladium* and *Schoenus* communities. In a *Cladium* community, *Phragmites* has its normal form and frequency, in contrast to the cramped changes that occur in the *Schoenus* habitat (Chapter 9).

The interactions between *Cladium* and *Schoenus* depend on the form, the habit, of the *Cladium*. The habit is partly but not entirely related to present habitat (related to past mowing, perhaps?).

Dwarf *Cladium* in East Anglia is mostly confined to strongly calcium-influenced fens, that is, to areas in fens just away from the most limestone spring-dominated parts. The shoots may be dense – denser than in normal *Cladium* – but they are only 1 m or so high, narrow leaved and do not bend over together to form a tall sward as in normal *Cladium*. Because there is no shading canopy, *Schoenus* and even *Juncus subnodulosus* can co-dominate. A co-dominated community, stable in the short term, results. All three species are larger in wetter places than in the drier.

Normal *Cladium* can occur with *Schoenus* also. Where normal and dwarf meet, each can have rhizomes penetrating the other, bearing dead shoots only. Normal *Cladium* is at least 1.5 m tall, with a shading canopy of wide, bent-over leaves. These are large enough to form a litter mat, which may become thick enough to harm the plant (until, of course,

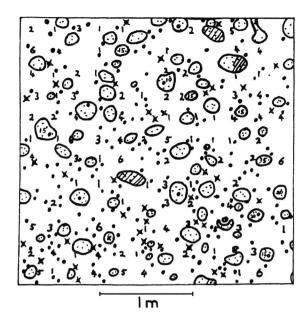

Figure 10.16 Co-dominant short *Cladium, Schoenus* and *Juncus subnodulosus*, plan, Thelnetham Old Fen, Norfolk. *Cladium* shoots shown as dots (leaf) and crosses (flowering), *J. subnodulosus* shoots in each 25 cm² shown as numbers. Living *Schoenus* tussocks dotted, dead tussocks shown dark (Haslam 1960).

the litter mat breaks down). Most young shoots are unable to grow through this, and die young (the crop may be down to only a quarter).

Advancing normal *Cladium* kills *Schoenus*. Shoots invade sparsely (see Figure 10.15), build up density, shading power and litter mat: and all that remains of the *Schoenus* is dead tussocks underneath. Some dead *Cladium* shoots may be found below tussocks. *Cladium* shoots started to come up all over the habitat. Those developing under tussocks are likely to die. Just a few come through very short tussocks (tufts), or through the sides of dead tussocks.

The advancing wave, some 2 m wide (where observed), has more young shoots, taller adult ones and more flowering heads than the stand behind: marginal vigour (Watt 1955). All smaller plants, bryophytes, *Parnassia palustris*, *Drosera* spp., orchids, etc., are killed by the shade likewise. Only the tallest plants survive (e.g. *Eupatorium cannabinum*, *Phragmites*). The litter kills, even more than the living shoots do, and remains thick for some time after the advancing wave has passed.

There can also be succession of *Schoenus* without an advancing wave of *Cladium*. Shoots slowly infiltrate into the *Schoenus*. The result is the same, but slower.

A litter mat accumulates only in unharvested places. For centuries, almost all *Cladium* was harvested, so a litter mat is effectively a new phenomenon.

Phragmites made sparse in three other vegetation types

Allowing grazing into a valley bog can, over a few years, turn near-dominant read into sparse short reed. Tall shoots (dominant because there are no other tall species) can

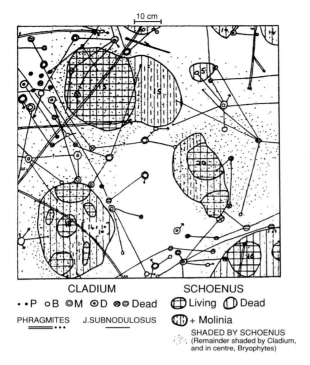

CLADIUM   SCHOENUS
•·P oB ⊙M ⊙D ⊗⊝ Dead   ⊞ Living ⊓ Dead
PHRAGMITES   J.SUBNODULOSUS   ⊞ + Molinia
━━━···━━━   ━━━━━━   SHADED BY SCHOENUS
                               (Remainder shaded by Cladium,
                               and in centre, Bryophytes)

Figure 10.17 Wave advance of tall *Cladium* into *Schoenus*, below-ground plan, Eriswell lode head fens, Suffolk. *Cladium* shoots marked as: P, pioneer; B, building; M, mature; D, decaying. Dark shoots are dead. *Cladium* is starting to shade left and below (Haslam 1960).

decrease in (modal) height from 150 to 60 cm, as a result of buds decreasing from a (modal) width of 5.5 to 3 mm. Because of the low-nutrient status, the rhizomes could not replace grazed shoots – even with light grazing – with ones of equal width (see Chapter 9). Therefore, *Eriophorum angustifolium* and small *Carex* spp. were able to take over. In a more nutrient-rich soil, with similar grazing, reed would remain abundant. (Heavier grazing would reduce and eliminate it.)

In the New Forest, continued heavy grazing on such bog stands eliminated the *Phragmites*: but only after the reedswamp had been drained so livestock could get access. Earlier, *Phragmites* was protected by the wet ground preventing livestock access. The plant interactions here are determined by animal (including human) impact, aided by the lack of resilience of *Phragmites* in low-nutrient places.

Juncus subnodulosus (see Chapter 9), when in a dense mat, prevents *Phragmites* from growing tall enough to shade it. Its rhizomes are well separated from those of *Phragmites* (they are higher). They form a thick weft, in which *Phragmites* emerging buds are narrowed greatly. This, together with the *Phragmites* internodes growing much less than normal, means that the reeds stay short. (This is toxin production: harming the *J. subnodulosus* rhizomes allows normal *Phragmites* growth.) A simple way for a short plant to avoid being shaded and taken over by a (potentially) taller one! Various species have this ability.

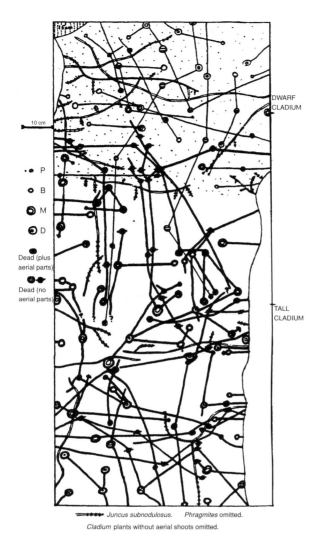

Within the figure:
DWARF
CLADIUM

10 cm

• P
○ B
◑ M
◐ D

Dead (plus
aerial parts)

Dead (no
aerial parts)

TALL
CLADIUM

Juncus subnodulosus.　Phragmites omitted.
Cladium plants without aerial shoots omitted.

Figure 10.18 Tall and short *Cladium mariscus*, below-ground plan, (other species omitted), Eriswell lode head fens, Suffolk, showing the two different clones, in the same habitat (Haslam 1960).

Slightly brackish, vigorously growing *Agrostis stolonifera* has the same ability (see Chapter 9). *Phragmites* vigorously advancing into thin *Agrostis* had about 50 shoots/m^2 in the first year, and about 70/m^2, which was a shading density, in the second year. In contrast, in thick *Agrostis* there were about 12 shoots/m^2 in both first and second years. Their emerging buds were much narrowed (again by 2–5 mm each) and so their height reduced. The approaching *Phragmites* was an advancing wave, so with much extra food. Yet its shoots were unable to become large and dense enough to shade – in just short grass! Mature (hinterland) stands are even less able to cope with large *Agrostis*. If it is partly reduced by an advancing wave of *Phragmites*, *Agrostis* can increase again after the advancing wave is past, and there is less vigour and impetus in the *Phragmites*.

The interaction between *Phragmites* and *Agrostis* depends on the ability of the *Phragmites* to shade, and the ability of the *Agrostis* to prevent such shade developing. Habitat factors control both. In a drying stand, for instance, *Agrostis* grows better: and therefore *Phragmites* grows worse.

Reedswamp invasion of open water

By zonation or by succession (see Chapter 3), reedswamp on shallow lake shores ordinarily lies between, on the deep water side, submerged and floating plants, and, on the landward side, carr or (carr-prevented) marsh or fen communities. Most dominant reedswamp plants are sward species, such as *Scirpus lacustris* (deepest), *Equisetum fluviatile* and *Cladium mariscus*. Some are tussocks, e.g. *Carex elata* and *C. paniculata*. *Phragmites australis* is the main peat-building species, but far from the only one (see Chapter 3). Even *Phragmites*, however, can grow (either dominant or sparse) where it does not form peat. Peat is not formed (1) when *Phragmites* is sparse, owing to habitat limitation, e.g. unsuitable water, nutrients, competition or management; (2) when current or waves remove litter, so no material can accumulate; or (3) where there is much incoming mineral sediment. With such deposition, even with peat-building capacity, a peat-rich mineral soil results; and finally (4) in dry places oxidation and decomposition prevent any accumulation of organic material.

Each species has a different inherent maximum water depth. In addition to this internal control, there is external habitat control. In lakes, part of the substrate may be exposed rock, unstable shingle or other unsuitable substrate (see Figure 10.19) and complex patterns may result in consequence. Plants need a substrate to root in and anchor to, one in which roots can grow, in which they can live in without being squashed, and from which they can get enough nutrients.

In rivers, the reedswamp fringe can grow where the depth and substrate are suitable, and where storm washout is suitable: this varies with species. In ordinary hill rivers, *Sparganium erectum* (the commonest river reedswamp species) will grow 0.5 to 1 m per year down from the bank into the shallow edge. At least when this band is 2–4 m wide, it will be swept away by storm flows. In a band this wide, the plant is no longer anchored to the firm soil of the bank. It is encroaching into over-deep water. It is quite possibly constricting the flow in the river, so making flow even fiercer and more likely to wash the vegetation away. Unusually severe storm flows will wash away even narrow fringes. In lowland rivers the engineers, worried about ponding in the river and waterlogging on the land, usually arrive before the reedswamp has reached 3 m wide.

Boats, swimmers and other recreation activities can limit or indeed eliminate reedswamp, restricting it to quieter places or even just quieter places up the bank, above the disturbance in the water.

Each species has its own range of nutrient status. *Carex rostrata* is perhaps the most nutrient-poor, with *Equisetum fluviatile* and *Eleocharis palustris* also nutrient-low. *Schoenus nigricans* and *Cladium mariscus* can do well in calcium dominance. *Glyceria maxima* needs high silt.

With the different ranges of water and nutrients, many patterns can occur. These species are usually monodominant; there is only one dominant in a given place, even if there is a mosaic overall. The habitat ranges of species overlap. Chance and past history as well as present habitat determine which species will dominate where.

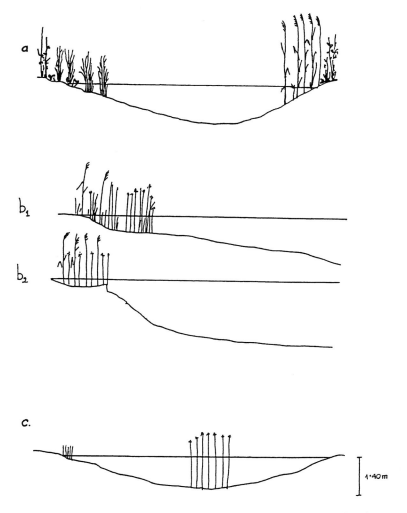

Figure 10.19 Lake edge reedswamp profiles. (a) Large sedges to left, *Phragmites* grading to tall herbs, right; (b) *Phragmites australis* and *Scirpus lacustris* varying with shore type; (c) *Scirpus lacustris* in centre of loch, where alone there is mud. (a, b, redrawn from Tansley 1949; c, redrawn from Spence in Burnett 1964.)

A native and an introduced grass in the Camargue, France (Mesleard *et al.* 1993)

Both *Aeluropus littoralis* (native) and *Paspalum paspalodes* (South American) grow in monodominant stands in the marshes of the Rhone delta. *P. paspalodes* is good for pasture and for wintering waterfowl, and is abundant on abandoned agricultural land. *A. litoralis*, with the same growth form, is a member of seasonally flooded brackish marsh margins. Both species tolerate the winter flooding, *P. paspalodes* is the more tolerant, so has a competitive advantage. However, *P. paspalodes* is not salt-tolerant, and where there is salt, *A. littoralis* has the advantage. Knowing the salt content means knowing the grass (the two are not found together, except as separate patches in multi-dominant communities).

Phalaris arundinacea and *Urtica dioica* in flood meadows, River Luznice, Czech Republic (Prach 1992, 1993) (Figure 7.6)

Both species are common in flood meadows abandoned after the Second World War. Both grow well here by seed and by rhizome. There are plenty of young plants in places opened by floods, etc. *Urtica* is less tolerant of flood (though more so, of shade).

The water level in the meadows varies widely and unpredictably, and this controls the plant pattern. Drier places have more *Urtica*, wetter ones, more *Phalaris*. As, overall, the meadows are getting wetter, *Urtica* is declining and may end as an ephemeral in disturbed places. However, part of the meadows is more managed and drier. Here both *Urtica* and *Phalaris* decrease in favour of short sward species that have their growing points below mowing level.

Typha spp. in North America

There are four Typhas in the USA, *T. latifolia* and *T. angustifolia* (British and European also), their hybrid, *T. glauca*, and *T. domingensis* (Mediterranean but not British). *T. angustifolia* can grow into deeper water than *T. latifolia*, so has a competitive advantage there, ascribed to its tall narrow leaves and rhizomes storing much food. *T. latifolia* grows better in shallow water, ascribed to its greater leaf area. Water depth, therefore, determines which of the two, in the USA, will succeed and form a monodominant stand. Where both have vigour, *T. angustifolia* has much early growth, while *T. latifolia* grows slowly but well throughout the summer. *T. angustifolia* would therefore do better in the absence of interference. If the stand was traumatised, however, *T. latifolia* has the advantage. *T. latifolia* is also more tolerant to shade. Where the competition is between *T. latifolia* and the slightly narrower leaved *T. domingensis*, *T. latifolia* competes best in drier habitats (Grace 1985, 1989; Grace and Wetzel 1981a,b,c).

Vigour and hence competitive advantage between species of this one genus are determined by variation in water depth, shading, growth pattern, and proportion of food held in leaves and in rhizomes.

T. glauca probably does better than either parent in nutrient- or organic-higher places with much water level fluctuation (S. G. Smith, personal communication). This is like *T. domingensis* in the Mediterranean, where it spreads well with increasing pollution.

Cladium mariscoides and *Typha jamaicense* in the Everglades, Florida (Davis, in Davis and Ogden 1994) (Table 10.1)

Typha is the species of greater nutrient range, an early coloniser of naturally disturbed nutrient-low ones, and a persistent dominant of nutrient-polluted sites where *Cladium* used to dominate. Pollution alters species dominance, here.

Myrica gale, *Cladium mariscus* and the Keeper (Parmenter 1995)

Myrica is a characteristic sub-shrub of *Cladium* communities, and in harvested beds increases in the open habitat, unless cleared.

Table 10.1 The behaviour of a low-nutrient species (*Cladium jamaicense*, sawgrass) and a high-nutrient one (*Typha domingensis*) in relation to nutrient pollution in the Florida Everglades (after Davis, in Davis and Ogden 1994)

Sawgrass	Cattail
Stress-tolerant species from infertile habitats	Competitive species from fertile habitats
Relatively small leaf growth response to short-term variations in nutrient pollution	Relatively large leaf growth response to short-term variations in nutrient pollution
Relatively low uptake capacity under polluted conditions	Relatively high uptake capacity under polluted conditions
Lower rate of phosphorus loss from senescing leaves	Higher rate of phosphorus loss from senescing leaves
Longer leaf longevity	Shorter leaf longevity
Lower leaf growth rate	Higher leaf growth rate
Slower leaf turnover rate	More rapid leaf turnover rate
Well-developed leaf cuticle	Poorly developed leaf cuticle
Dominates in typical unpolluted areas	Dominates in eutrophic pollution (and any naturally eutrophic places)

One keeper altered the balance of dominance more. He objected to a naturalist regularly visiting the *Myrica* (it bore a rare invertebrate): and there was no more *Myrica*.

Combined ills

It often happens that a species can withstand one unsatisfactory factor, only to succumb when a second, and different one is added. *Phragmites* can grow and dominate in nutrient-low places: provided grazing is negligible. It can survive the weakening from two unusually dry winters: provided no invader takes advantage of the window of opportunity to invade. *Myrica gale* tolerates routine burning: but not the personal attentions of the keeper. *Urtica dioica* holds its own against *Phalaris arundinacea*: but not if the habitat is regularly flooded. In the field, many factors may be unsatisfactory, and a slight increase in one of them may weaken the plant sufficiently for another species to invade and take over. Conversely, a plant in an otherwise satisfactory habitat may be able to remain robust despite some (moderately) unsuitable features.

Conclusions

Changes occur along a continuum, but the extremes can be separated. A long-term trend, e.g. drying, whether in space (e.g. along a slope) or in time, will impose itself on the community. Changes in dominance will (except in the very short term) be in conformity with that trend. Superimposed on this is the other extreme. There are changes that happen only because of a historical accident: because the flock of geese or the flood just happened to come to this place and open up the habitat, *and* other species were in waiting.

When one species or community takes over another, therefore, it is easy, true but too facile, to say that one grows better than the other in a given set of conditions.

There is a definite pattern, varying with the species, of how the takeover mechanism works. It can only work if both the species and the takeover conditions are present. 'Weaker' vegetation may persist for decades or centuries if no 'stronger' type arrives in the given habitat.

Species balance   The ability of species to invade or to resist invasion can be altered by altering the vigour (or robustness) of one of the protagonists. If one is weakened or the other strengthened, competitive balance may be altered. Habitat factors likely to alter robustness include: water regime; chemical regime (including nutrients, pollution, salinity); presence of propagules of species able to colonise; grazing and cutting; disease; aeration; erosion; substrate texture; factors lodging shoots (storm, trampling, rambling or climbing species, etc.); burning; other disturbance and historical accidents.

Fitted to these are the ability of a plant to (1) germinate and establish in conditions occurring fairly frequently, both for the first colonisation and for later recruitment (except, for the latter, when this can be done by rhizome); (2) shade other species; (3) tolerate the shade cast by other species; (4) form (in dry places) a litter mat; (5) tolerate a litter mat formed by other species; (6) produce root toxins that dwarf, restrict or weaken other species; (7) produce an advancing margin of much vigour that can suppress vegetation in its way; (8) tolerate a wide range of physical and biotic conditions; and (9) grow, occupy space, and shade the ground early in the year.

Vigour and resilience   In *Phragmites*, lateral rhizome growth and hence shoot density are easily stopped by other species (root toxins), but horizontal growth and hence *Phragmites*, present sparsely, are very difficult to stop. *Phragmites* can invade and remain where it cannot dominate.

Rhizomatous species such as *Glyceria maxima* or *Typha* sp., which can cover all the ground (with a sward), can be contrasted with those such as *Schoenus nigricans*, equally dominant, equally controlling the community, but with open spaces and a rich and rare associated flora.

Annuals are dependent on frequent repeated establishment by seed. Trees, tussocks and others unable to move far vegetatively also need establishment from seed, although of course less frequently. Windows of opportunity for establishment must occur at proper intervals for species to survive.

However narrow the habitat range, a species, to live in wetlands, must find this range at suitable seasons, sufficiently often. Even rhizomatous plants such as *Phragmites* and *Typha* spp., doing well by rhizomes for centuries, spread well by seed when bare places are created (e.g. gravel pits). With the general tendency to regulate, bare spaces on the ground are getting fewer. They can be created by stock, domestic or wild, by grazing, trampling, other disturbance, faecal deposits, etc. They can be created by other forms of human impact, for example mowing, burning, trampling and such like and by construction, excavation and such like.

In some Swedish lakes, shores without emergents have dried out only a few times this century (e.g. 1932–3). Species having (at best) very large numbers of small, wind-dispersed seeds, such as *Phragmites* and *Typha* spp., can invade rapidly in response to the sudden appearance of suitable conditions (conditions *and* the seed). Episodes of seeding may spread species rapidly, and indeed alter dominance patterns (Ekston and Weisner, in Finlayson and Larson 1991).

Some examples of response are:

Aeration. Alder invades fens better where the soil is better aerated, sallow tolerates a wider range, so in early carr, sallow is often in more stagnant places.
Disease. A leaf virus killed *Epilobium hirsutum* in a tall-herb community: four years later *Epilobium* was still absent. *Glyceria maxima* died of rhizome rot in a mixed *Glyceria–Phragmites* stand. *Phragmites* became monodominant.

Height: shading of shorter by taller plants. This is the most common, important and wide-spread means of competition.

Lodging. Species such as *Calystegia sepium*, *Galium aparine* and *Glyceria maxima* can bring down *Phragmites*, for example, in late summer, causing poor maturation of the *Phragmites* shoots.

Nutrient regime. Species of higher nutrient status are generally (not always) more able to shade and suppress those of lower nutrient status, when habitat allows them entry.

Toxins. For *Phragmites*, *Oxalis pes-caprae* has none, *Epilobium hirsutum* some, *Juncus subnodulosus* much.

Water level. Saturated soil and flooding depress, for example, *Calamagrostis canescens* and its more nutrient-poor counterpart, *Molinia caerulea*, allowing reedswamp species to flourish.

Weather. The spring start (so important for establishing dominance or even presence in mixed stands) is dependent on two factors. First, the inherent behaviour of the plant for germination or emergence. Second, the weather, which, through temperature and rainfall, modifies this date, and does so differently for different species, altering balance by the year. Summer growth, viable seed production, etc. are modified similarly.

At first sight it would seem that, given equal height, rhizomatous species would do best as they are not limited by the necessity of having new plants from seed. This is untrue, however. *Cladium* can invade and succeed *Schoenus*, but only if present (in tall form) and if the soil is firm (dry) enough, and then because of its superior shading ability, not its rhizome. *Phragmites* can succeed *Carex paniculata* only if the latter has no windows of opportunity for seed regeneration. That is, *Phragmites* cannot weaken *C. paniculata*, but in theory could wait it out. These windows are likely to occur, given the tolerance of young *C. paniculata* to shade, although a change of habitat could be unsatisfactory for the more restricted *C. paniculata*. *Phragmites* cannot succeed *Schoenus*, the nutrient status is wrong. It can only succeed *Cladium* if the nutrient and water regime suits them both, and management (or similar) favours *Phragmites*: a narrow window. *Phragmites* cannot even succeed vigorous (brackish) *Agrostis stolonifera*, it is inhibited by it. Relations are far more complex than just growth form.

Plant vigour is a phenomenon needing further study. It is partly genetic, partly environmental and partly due to plant phase (advancing vigour). It gives a greater competitive advantage.

The quiet, remote wetland, so peaceful for people: truly, a silent battlefield.

11 Threats and losses, past and present

(of Great Ouse riverside)
The poplars are felled, farewell to the shade,
And the whispering sound of the cool colonnade!
The winds play no longer and sing in the leaves,
Nor Ouse on his bosom their image receives.

The blackbird has fled to another retreat,
Where the hazels afford him a screen from the heat,
And the scene where his melody charmed me before
Resounds with his sweet flowing ditty no more.

(W. Cowper)

(The trouble now is that too often the 'other retreat' is small and far and as full as it can hold, and this means the blackbird, etc. not moving, but dying out from habitat loss.)

The fen and quagmire, so marish by kind
Are to be drained

(Tussor)

The ways being foul, twenty to one
He's stuck in a slough, and overthrown

(Milton)

Dutch polder areas are losing much of their characteristic biotic diversity because many wetland species are decreasing or have disappeared.

(Barendregt *et al.*, in Vos and Opdam 1993)

One man's mess is another's opportunity.
Not all is irretrievable.

The major dangers

Most wetlands have gone. They were in danger, and the dangers materialised (Table 11.1). The Basic Figures (1–22) illustrate some of the more frequent of these.

Loss of water (Basic Figure 8)

Drainage is widespread. The first drainage shown here is open channels. This was used anciently on marsh and fen (and raised bogs on them), and has newly spread, from raised to blanket bog. Blanket bog is in the least populated areas. After surface drainage comes,

Table 11.1 Loss of European mires (from Heathwaite, in Hughes and Heathwaite 1995)

Country	Loss
The Netherlands and Poland	All exploited
Switzerland and Germany	500 ha remain
Ireland	80 000 ha drained since 1946
Britain	90% loss of blanket bog
	98% loss of raised bog: 445 ha remain

in agricultural areas, under-drainage; alluvial flood plains may be in this next stage, with under-drainage, carrying water from the slope above direct to the stream. If much water comes from the slope above, there may be a catchwater drain running along the outside of the wetland, receiving run-off and again taking it direct to the stream, removing this 'wetland water' from it. The river is embanked, removing that source of 'wetland water' from the wetland. Both processes alter the water quality in the wetland: POLLUTION. The river hydrological system is by now separated, decoupled as it is termed, from the flood plain. This disrupts hydrological cycling and the cleaning function of the wetland.

High embanking and very low river levels are seldom needed together, to prevent flood. The embanking, however, may have been done before the drainage, or both may have been planned together.

Then there is abstraction, which also lowers water level, and, because groundwater is usually of different chemical composition, causes pollution by altering the chemical status of the previously spring-fed area. If that area is now arable, or housing estate, this is irrelevant to the 'wild' vegetation, which has gone. Where 'wild' vegetation is left, however, it is much changed.

More intensive farming (Basic Figure 16)

In near-natural habitats, the harvest is of wetland products. The next stage is rough and wet grassland (or bog grassland or conifer forestry). Arable is the final stage. This succession proceeds at different rates in different wetlands, and even in different parts of the same wetland. It started in pre-Roman times. The general trend, with many ups and downs, has been accelerating in recent centuries, with intensification of yield much increasing in the past half-century. In the most intensive farmland, there may be few drainage dykes. These may also be used for irrigation in summer: such a waste of the little water left! The dykes probably bear a poor or non-existent aquatic and marsh community and, when drier, vanish. With the low-nutrient and dystrophic nature of bogs, it takes much more nutrient, etc. addition to get fertile land.

Losing the functioning of a bog is more destructive than losing that of fen or marsh. With suitable and adequate water, and a source of plants, and animals, wetland functioning can return fairly quickly (decades). Bogs, having a more active functioning, take much longer, and need more restricted conditions (discussed later).

Destructive change in products (Basic Figure 18)

Changing a lowland wetland from, for example, reedbed to sedgebed maintains 'wild' vegetation, and – as long as the proportions of each remain satisfactory – is not destruction.

(The reverse change, from sedgebed to reedbed, it should be pointed out, may be destructive in some conditions. If the sedgebed is on calcium-dominated peat, it is necessary to raise the nutrient status, and destroy the calcium dominance to produce reedbed. Where sedgebed and reedbed habitat overlap, however, there is no destruction.)

Changing a reedbed to a gravel pit is destructive. However, most such gravel pits are on riverside grassland or arable, developed by drainage and agrochemicals. Changing intensively managed farmland to – after the extraction – new wetland and water habitat could be regarded as a gain.

Changing a living bog to a milled peatland after peat extraction is much worse, is serious destruction and over a large area. This is further discussed later.

Changes in construction (Basic Figures 7, 14, 16)

Wetland that has vanished under housing estates, industry and roads is even more destroyed than that under arable. Once drainage is adequate, roads and farm buildings start. As drainage increases, the denser construction moves in, and may come to cover much of the flood plain.

Pollution (Basic Figure 14, Tables 11.2 and 11.3)

Vegetation, in comparison to other groups of animals and plants, has the advantage – to researchers – of being large, stationary and consisting of a relatively small number of species. It is also more sensitive to chemicals than is often realised. Therefore, species assemblages, communities, can be used for assessment. To identify change, however, the pre-change biota must be known: and seldom is. An additional difficulty is the presence of time-lag communities (see Chapters 6 and 7), where long-lived dominants persist, but do not reproduce, in conditions no longer suitable for them.

Birds are the most mobile group, they use large units of habitat, and – at present – appear to be, as a group, less sensitive to minor habitat features. Mammals, again on inadequate evidence, are perhaps even less sensitive. However, these two can be the best monitors for some purposes, such as monitoring organochlorine and other pesticides by their distribution in otters and birds of prey.

Table 11.2 Difficulties in controlling diffuse (non-point-source) pollution (after Novotny and Olem 1994)

1	Farmers, agricultural, industrial and mercantile groups, planners and the general public find it difficult (or impossible) to believe in it (NOT *our* supermarket car park! Why, they have a Green policy!)
2	The environmental effects often occur downstream, not seen by the land user or owner
3	The effects are cumulative. One form of pollution may be tolerated by the river, while the total effects of four busy roads, three large car parks, two large farms and a golf course are not
4	There are often no historical data. 'The river has always been like this', however untrue, may be difficult to disprove.
5	Relationships between cause and effect can be difficult to establish: the busy road is far away, the soil on my farm does not release that much fertiliser, etc.
6	When firm data documenting decline in water quality (or, e.g., fish) are available, even here the reason may be unclear – straightening the river may have decreased its cleaning power, so the agrochemicals have not increased, but are cleaned less

Table 11.3 Wetland habitat sensitivity to and relationships with different types of pollution (Haslam 1994)

Pollution source	Sensitive habitats
Aerial deposition	Wetlands with few solutes are the most sensitive: bogs, some lakes, and to a lesser extent nutrient-poor mires. The concentration of pollutants varies regionally with sources and wind patterns
All pollution entering from the side	The outer roughly 50 m (from the outer boundary or a polluted watercourse) is the most vulnerable as the boundary receives the full strength of the pollution. Pollutants in low concentrations and easily transformed to harmless substances are effectively cleansed within this distance. Those in low concentrations saturating the habitat may be held within the 50 m for decades or longer. (Gross pollutions spread and damage further away)
Effluents (sewage treatment works, other sewage, factories, farm and other spills, etc.)	The most vulnerable wetlands are those with (1) water draining in from effluent sources and (2) low solutes, buffering capacity and transforming capacity. These are nutrient-lower swamps, fens and mires, especially areas too small to be protected by a polluted outer buffer strip. River water seriously polluted with effluents and run-off is seldom permitted to reach wetlands, but where it does, it is harmful
Run-off	Similar to effluents, where dirty run-off enters wetlands (e.g. from busy roads). This may enter over a wider area than effluent pollution
Agrochemicals	Compared to effluents and run-off, this pollution is much more widespread and, per km, less concentrated. Fertiliser run-off is particularly high in nitrogen. Pesticide concentrations are usually low and mainly in the sediments, and so far these are rarely troublesome in wetlands

More nutrient-poor, and drying wetlands are the most sensitive: nutrient-medium fens are the most vulnerable, as agricultural run-off does not often reach bogs |
| Dumping | Council tips or hazardous waste dumps may be in or beside wetlands and some may not be lined. Leachate reaching wetlands is probably initially concentrated enough to damage all habitat types. After some purification, the principles described for all pollution sources above apply |

Table 11.3 (cont'd)

Pollution source	Sensitive habitats
Drainage	Severe drainage totally removes wetlands, eroding peat from Highlands, creating, e.g., sugar beet fields in lowlands
	Lesser drainage alters wetland chemical status, and is therefore polluting. Sensitive habitats are (1) where there are mosaics and patterns of different water types (flushes, run-off hollows, etc.) that are amalgamated or lost on drainage, and (2) organic-rich and peat soils that become dried and suffer irreversible chemical changes, including a higher nutrient status
Disturbance	Disturbance of drier fen, etc., soils aids the irreversible chemical changes, including the higher nutrient status. (Trampling occurring with regular grazing, however, prevents or lessens this damage, by keeping the soil compact and pressed down)
Abstraction (enough for water loss to affect wetland groundwater)	Abstraction lowers ground water table. Drying alone acts as in above. Abstraction, however, may be even more damaging chemically. Wetland habitats develop in equilibrium with the incident water. Groundwater may be a major, minor or absent component of this water. Groundwater usually differs in composition to the run-off or other surface water. Consequently if groundwater is decreased or removed, the chemical status of relevant wetlands will change
	Many fen-type wetlands, e.g. in East Anglia, are particularly vulnerable, as their character depends on flooding with calcium-rich groundwater
Cutting, mowing, etc.	Removing vegetation from a wetland removes all the nutrients in that vegetation, and can help to reverse the effects of eutrophication
	Traditionally harvested wetlands have developed in equilibrium with this loss
Faecal material encouraged by man's activities	Herds of livestock held over-long in a small area, and large flocks of birds encouraged into a wetland by man's activities (e.g. making the site more attractive, or destroying alternative sites) can much raise nutrient status, and so alter vegetation and habitat

More intensive farming, changes of products, and change in constructions all increase pollution within wetlands. Changes elsewhere bring pollution to the wetland from outside, whether from the air (with industrial, agrochemical and vehicle, etc. chemicals), the river (with sewage treatment works run-off and other effluents) or the slopes around (farming wastes, road and urban run-off, etc.).

Air pollution has damaged large tracts of the south Pennines and lesser areas elsewhere, producing, with acid rain on bog peat, eroding, species-poor cotton grass moor (*Eriophorum vaginatum*). Grazing, burning and drainage have contributed to the change (an example of combined ills, see Chapter 10). The original communities were *Calluna–Eriophorum*, *Erica–Sphagnum* with *Calluna* and *Erica* heaths and *Nardus stricta* grasslands (Rodwell 1995). Sphagna, other mosses, lichens and associated flowering plants have been much reduced, and erosion may be widespread. Bogs, being solute-poor, are the most susceptible to outside pollution.

Air pollution (without or with eroded peat in run-off) has acidified over a quarter (over a million acres) of Welsh lakes. Some of these have reedswamp: therefore acidified wetland. Conifer planting increases vulnerability to acidification. Acid conditions mean aluminium released (see Chapter 7) and toxic. With stress, in drier parts birch wood can result. In places with 'never acid' waters these buffer present acid deposition, so are not likely to change in the short term. The 'occasionally acid' water may well deteriorate and add to the 'permanently acid' ones. Affected areas are more in the North.

Nutrient-rich lakes such as Llyn Chwythlin, with *Phragmites australis, Potamogeton pectinatus, Nuphar lutea*, reed bunting, moorhen and probably breeding dabchick, are not likely to change (Rimes 1992a,b; Sketch and Bareham 1993).

Near York, Askham Bog was a raised bog. It is now extraordinarily rich in species (over 300 plants, nationally and regionally rare invertebrates). Bog peat was cut at least from Roman times to the eighteenth century: by which time the usable peat had almost gone. Now, acid peat is back in the (again-raised) centre, with *Sphagnum, Molinia*, etc., but the lime-rich run-off water from the moraine around brings fen at the edges. There is open vegetation and woodland, bog, fen and water: most diverse, and the diversity is due to the removal of the peat! There is now active management. A new threat has come to Askham Bog: air pollution. The bog is downwind of industry, and emissions from that industry and nitrogen and other fertilisers are dumped on it. Sphagnum and low-nutrient species from both bog and fen are getting less, and nettle and other ruderals are spreading. Tall-herb communities threaten. North-central England combines high industry and low nutrients, so wetland buffering is low and this effect is shown sooner than many other parts. It is an unpleasant pollution, as nothing can be done on site to protect a site (Fitter and Hogg 1996).

A very frequent pollution, in the sense of man-made chemical change, is the removal of the water type with which a habitat developed, and without which it cannot, long term, continue. This is removing groundwater discharge by abstraction, and amalgamating or removing surface-water types by drainage.

In the Netherlands, water from the River Rhine has got into the general (complex and inter-related) watercourse system. Where that feeds wetlands (see Chapters 6 and 7), this water is beginning to have an effect. The Rhine itself, of course, contains a mixed bag of contaminants, many very poisonous.

In Norfolk fens, decreasing the input of calcium-rich groundwater allowed phosphate to rise and vegetation to change (from calcium dominance). Rare species in the *Schoenus nigricans* community are inversely proportional to phosphorus, and directly proportional to calcium (Shaw and Wheeler 1990; Boyer and Wheeler 1989).

Lowered pH damages roots and soil organisms (Barth and l'Hermite 1987). Acidity in Norfolk ditches alters flora and fauna (George 1992).

Iron in Broadland (with low redox) was associated with low diversities (Wheeler and Shaw 1987). Furthermore, in a spring-fed, base-rich fen, *Juncus subnodulosus* was tolerant to iron toxicity, and *Epilobium hirsutum* was sensitive. (Tolerance improved with added phosphate or drying.) Iron toxicity patterns were probably determining species distribution.

Ammonia pollutes via water and air. It is acidic and damages plants (e.g. Reddy and Smith 1987). A Polish nitrogen-processing factory emitting ammonia (as the principal pollutant) had vegetation destroyed nearby, then the reedbeds, alder carr and nitrogen-favoured species, and finally the near-clean vegetation beyond (Wassen *et al.* 1990).

Even Wicken Fen, isolated amid farmland, has to have polluted river water, that is now too polluted and is seriously damaging the fen (the alternative was to let it dry). At least British catchments are separated, and East Anglian pollution does not reach Wales. In the Netherlands, as noted earlier, Rhine water permeates the land through rivers, except for discharge areas.

Wetlands with lateral (sideways) flow are the most susceptible to agrochemical pollution, as the pollutants can enter though a pre-existing and fast pathway (Shaw and Wheeler 1990). With (Dutch) groundwater flow 0.1–1 m/day, it may be decades before pollution starts, but once started, it will continue for a very long time (Verhoeven 1992).

Two small Dutch fens in fertilised pasture were contrasted, one is a discharge fen with much spring water. The other is a recharge fen, its aquifer refilled by run-off from the land around. This run-off now contains many agrochemicals. Both this fen and this aquifer were much more polluted (Koerselman *et al.* 1990).

The Trent plain had, until the 1980s, acid flushes with good populations of, for example, *Scirpus fluitans* and *Littorella uniflora*. Drainage removed both the flushes and the populations (Mountford and Sheail 1987).

Farm wastes are increasingly polluting, with 200 million tonnes of slurry, silage, etc. produced each year, and no overall disposal strategy. The proportion of a total of 78 lakes in Sites of Special Scientific Interest that are polluted is horrifying: these are supposedly protected!

Polluted by wastes of:	*Certain*	*Probable*
Farm	44	27
Sewage	36	24
Others	8	29

(*English Nature* 1991, 1992). Lakes should have fringing wetland.

Wynbunbury Moss, Cheshire, was a nutrient-low basin bog, with floating peat, etc. In the 1930s, road run-off was diverted to the bog, which probably then purified all. In the 1940s, septic tanks were added. By 1981 alder–sallow carr and fen had spread. *Typha* and *Phragmites* were growing on ditch-sides (where, without pollution, it was *Osmunda regalis* and fen scrub). Drains were then looked for, and those found were blocked. Bog vegetation continued to decrease (due to lost drains or slow input of earlier pollutants?). By 1993 the pollution might have been halted (C. Hayer, personal communication; Page and Rieley 1992). As there were no chemical and hydrological analyses, diagnosis was by vegetation, and remedy by 'guesstimate'. This is a fairly typical exercise.

An American cedar wetland receiving waste water had higher levels of the main nutrients, heavy metals, organic materials, etc. The mature trees remained, but down on the ground ruderals and exotics spread (Eisler 1992).

Organic and mixed-organic pollution tends to increase invertebrate populations and skew their populations. In the Clyde, the effect of sewage outfalls on mudflats is observable for less than 2 km, and is (Furness 1989):

Very low oxygen, toxic	Low oxygen	Moderate oxygen	High oxygen (clean)
No fish	No fish	Many fish	Many fish
No invertebrates	Much *Nereis, Corophium*	High invertebrate biomass	Diverse invertebrates of moderate biomass
No waterfowl	Many waterfowl	More waterfowl	Moderate waterfowl

The birds changing in this sequence include goldeneye, redshank, dunlin, pochard, grebes, other waders and ducks. There were no recorded harmful effects to the birds from eating polluted mud or invertebrates, despite a high concentration of heavy metals and other poisons.

Organochlorine pesticides and other halogenated hydrocarbons and related compounds are liable to lessen bird body weight, make eggshells thin (reducing breeding success) and generally affect reproduction. If levels are high enough, they can kill. Pesticide harm is more likely to come from outside than from vegetation or prey within the wetlands.

Heavy metals are also damaging to birds, although these do have protective mechanisms. Cadmium damages the kidneys first, and this happens even in healthy breeders. High lead also kills (Furness 1989). High lead is more local; but lead shot eaten in place of gravel is dangerous. In Britain, swans died by the thousand – as anglers were unwilling to retrieve the lead shot they used – until non-lead weights were introduced instead. In Broadland, the lead did little damage until the vegetation mat was lost, allowing lead and swans to meet (Linsell 1990). Lead pellets in the Lindisfarne saltmarsh led to increased levels in the plants (highest in *Enteromorpha* spp.), but these seemed unaffected. Wigeon feeding on the pellets, seeds and shoots, however, had lead at levels (3 to 14 ppm) probably affecting reproduction (Palmer and Evans 1991). Mercury causes embryo death, infertility, and nerve and behaviour disorders. Game birds are more sensitive than sea birds. Copper and iron are probably regulated by the bird's metabolism and cause little problem (Furness 1989). Selenium and boron from agrochemicals can impair reproduction in wetland birds (Pavaglio *et al.* 1992).

Where grazing (or mowing) is needed to maintain a vegetation type, bird grazing may be the more desirable: although livestock are more predictable. Bird grazing can lead to overkill, however, and wetlands may need protection from them (Hik and Jefferies 1990; Hik *et al.* 1991).

Birds move, some over areas of kilometres, some over thousands of kilometres. Protecting a small valley bog from pollution is therefore very different from protecting migratory waders. Loss of habitat (structural damage, not the chemical damage of pollution) is by far the greater hazard to wetland birds.

(Pollutant interactions in wetlands are well summarised in Novotny and Olom 1994.)

Peat loss

Where peat is being lost, through drainage and erosion in bog, through wastage following drainage in fen, chemistry changes. This is of little conservation importance where fen peat soils and their clay soil successors are heavily fertilised anyway. Where blanket bog is removed, however, leaving mineral soil, there is both bog loss and a change in the run-off that reaches other wetlands. Run-off before, during, and after erosion is very different. Bogs being eroded differ chemically as well as biologically to those not eroded.

The peats of Europe (bogs and fens) are being mapped (for the Vegetation Map of Europe): a small-scale map is already available, a large-scale one (1:2 500 000) is in progress (Rykincek and Yurkovskaya, in Moen 1995). Britain has a mire and wetland inventory. Such exercises look at the remnants. In Britain, Lowland raised bogs have only 2 per cent left in good condition, blanket bogs 10 per cent (Heathwaite, in Hughes and Heathwaite 1995) (Table 11.1).

With more drainage, losses to agriculture, and, in the twentieth century, forestry have been great. Until quite recently it was considered: bog is waste, cover it with trees and make it valuable. (The same as fen, earlier: fen is waste, cover it with crops and make it valuable.)

Export for power stations and horticulture is responsible for a much quicker and – given the small amount left – more damaging loss in the late twentieth century. Machine extraction of peat can remove up to 50 cm a year. The peat growth in undisturbed bogs is hardly over 1 mm!

In Canada, where peatlands are still vast, it can be argued that extraction is irrelevant as the losses to agriculture and settlement are so much greater (Rubec, in Moen 1995). Elsewhere, extraction is (non-arguably) destruction.

There are three main machine extraction methods. All require drainage first. The most damaging (and so most widespread) is milling. This involves scraping off and vacuuming thin layers of peat over large areas, leaving areas of just about flat (or gently crowned) bare peat, with a few ditches. Some 3000 ha of British raised bog is being milled as this book is written (65 per cent of the British and 95 per cent of the world production is by milling).

Machine block cutting removes peat in strips, leaving troughs in which rain can accumulate, 10-yard strips in which peat may recover, which is better.

The third method is extrusion, removing peat by tubes, which occurs in, for example, Hatfield Moors (Moen 1995).

Unfortunately, many permits to extract were given up to the 1950s (government being slower than ecologists to recognise the danger) and permits given need legislation and compensation to reverse. Thorne Moors has 73 ha of National Nature Reserve in 1918 ha of bog. Fortunately, English Nature has been able to buy large areas, resulting in extraction being phased out at these locations.

Given the sensitive water balance in bogs, taking off the surface, removing the active acrotelm, going down to the compacted catotelm, gives a drastically altered habitat. Active bogs retain rain (see Chapter 3). Damaged bogs have much less water storage. They have ditches that drain. They have greater water loss from evapotranspiration. They have less (acrotelm) peat for wetter storage. Downwards seepage is increased with the loss of the active peat and exposing cracks, etc. The catotelm peat is much more humified. Having a dry surface exposed means mineralisation and much change. The catotelm peat is chemically different too. To add to all this water loss, abstraction is, in some bogs, taking place underneath. (Money, Heathwaite, in Hughes and Heathwaite 1995). Water loss can lead to subsidence, to add to the trouble.

With just ditch drainage, there is colonisation, in a Dutch example, of *Erica tetralix, Eriophorum vaginatum, Scirpus caespitosum, Molinia caerulea* and *Cladonia* spp. More drainage means the loss of *Oxycoccus palustris* and many *Sphagnum* spp. and the increase of *Calluna vulgaris, Vaccinium* spp. and *Betula* spp. (Barkman, in Verhoeven 1992).

Many blanket bogs have had recent degradation after earlier peat accumulation. Degradation comes from drainage, also from too much grazing, burning, air pollution, storm, etc. It leads to bare peat.

Removing plant cover means loss of plant community, loss of animal cover and shelter, and loss of animal food (whether invertebrates and other 'wild' fauna or farmed grouse, deer and domestic livestock).

Gullying following bare peat has taken place in the past as well as the present, occurring, in dated sites, at around AD 450, 750–1050 BC and 1350–1700 BC. Small gullies of cross-section 1–1.5 m² are dated around 200–250 years. Larger gullies, of 4–5 m², are dated to over 500 years. Gullies have therefore been around for a long time, despite the greater recent impact. It is uncertain how far all are due to impact (e.g. Roman forest clearance, post-mediaeval sheep prolongation of the effects of the mediaeval dry phase), or whether such bogs reach a stage in which they are unstable (Tallis, in Wheeler *et al.* 1995).

On dry bare peat, whether bare from extraction, disturbance or pollution, birch–bracken–heather often invade (e.g. Thorne Moors) (Heathwaite, in Hughes and Heathwaite 1995).

Forestry degrades bogs. This is not just through draining. First, land is fenced off, so grazing is lessened, and perhaps burning is too. Management is altered, and so therefore is community. Ploughing changes the vegetation even more. A Caithness site altered from bog with *Trichophorum caespitosum* to moor with heath and heather. Fertiliser led to more change: to *Eriophorum vaginatum*. By now the site was of course too dry to regenerate without re-wetting. Wetting would usually be impossible as the cracking of the drying peat prevents water being held on the surface. This cracking is due more to evapotranspiration by planted trees than to earlier drainage: and it prevents any easy rehabilitation (Anderson *et al.*, in Wheeler *et al.* 1995).

Draining peatlands for forestry is damaging, usually irreversibly so. Trees grow better when the upper peat is more aerobic: more like that on which woodland 'naturally' occurs.

Recreation (Basic Figures 19, 20)

Damage from recreation now causes much concern. It must be kept in proportion. It rightly causes concern for the few, mostly small wetlands that are left: but is nothing compared to the large-scale and massive destruction and damage from drainage, farming and product removal.

It has been said that peat is buried history. Homes, causeways, pottery, jewellery – all and more come from wetlands, either because of settlement in them (e.g. Somerset Levels, The Fenland), because the wetland developed over them, or artefacts were disposed of in them. The wetland surface is the accumulation of recent history and management: or lack of management.

Damage comes primarily from disturbance (sport-killing, in wetlands, is hardly destructive, as it is mainly of birds grown for the purpose). Table 11.3 summarises pollution sources and the sensitivity (vulnerability) of different wetland type. Table 11.2 summarises run-off, etc., non-point source, pollution.

While recreation has always existed, recreation pressure outside towns has increased at an ever-increasing rate in the past two centuries: with the overall increase in population, with the concentration of population in towns, and the vast increase in wealth, mobility and leisure. The increasing recreational damage to the countryside, including wetlands, is the result. Litter – some dangerous – is left, paths are turned into mud tracks, vegetation is trampled, birds are disturbed, private land is invaded, livestock and crops are damaged, rare plants and eggs of rare birds are stolen, land is taken up by car parks, roads by noisy, polluting traffic: the list is endless.

Yet, considered rightly, recreation can be a lifeblood of conservation. There is a demand, a proper demand for recreation: and those making that demand are prepared to pay for its supply. This also means paying for the upkeep of wetlands. Only a few can be kept up by charitable or government organisations. Expecting farmers and other landholders – who own most wetland – to put large sums in for management with no financial return is unreasonable and often impossible. Recreation money wisely planned can help.

Those visiting wetlands are a small proportion of the whole, although in absolute numbers, ever increasing. Wicken Fen, the first British Reserve (1894), the British fen with most research and with the highest international reputation, has, in the past few decades, increased its recreational visitors from a handful to near 30 000, despite being away in the Fenland.

This is worthy. People should be able to enjoy the wetlands of their country. However, only if people do enjoy them can there be public will to preserve them: ecologists have but few votes. Wetlands are endangered: but who would be prepared to, for example, halve their water consumption for the sake of wetlands they have never seen or heard of?

Seeing wetlands may vary from pottering round a school or village pond with a wetland fringe, visiting Broadland for a summer holiday (on or off a boat), tramping over the bogs of Dartmoor or Wester Ross, or visiting a nature or recreational reserve. These last range from Wicken Fen at one extreme to, at the other, say the Melton Mowbray Country Park recently constructed from a (man-made) flood balancing lake, a former railway line and their surroundings.

Neglect or mismanagement

Active little-impacted wetlands such as blanket bogs or northern lake shores (some!) need only to be left alone. Wetlands managed for their products for centuries or millennia continue to need that management, if the crop community is to continue. Otherwise it will change in accordance with the underlying habitat trend: generally, now, to drier and more nutrient-rich conditions. Accidental damage, damage done unintentionally and in pursuit of another purpose, can come under this heading of neglect.

Management and loss of Broadland over time in East Anglia (Table 11.4)

Over the centuries, Broadland has dried. The question arises: at what point does this become loss of conservation value? When peat was cut and the Broads were made, any conservationist in the 1990s would have been horrified: yet these became of extreme conservation value! These and later peat cuttings have been infilled in shallow places. Is peat infilling damage? Surely not: yet if it leads to woodland, species-poor tall-herb community, arable or other dryland community, surely those are of low conservation value? An insoluble question: although management to prevent the development of these last dry communities of low value is considered to be correct. To prevent drying to terrestrial habitat, to prevent woody plant invasion, to prevent the type of disturbance ruining 'wild' vegetation (whether farmers for arable, or coypu for food) is proper.

Table 11.4 Damage to some East Anglian fens (selected from Wheeler and Shaw 1992)

Site	Flora	Dehydration	Drains	Abstraction	Dereliction	Tall herb	Scrub	Peat extraction	Reclamation	Other
Beeston bog	★★★★					★	★		Agriculture	Nutrients added?
Beetley and Hoe meadows							★			
Bressingham fen	★★★★		★★★					✓		
Briston common	★★	★?					★★	✓?		
Brock's watering			★			★				
Buxton Heath	★★	★?		★?	★		★★	✓	Fires/Sphagnum pulling	
Crostwick marsh	★★	★★	★?	★?	★★		★			
Ducan's marsh	★	★		★	★★	★★		✓	Agriculture	
East Ruston common	★★★★	★★★	★	★★★	★★★★	★★	★★	✓		Tip
Felthorpe bogs	★★★★	★★★?	★★★?		★★★★	★★		✓	Forest	
Forncett meadows	?	★★★		★★	★★	★★	★★			
Lopham fens	★★★	★★★	★★	★★	★★★	★	★★	✓		
Cavenham Poor's fen	★★★	★★	★★		★★★★	★★	★★	✓		
Icklingham Poor's fen	★★★	★★	★★	?.	★★★★	★★★★	★★	✓		
Redgrave and Lopham fens	★★★★	★★★	★★	★★★	★★	★★	★★	✓		
Fowlmere	★★	★		★?	★★	★★	★	✓?		Watercress

Damage has come from the usual variety of causes: drying, neglect, too many fires, pollution, new excavations, more intensive agriculture, and constructions. Neglect can bring wetter conditions (failed drainage) so better wet vegetation. It can also bring drier conditions (failed sluices, etc., again). It can bring woodland, it can bring species-poor instead of species-rich vegetation, ruderals instead of wetland species. And it can bring a more uniform fen type rather than the mosaics that come from management when every community is of value to farm or household.

It is important to lay responsibility where it belongs: blaming the water company for changes due to neglect of management will not, in the long run, advance conservation. (It is easy to blame the water company, far more of a nuisance to go out and remove shrubs and saplings. Make sure the water company really *is* responsible before saying so!)

The examples also show the loss of small water types, both calcium-dominated and acid waters (the acid being shallow surface water under acid sands), and therefore the loss of the associated vegetation.

Where solutes are sparse, adding even just a little of these same solutes or similar ones will have a great effect as the total has gone up greatly. Adding the same quantities to solute-rich wetlands, however, will have little or no effect, as there is negligible increase in the totals. The capacity of the wetland to buffer pollution, to balance or exchange chemicals, is important. Wetlands clean (see Chapter 8).

England is indeed fortunate to have the detailed historical work of Parmenter *et al.* (1995), documenting the management and descriptions of Broadland marshes over 200 years. This book should be consulted in full to understand the value of wetlands, and their vegetation types down the ages, and the terrible loss that has occurred. The following extracts are representative.

1   *East Ruston Common – Kings Fen. Borehole 1974, fully working 1985*
   1797 Rough pasture.
   1810 Well-drained (for 1810!), well fenced.
   1832 Controlled peat cutting, rush cutting, grazing.
   1840 Gravel and sand extraction.
   1885 Marsh.
   1909 Litter, rushes and mixed fodder (bullocks), peat cutting almost finished. Soft peat and mossy beds, usually with a hard bottom 1 m down, but also dangerous holes. Sphagnum beds usually safe to walk on (note: bog developing above, not inside, fen). Sedges have firmer bottom than reed.
   1919 Floating carpets over unknown depths, and 60 cm hidden drops into peat cuttings. Slippery putrid mud.
   1921 Drought. Deeper holes remained wet.
   1929 Wet marsh with mossy bottom (i.e. drying).
   1958 Attractive area of mixed fen.
   1970 Abstraction started beneath.
   1975 Probably too dry and burnt for *Carex limosa*.
   1984 At least 2 m of water on marsh in June (no longer permanent).
   1985 Very wet summer, but abstraction at full extent, and can walk over fens and flushes in ordinary shoes. Sixty cm drop in water in two months, up to 30 per cent of former water intercepted.
   1989 *Juncus–Molinia*, invasion of acidophilous and tall-herb communities (*Urtica dioica, Rubus fruticosus* agg., *Calamagrostis canescens, Epilobium hirsutum*, sparse *Phragmites australis*). Peat oxidation. Birch invasion. Catastrophic deterioration.

1992 *Sphagnum* gone, numerous birch seedlings, *Carex acuta* still present along seepage lines.

1993 More drastic change. Very dry, severe burns. Gorse (a completely dry land species) invades. Winter pools, however, killed birch seedlings.

2 *Mown Fen, East Ruston, affected by same borehole*

1797–1920s As in Kings Fen.

1957 Lost: *Anagallis tenella, Drosera anglica, Epipactis palustris, Eriophorum angustifolium, Liparis loeselii, Ophioglossum vulgare, Parnassia palustris, Pinguicula vulgaris, Sphagnum* sp., *Stellaria alsine, Thelypteris thelypteroides, Utricularia intermedia, U. minor.*

Still present: *Caltha palustris, Carex rostrata, Hydrocharis morsus-ranae, Menyanthes trifoliata, Osmunda regalis, Peucedanum palustre, Ranunculus flammula, R. lingua.*

1958 Mostly alder (sallow carr, herb fen). No more mowing and grazing.

1973 A little sedge mowing.

1983 Sedge beds abandoned, scrub invasion.

1985 Rapid scrub invasion, very dry. Narrow band of wet heath above fen. Still present: *Eriophorum* sp., *Peucedenaum palustre, Phragmites australis* (a fen or marsh where this species is worth mentioning, is in a bad way!), *Potentilla palustris, Sphagnum* sp.

1989 Too dry to regenerate *Cladium*. Water 75 cm below surface instead of 45 cm above.

1993 Botanical interest (i.e. rare or specialised species) lost.

3 *Dilham Broad. Representative of many*

Late-nineteenth century: drying (early).

Mid-twentieth century: drained for agriculture.

1990sa Well-drained grazing marsh.

4 *Barton Fen. Scrub invasion serious and damaging*

5 *Sutton Broad Fen*

1797 Broad (lake).

1826 Broad.

1840 Open water of more regular shape.

1884 Discontinuous swamp.

1903 Harvested *Phragmites, Typha angustifolia* near edge. Still some quaking fen.

1909 Quaking bog and mowing marshes, unreclaimed, usually flooded in winter.

1962 *Nymphaea* pools gone since 1955. Reed.

1963 'Early' herbaceous fen.

1970 *Liparis loeselii* all right.

1980 *L. loeselii* on tussocks of *Carex appropinquata* and *Schoenus nigricans*. Bryophyte carpet.

1990s Drying, deteriorating through lack of management. Still one of the most floristically rich areas.

6 *Ormsby Broad area (Ormsby Common). 1850s waterworks*

Eighteenth century: some arable.

Nineteenth century: (pump drainage) more arable. 1850s waterworks.

Circa 1900 Conifers from before 1884, more by 1907. Broad leaved forest on broad edges prevented good marginal vegetation.

1940s Grazing decreases. Rich fen sward from meadow (including *Agrostis stolonifera, Anagallis tenella, Carex echinata, C. pulicaris, Eriophorum angustifolium, Holcus lanatus, Juncus articulatus, J. subnodulosus, Menyanthes trifoliata, Potentilla palustris, Salix repens, Triglochin palustris.* Alder carr invaded (management less). Most interesting area lost, presumed due to waterworks.

1990s No trace of fen vegetation.

7 Hoveton Marsh

1908 Active drainage, rough grazing.

1940s Drained and ploughed (much), grazing marsh, cut yearly.

1979 Stopped mowing with decreased drainage.

8 Ranworth Broad Marshes

Thirteenth, fourteenth and nineteenth centuries: peat-cutting.

1797 Fen and wood.

1838 Fen (and wood). Peat-cutting.

1879 With a duck decoy (a pond designed to attract and capture ducks, usually using tame decoy ducks).

1883 In a dry year and (if walk quickly) cross with water boots.

1885 Open fen.

1902 Dangerous bogs and swamps.

After 1940s, loss of management and, e.g., *Liparis loeselii.*

9 Decoy Carr, Acle

Calcium-dominated 'good' fens deteriorated because of greater drainage, drying out principally between 1969 and 1980s. In the 1990s, it was quite dry, with too much carr. Dams were put in the dyke network in 1992, putting water level back to around ground level, and results are awaited.

10 Surlingham Marsh

1797 Rough grazing.

1908 Fairly dry, grazing.

Late twentieth century: carr invasion from lack of management. Sewage treatment works pollution increased *Glyceria maxima, Phalaris arundinacea, Phragmites australis.* These two have led to the loss of, e.g., *Carex diandra, C. dioica, C. limosa, C. vesiculosa, Dryopteris cristatus.*

11 Thorpe Marshes

Circa 1830: Norwich (Whitlingham) sewage treatment works (STW) brought effluent in River Yare flood water. Constructed after a cholera epidemic.

1955–75 Works expansion increased the pollution.

1990s Bypass construction altered the drainage. Gravel extraction both sides of the river. Damaged by Sewage Treatment Works pollution, construction and extraction.

12 Reedham Marshes

1838 Grazing marsh, wind-pump drained.

1855 Often flooded.

1907 Semi-improved pasture.

1960s Part flooded from abandoned drainage, developed to good reedbed.

1990s Much reedbed, some sedge beds and others (including *Juncus subnodulosus*), the highest number of fen communities of any Broadland fen.

13 Flixton Decoy area

Acid water seepage from the crag above lost with local water abstraction and carr invasion.

14 Beccles Marshes

Most grazed since the eighteenth century; twentieth century part to arable, grazing abandoned. Effective drainage from mid-nineteenth century. No fen communities.

15   *Ashby Warren*
    Acid valley bog lost by abstraction. (A little drying *Sphagnum* in conifer plantation, etc.)
16   *Halvergate Marshes*
    Late eighteenth-century drainage and grazing. Nineteenth-century under-drainage and part to arable. No fen, but much wet grassland. (The famous wet grassland.)
17   *Horsey-Braydon*
    Dumping has destroyed most of the acid (nutrient-low) nuclei.
18   *Barton Broad Fens*
    Semi-floating swamp, surface now stable.

Changes in Wicken Fen in the past century (Calston and Friday, in Friday 1999) include those due to a serious loss of calcium-rich water, and of wet banks (35 macrophytes extinct), an increase of oligotrophic plants, and a net loss of dragonflies, butterflies, moths, beetles, bees and crayfish.

A summary of the known, probable (or possible) causes of damage in a greater number of East Anglian fens is shown in Table 11.4.

Deterioration of waterfowl and wet grassland

When farming is intensified, stock can be put onto the land more densely, and earlier: by two weeks in the Netherlands over the last century. Species such as meadow pipit, skylark and breeding gargeny decline. Drying means more predation of the young, because they are unprotected for longer, as the parents have to go to now-distant wet grounds to feed. Waders replace destroyed early nests, but species vary in their willingness to replace nests formed early. However, in drier places there is less replacement in all: another reason for decline. Larger birds have better tolerance to drying than smaller ones: smaller can go into decline from intensification while large ones are satisfactory.

Conversely, if rough grazing is abandoned, this also can trigger the decline for some species. If carr results, all waders decline.

Population losses from farming are therefore great. There is a decline from wet grassland, to dry grassland, and to wood and/or arable.

Ruff is a good indicator, and habitat management for ruff gives the most diverse bird community. This is least impact, with the strongest restrictions on agrochemicals, drying and grazing (RSPB no date, 1983; Green and Robins 1993; Grieve *et al.* 1993).

Reedswamp dieback

This has been primarily of *Phragmites australis*, much the commonest European reedswamp species. In the Broads, increased sedimentation and presumed increased nutrients led, in the nineteenth century, to *Cladium* decline, and, often via *Scirpus lacustris* and *Typha angustifolia*, *Phragmites* increase. *Cladium* is now restricted to nutrient-medium places, *Scirpus* and *Typha* are sparse. Their decline was hastened by coypu damage in the 1950s. *Phragmites* die-back, though noticed in the 1940s, caused little concern until the 1970s. The die-back also halted succession from reedswamp to (on softer ground) tussock sedge fen, then alder, and on drier, sallow carr (see Chapter 10).

In the nineteenth and early twentieth centuries, commercial boats (the Norfolk wherry) slowly went into decline, and leisure boats were few. The die-back started later on clay and gravel floors, whose bed is less easy to disturb than peat. Rollesby Broad did not start die-back until the 1970s, Filbey Broad until the 1980s. In general, the amount of reedswamp in 1968 was the same as in 1881. Even for 1948–88 Rockford Broad had reedswamp gains equalling losses. More die-back means less reedbed. Less reedbed means less bird habitat: so more pressure from bird flocks, so more damage from these (George 1992).

Over Europe reedswamp has declined: but only in reed fronting open water. Inland, loss is mainly with arablisation or drying.

Typical impacts:

Lakes. (a) Wave scour, stressing the shoots. The force is enhanced by drift, algal mats, boats, etc., and may cause soil erosion. (b) Disturbance from recreation (other than the last). (c) Embankments, roads, car parks, etc. (d) Pollution from settlements, factories, farms, motorways, agrochemicals, etc., sometimes severe. (e) Disease is possible, but has not been recorded from Britain.

Marshes. (a) Loss of water by drainage, abstraction or failure to maintain high water levels by management. (b) Ploughing or other conversion to different land use. (c) Recreation, usually minor and controlled. (d) Pollution, usually minor and agrochemical. (Haslam 1994.)

Intentional loss is more likely to be in marshes, unintended in rivers and lakes.

In Britain reedswamp die-back has only been studied as a commercial or ecological problem in Broadland. Counter measures found to be satisfactory on the continent include (Ostendorp *et al.* 1995): underwater fences to break waves; other constructions to dissipate waves; planting; stopping recreation; stopping gravel extraction by and near flooded reed; stopping damaging construction on the inland side of the reedbed (embankments, roads, shore works, etc.); protection to banks by brushwood, weir, etc.

It is not, unfortunately, socially practical to stop recreation on Broadland rivers for a decade or two to replace and protect fringing reed.

12 Conservation

Wetlands are wet lands.

That form of destruction which is called restoration.

(Anthony Trollope)

Good quality mire vegetation can only be restored where suitably wet conditions exist.

(Hughes and Johns, in Hughes and Heathwaite 1995)

To be successful in wetland management you need not only a lot of biological knowledge, you need a lot of money to plan and perform the work.

(Cronert, in Finlayson and Larson 1990)

The main aim of this Directive being to promote the maintenance of biodiversity, taking account of economic, social, cultural and regional requirements, this Directive makes a contribution to the general objective of sustainable development; whereas the maintenance of such biodiversity may in certain cases require the maintenance, or indeed the encouragement, of human activities.

(*The (EEC) Habitats Directive* 1992; Council Directive 92/43/EEC)

Scottish Natural Heritage's vision is one of an accessible and welcoming countryside, but one in which access is arranged so that it does not place an unreasonable burden on rural land or those who live and work in the countryside. We have a responsibility for managing the natural heritage carefully so that we can pass it on unblemished or, where possible, enhanced.

(Scottish Natural Heritage)

Market forces need not lead to social goods.
A sound hydrological regime permits the persistence of a wetland as a self-perpetuating oscillating system.
Commercial values are finite, wetlands provide values in perpetuity.

(Mitsch and Gosselink 1993)

Introduction

While wetlands were wet and being managed traditionally there was no thought for conservation: also little need. A general interest in wild flora and fauna for their own sake is a recent phenomenon. Medicinal plants, of course, have been long known, and herbals described their use. Increased impetus for naming and understanding the uses of plants came with the great voyages of discovery and trade, in the seventeenth and eighteenth centuries. Pioneer natural historians included the Rev. Gilbert White in the eighteenth century (The Natural History of Selborne 1789).

In the nineteenth century, natural history reached the educated British public, who investigated and discovered the flora and fauna of British habitats, without yet calling them habitats. Those who, in the late nineteenth century, would have belonged to the Cambridge Natural History Society, belonged, a century later, to the Cambridgeshire (etc.) Wildlife Trust, dedicated, not to lists and collections, but to the preservation of the habitats in which the listed species can grow.

Wicken Fen was the first British nature reserve (1894). After that, with a slow start, reserves and sanctuaries snowballed. They could be in private hands, owned by a government organisation (e.g. English Nature, Scottish Natural Heritage, Countryside Council for Wales and their predecessor bodies), or voluntary organisations (e.g. the Wildlife Trusts, Royal Society for the Protection of Birds (RSPB), or the Woodland Trust), or private companies (e.g. a water company – reservoirs, etc.). Much land has been designated and, to some extent, conserved. 'To some extent' – Why?

First, conservation means more than putting a fence around, it means protection and, often, management. This takes knowledge and probably money, often very large sums of money.

Although English Nature, etc. can designate a place as a Site of Special Scientific Interest (SSSI), whether in or out of a reserve, this does not mean protection. Making a wetland a reserve does not necessarily make that reserve wet, or calcium-dominated, or keep it free from pollution, or keep the mosaic of open vegetation for whose preservation the reserve was established. Second, designation for conservation does not mean exemption from destruction by either government or owner. An owner can announce change of land use that will cause damage: it is difficult to stop it (without inexhaustible money). Government can announce the same, and is even more difficult to stop. SSSI status did not require diversion of sewage treatment works effluent from Slapton Ley reserve (Sullivan and Wilson, in Finlayson 1992), or diversion of the Newbury bypass.

There are international reserves schemes. Ramsar sites are wetlands of waterfowl importance, designated after the Ramsar Convention in 1971. The EEC Habitats Directive (92/43/EEC) set up a network of special areas of conservation called Natura 2000. These include the 1979 Bird Directive sites. Directives, unlike UN charters, carry penalties if too obviously seen to be ignored: but where information is unavailable, or poorly applied, what then? The sites here include linear ones such as valleys, and intermittent ones such as fens.

There are plenty of international conventions and directives, and organisations, that could help wetland conservation. These include:

Agenda 21 (UNCED)
Berne Convention
Birds Directive (EEC)
Bonn Convention
Commission on Sustainable Development (UNCED)
Convention on International Trade in Endangered Species of Wild Fauna and Flora (CITES)
Global Convention of Climate Change
Habitats Directive (CEC)
International Biodiversity Convention (Rio Convention) (UNCED)
International Peat Society Commission on Land Use Planning and Environment
International Waterfowl and Wetlands Research Bureau

IUCN Wetlands Programme
Precautionary Approach (UNCED)
Ramsar Convention
UN World Charter for Nature
UNESCO Biosphere Reserve Programme
UNESCO Convention for the Protection of World Cultural and Natural Heritage
World Conservation Union (IUCN) (Red Data Lists)
World Wide Fund for Nature

There are ever-increasing schemes for conserving wetland. It would be nice to think their rate of deterioration is slowing! Schemes fall into two groups. One comprises reserves (taken out of modern production or never in it), which here include land (reasonably) traditionally managed for non-grass crops, e.g. reedbeds, osier beds, coverts for game birds, etc. The other is land under 'ordinary' farming, where some wetland value can be maintained, or rehabilitated. Necessarily the first are only a few of the best examples because of the expense of upkeep. Earlier there was no need to 'conserve' agricultural land as the traditional management of, say, wet grassland was, in itself, its own conservation. With intensification of agriculture, management was no longer conservation management. Changing to management practice of wet grassland value is of extreme importance for its survival.

Agricultural schemes include:

1   The Environmentally Sensitive Areas Scheme pays farmers for carrying out management aimed at maintaining landscape and grassland in river valleys. It is funded by the Ministry of Agriculture, Fisheries and Food (MAFF, now DEFRA).
2   Water Fringe Scheme (also funded by MAFF, now DEFRA).
3   Countryside Stewardship (started under the Countryside Commission, then MAFF, now DEFRA) includes waterside landscapes. This has three aims: to support and reintroduce traditional management for meadow, pasture and the wildlife they support; to restore and protect characteristic waterside features; and to foster enjoyment of landscape and wildlife. This, if used widely, would be most valuable. The best examples of habitat go, in theory, into reserves. The working landscape works, and having traditional management is excellent.
4   The 'set-aside' scheme is also administered by MAFF, now DEFRA, and is designed to reduce crop production rather than aid wetlands. Buffer zones at least 15 m wide can be created under this scheme; they help to clean agrochemicals and might carry some form of wet grassland, etc.
5   The Habitat Improvement Scheme is a variant of the ESA Scheme, taking land out of cultivation for 20 years.
6   English Nature has a Wildlife Enhancement Scheme, paying farmers for beneficial management.
7   The Internal Drainage Boards have as their effective aim to increase crop yields through drainage, and in so doing to please the main landowners, who pay the main drainage rates. When the RSPB acquires land and becomes one of these largest landowners, wanting decreased drainage, some adjustment is needed.

Farmers are in charge of land. To work, agri-conservation schemes must be what farmers like: simple, with adequate funding to cover loss of income, etc. Income is not the only consideration: should farm workers be put out of jobs to get more marsh hay?

Principles of conservation

The primary requirement for the conservation of wetlands, 'reserve wetlands' or 'agri-wetlands' is to have **WATER** ample in both quantity and quality. Basic Figures 5–8 and 21 show this: hold water in, do not speed it away. And, keep water types separate, to keep the communities depending on small (e.g. flush) as well as large (e.g. flood plain) water types.

Other measures are shown in Basic Figure 22.

1   preventing damage;
2   continuing or mimicking the treatment for former (or indeed present) use, the treatment under which the vegetation type developed;
3   study and consideration before making other changes;
4   consideration of possible other changes for enhancement.

The historic and archaeological value of wetlands is seldom obvious (Basic Figure 20). No bridges, water mills or chapels grace wetlands as they do rivers. But settlements can be buried in peat or flood plain alluvium (see Chapter 2), and peat is a good preserver of pollen grains, fruit and other plant and animal remains. The great detail now known of British post-Ice Age vegetation comes largely from pollen analysis.

Looking at Basic Figures 1 and 2, and once more starting from the left, the blanket bog and heather moor here are, in the short term, stable. They occur under light grazing, by sheep, cattle, perhaps other livestock, deer, birds, etc. Fencing off areas, e.g. for forestry, alters the species composition and pattern (see Chapter 11). It is well known that ungrazed grassland grows tall and changes in composition. It is easy to forget that the 'wild' moors with – at first sight – so little interference are grazing-determined too. They also need protection from tree invasion (drier places), denser stock, recreation, drainage, (commercial) peat extraction, other disturbance and water abstraction except from the run-off streams. The small pool (a tarn in northern England, a llyn in Wales, lochan in Scotland) also needs protection: from excess trampling and erosion and from drying up.

The variety of flood plain communities require the same care: maintenance management, and protection from disruption from outside. It takes study to work out under what conditions each community developed, and how these can be continued or mimicked. For years, or even decades, vegetation may have inertia, and continue with little change and little management. But changes build up, and communities are lost (Chapter 10).

Changes made outside the wetland may also influence it. Downstream river drainage and channelling, for instance, may act upstream too, and drain and dry the land beside the stream. River water polluted from towns and roads upstream can enter and alter the wetland. The more solute-rich, and the greater the buffering capacity and purifying power of the wetland, the more pollution it can tolerate without change. Pollution able to destroy a bog may be hardly noticed in the most nutrient-rich reedbeds.

Of course changes may be wanted, the vegetation may have been neglected, and reversion to mowing or grazing be needed. It may have been dried and need the water back. It may be too uniform, and a mosaic be needed.

The most important and primary task of conservation is to prevent the wetland being converted to arable, towns, roads, car parks and other such habitats. It may be difficult to re-create a diverse wetland from a carr or dryish grassland. It is much more difficult to re-create it from a housing estate (where, too, native species are probably lost).

Conservation can therefore be summed up as follows.

The two principles of conservation:
1   Defend from disruptions from outside, the primary three being water loss, pollution and disturbance.
2   Manage in ways that either are, or simulate, traditional practices for sustainable exploitation.

Principle, however, does not indicate which of the possible vegetation (and animal) types 'should' be in a given site of, for example, carr, meadow, small sedge fen, rush fen, or rush pasture.

The aims of conservation must be clear when a wetland is made a Reserve or the subject of a Scheme. If the (ultimate and unattainable) aim is to restore the intact, unimpacted wetland, this is looking at conditions between about ten thousand and one thousand years ago, with: whole and un-split water systems (see Chapter 6); peat growing or degrading from climatic and geomorphic factors alone; no outside sources of chemicals (including only background levels from the air).

Unfortunately for general wetland conservation, waterfowl do well in wet places unsatisfactory for other groups. It is possible to take abandoned farmland, dig pools with nice islands in, and have good birds: these being in places with negligible wetland habitat or community. That is an extreme, but there is a difference in philosophy between

- a high-quality wetland that nearly approaches a traditional one in the same environment and supports communities and species proper to that traditional wetland; and
- a high-quality wetland that supports as many birds as possible, even if it means feeding the birds by increasing nutrient or organic input.

Birds attract the most money for conservation: so this is not a theoretical but a practical issue of management.

For details of conservation practices such as methods of regulating water level, effective methods of cutting, removing trees, planning, etc., the publications of the Countryside Commission, English Nature, International Wildfowl Research Bureau, RSPB, Scottish Natural Heritage and similar organisations should be consulted. (They are not listed here individually, as valuable new ones come out each year.)

References

Anglo-Saxon Chronicle. Translated by G. M. Garmonsway (1953). Everyman, Dent, London.

Aquatic Botany (1999). Special issue on genetic diversity, ecophysiology and growth dynamics of the common reed (*Phragmites australis*).

Armstrong, J., Armstrong, W. and Beckett, P. M. (1992). *Phragmites australis*: Venturi- and humidity-induced pressure flows enhance rhizome aeration and rhizosphere oxidation. *New Phytologist* **120**, 197–207.

Ashfield, C. J. (1861). Norfolk and Suffolk botany. *Phytologist, New Series* **5**, 321–5.

Ashfield, C. J. (1862). Norfolk and Suffolk botany. *Phytologist, New Series* **6**, 321–2.

Athie, D. and Cerri, C. C. (eds) (1987). *The use of macrophytes in water pollution control.* Pergamon Press, Oxford.

Aubrecht, G., Dick, G. and Prentice, C. (1994). Monitoring of ecological change in wetlands. *Stapfia* **31** and *IWRB Publication* **30**.

Austen, J. (1811). *Sense and sensibility.* Egerton, London.

Banasova, V., Otahelooa, H., Jarolimek, I., Zaliberova, M. and Husak, S. (1994). Morava river flood plain vegetation in relation to limiting ecological factors. *Ecologia (Bratislava)* **13**, 247–62.

Bareham, S. and Smith, M. (1992). Acidification in North Wales – the implications for wildlife. *Nature* **86**, 8–9.

Barthe, H. and L'Hermite, P. (eds) (1987). *Scientific basis for soil protection in the European Community.* Commission of the European Communities, Elsevier Applied Science, London.

Beowulf. Translated by D. Wright (1957). The Penguin Classics.

Binz-Reist, H. (1989). Mechanische Belastbarkeit natürlicher Schilfbestände durch Waller, Wind und Treibzubug. *Veröffning Geobotanische Institut ETH Zurich* **101**, 1–536.

Bittman, E. (1953). Das Schilf. *Angewandte Pflanzensoziologie* **7**.

Blackmore, R. D. (1869). *Lorna Doone.* Blackie, London.

Boeye, D. (1992). Hydrologie, Hydrochemie en ecologie van een grundwater afhankelijk veen. Ph.D. Thesis, University of Antwerp.

Bond, C. J. (1981a). The marshlands of Malvern Chase. In *Evolution of marshland landscapes.* Oxford University Department for External Studies, Oxford, pp. 95–112.

Bond, C. J. (1981b). Otmoor. In *Evolution of marshland landscapes.* Oxford University Department of External Studies, Oxford, pp. 113–35.

Boon, P. J., Calow, P. and Petts, G. E. (eds) (1992). *River conservation and management.* John Wiley & Sons, Chichester.

Bosman, M. T. M. (1985). Some effects of decay and weathering on the anatomical structure of the stem of *Phragmites australis.* Trin. ex Steud. *AWA Bulletin* NS **6**, 165–70.

Boyer, M. L. H. and Wheeler, B. D. (1989). Vegetation patterns in spring-fed calcareous fens: calcite precipitation and constraints on fertility. *Journal of Ecology* **77**, 597–609.

Brändle, R., Cizková, H. and Pokosny, J. (1996). Adaptation strategies in wetland plants. *Special Features in Vegetation Science* **10**. Opulus, Uppsala.

Braun-Blanquet, J. (1932). *Plant sociology.* McGraw Hill, New York.

Bravard, J. P. (1987). *Le Rhône, du Léman à Lyon.* L'homme et la nature, Lyon.

British Crop Protection Council (1991). *Pesticides in soils and water: current perspectives.* BCPC Monograph **47**, Farnham, Surrey.

Brooks, N. P. (1981). Romney Marsh in the early Middle Ages. In *Evolution of marshland landscapes.* Oxford University Department for External Studies, Oxford, pp. 74–94.

Brown, A., Mather, S. P. and Kushner, D. J. (1989). An ombrotrophic bog as a methane reservoir. *Global Biogeochemistry Cycles* **3**, 205–14.

Bunbury, C. J. F. (1889). *Botanical notes at Barton and Mildenhall, Suffolk.* Mildenhall.

Bunyan, J. (1682). *The pilgrim's progress.* Book Society, London.

Burnett, J. H. (ed.) (1964). *The vegetation of Scotland.* Oliver & Boyd, Edinburgh.

Burt, T. P. and Haycock, N. E. (in press). Factors affecting the nutrient retention dynamics of near stream environments. In M. Anderson, *Floodplain processes.* John Wiley & Sons, Chichester.

Burt, T. P., Heathwaite, L. and Trudgill, S. (eds) (1993). *Nitrate.* John Wiley & Sons, Chichester.

Carbiener, R. and Ortscheit, A. (1987). Wasserpflanzengesellschafen als hilfe zur qualitätsüberwachsung eines der grössten grundwasser – vorkommens Europae (Oberrhinebene). *Vegetation ecology and creations of new environments.* Tokai University Press, Tokai, pp. 283–312.

Colgrave, B. (ed.) (1956). Felix, monk of Crowland (author) *Life of St. Guthlac.* Cambridge University Press, Cambridge.

Commission of the European Union (1992). *The CORINE Biotype Project.* Brussels.

Cooper, P. F. and Findlater, B. C. (eds) (1990). *Constructed wetlands in water pollution control.* Pergamon, Oxford.

Cosgrove, D. and Petts, G. (eds) (1990). *Water, engineering and landscape.* Belhaven, London.

Countryside Commission (1995). *Climate change, acidification and ozone.* Northampton.

Countryside Commission (no date). *The visitor welcome initiative.* Cheltenham.

Crowder, A. (1991). Acidification, metals and macrophytes. *Environmental Pollution* **71**, 171–203.

Curtis, D. J. (1977). *Lowland bogs spider survey.* CST Report **206**, Nature Conservancy Council, Scotland, South West Region.

Davies, M. (1977). A survey of the breeding birds of the Caorrunn Plateau, Loch Lomondside. Nature Conservancy Council, Scotland, South West Region, Internal Report.

Davis, S. M. and Ogden, J. C. (eds) (1994). *Everglades: the ecosystem and its restoration.* St Lucie Press, Delroy Beach, Florida.

Dawkins, C. J. (1939). Tussock formation by *Schoenus nigricans*: the action of fire and water erosion. *Journal of Ecology* **27**, 78–88.

Defoe, D. (1724). *A tour through the whole island of Great Britain.* Everyman, Dent, London (1959 edition).

Descy, J. P. (1976). Value of aquatic plants in the characterisation of water quality and principles of methods used. In R. Ameris and J. Smeats (eds), *Principles and methods for determining ecological criteria on hydrobiocinosis.* Pergamon Press for Commission of the European Communities.

Directorate General Environment, Nuclear Safety and Civil Protection (1992). *CORINE biotypes manual.* EUR 12587/1 EN. Commission of the European Communities, L-2920, Luxembourg.

Doyle, A. C. (1902). *Hound of the Baskervilles.* Worlds Classics edition, London.

Drake, C. M., Godfrey, A. and Sanderson, A. C. (1989). *A survey of the invertebrates of five lowland bogs in Cumbria.* England Field Unit **60**, Nature Conservancy Council, Peterborough.

Dunn, E. (1994). *Lowland and grasslands in The Netherlands and Germany.* Studies in European agriculture and environmental policy **10**, Royal Society for the Protection of Birds, Sandy.

Edwards, R. R. (2000). The potential for use of willow as components of practical buffer zones. *Annual Report 2000.* ACR-Centre for aquatic plant management, Reading, pp. 42–5.

Eisler, R. (1992). Diflubenzuron hazards to fish, wildlife and invertebrates: a synoptic review. *US Fisheries and Wildlife Services Biological Reports* **4**, 1–36.

Ellenburg, H. (1974). Zeigerwerte der Gefassplanzen Mitteleuropes. *Scripta Geobotanica* **9**, 1–97.

English Nature (1991). *Nature conservation and pollution from farm wastes.* Peterborough.

English Nature (1992). *River, lake and canal SSSIs in England subject to eutrophication.* Peterborough.

Fens Floodplain Project (no date). *Wet fens for the future.* EU LIFE Environment Project.

Finlayson, C. M. and Larsson, T. (eds) (1990). *Wetland management and restoration.* Swedish Environmental Protection Agency Report **3992**.

Finlayson, M. (ed.) (1986). *Integrated management and conservation of wetlands in agricultural and forested landscapes.* IWRB Special Publications **19**, Slimbridge.

Firth, F. M. (1984). *The natural history of Romney Marsh.* Meresborough, Rainham, Kent.

Fitter, A. and Hogg, P. (1996). Secrets of the bog. *Natural World,* Autumn 1996, 29–30.

Friday, L. (ed.) (1997). *Wicken Fen. The making of a wetland nature reserve.* Harley Books, Colchester.

Fuchsman, C. H. (ed.) (1936). *Peat and water.* Elsevier, Applied Science, London.

Furness, R. W. (1989). *Marine pollution and birds.* Report, Royal Society for the Protection of Birds, Sandy.

George, M. (1992). *The land use, ecology and conservation of Broadland.* Packard Publishing, Chichester.

Georges, A. (1994). Carabid beetle spatial patterns in cultivated wetlands. The effect of engineering works and agricultural management in Morais Poitevin (western France). In K. Desender, *et al.* (eds), *Carabid beetles: ecology and evolution.* Kluwer, The Netherlands, pp. 283–93.

Gerald of Wales (Giraldus Cambrensis) (Twelfth Century). *Travels through Wales.* Penguin.

Gilman, K. (1994). *Hydrology and wetland conservation.* John Wiley & Sons, Chichester.

Gilman, K. and Newman, M. D. (1980). *The Anglesey Wetland Study. First Annual Progress Report.* CST Report **290**, Peterborough.

Gilvear, D., Tellam, T. H., Lloyd, J. W. and Lerner, D. N. (1990). *The hydrodynamics of East Anglian fen systems.* Nature Conservancy Council **88**, Peterborough.

Gilvear, D. J., Andrews, R., Tellam, J. H., Lloyd, J. W. and Lerner, D. N. (1993). Quantification of the water balance and hydrogeological processes in the vicinity of a small groundwater-fed wetland, East Anglia, England. *Journal of Hydrology* **144**, 311–34.

Godwin, H. (1929). The 'sedge' and 'litter' of Wicken Fen. *Journal of Ecology* **17**, 148–60.

Godwin, H. (1978). *Fenland: its ancient past and uncertain future.* Cambridge University Press, Cambridge.

Gopal, B. and Junk, W. J. (2000). *Biodiversity in wetlands: assessment, function and conservation.* Backhuys, Leiden.

Gopal, B., Junk, W. J. and Davis, J. A. (eds) (2000). *Biodiversity in wetlands: assessment, function and conservation,* Vol. 1. Backhuys, Leiden.

Gopal, B., Junk, W. J. and Davis, J. A. (eds) (2001). *Biodiversity in wetlands: assessment, function and conservation,* Vol. 2. Backhuys, Leiden.

Grace, J. B. (1985). Juvenile vs. adult competitive abilities in plants: size-dependence in cattails (*Typha*). *Ecology* **66**, 1630–8.

Grace, J. B. (1989). Effects of water depth on *Typha latifolia* and *Typha domingensis. American Journal of Botany* **76**, 762–8.

Grace, J. B. and Wetzel, R. G. (1981a). Effects of size and growth rate on vegetative reproduction in *Typha. Oecologia* **50**, 158–61.

Grace, J. B. and Wetzel, R. G. (1981b). Habitat partitioning and competitive displacement in cattails (*Typha*): experimental field studies. *The American Naturalist* **118**, 463–74.

Grace, J. B. and Wetzel, R. G. (1982). Niche differentiation between two rhizomatous plant species: *Typha latifolia* and *Typha angustifolia. Canadian Journal of Botany* **60**, 46–57.

Green, R. E. and Robins, M. (1993). The decline of the ornithological importance of the Somerset levels and Moors, England, and changes in the management of water levels. *Biological Conservation* **66**, 95–106.

Greenland White-Fronted Goose Study (1988). *Greenland white-fronted geese in Britain, 1986/87.* CSD Report **814**, Nature Conservancy Council, Peterborough.

Grieve, I. C., Gilvear, D. G. and Bryant, R. G. (1993). *Hydrochemical and water source variations across a floodplain mire, Insh Marshes, Scotland.* Department of Environmental Science, University of Stirling, Stirling.

Grootjans, A. P. (1985). Changes of groundwater regime in wet meadows. Thesis, University of Groningen.

Gu, J. D., Berry, D. F., Tarabon, R. H., Martens, D. C., Walker, H. J. Jr. and Edmonds, W. J. (1992). Biodegradability of atrazine, cyanazine and dicamba in wetland soils. *Virginia Polytechnic Institute State University Water Resources Research Centre Bulletin* **0(172)**, I–XII, 1, 3–11, 13–17, 19, 21–4, 27–58, 61–72.

Gunnison, D. and Barko, J. W. (1989). The rhizosphere ecology of submersed macrophytes. *Water Resources Bulletin* **25**, 193–202.

Haldeman, M. (1993). Mycosociological research in hard wood alluvial forests near Brugg (Canton of Aorgau). English Summary. *Berichte Geobotanische Institut, ETH, Stiftung Rübel* **59**, 51–78.

Hall, D. N. (1981). The Cambridgeshire Fenland: an intensive archaeological fieldwork study. In *Evolution of marshland landscapes*. Oxford University Department of External Studies, Oxford, pp. 52–73.

Hammer, D. A. (ed.) (1989). *Constructed wetlands for wastewater treatment.* Lewis Publishers, Chelsea, Michigan.

Harding, M. (1992). *Redgrave and Lopham fens: a case study in change due to groundwater abstraction.* Suffolk Wildlife Trust, Report to English Nature.

Harold, R. (1990). *The birds of Woodwalton Fen National Nature Reserve.* EN/ER/8. English Nature, Peterborough.

Harrison, R. M. (ed.) (1992). *Understanding our environment.* The Royal Society of Chemistry, London.

Haslam, S. M. (1960). Vegetation of the Breck Fen margin. Thesis, University of Cambridge.

Haslam, S. M. (1965a). Ecological studies in the Breck Fens. I. Vegetation in relation to habitat. *Journal of Ecology* **53**, 599–619.

Haslam, S. M. (1965b). The Breck Fens. *Suffolk Naturalists Transactions* **13**, 137–46.

Haslam, S. M. (1969a). Stem types of *Phragmites communis* Trin. *Annals of Botany* **33**, 127–31.

Haslam, S. M. (1969b). The development and emergence of buds in *Phragmites communis* Trin. *Annals of Botany* **33**, 289–301.

Haslam, S. M. (1969c). The development of shoots in *Phragmites communis* Trin. *Annals of Botany* **33**, 695–709.

Haslam, S. M. (1970a). The development of the annual population in *Phragmites communis* Trin. *Annals of Botany* **34**, 571–91.

Haslam, S. M. (1970b). The performance of *Phragmites communis* Trin. in relation to water supply. *Annals of Botany* **34**, 867–77.

Haslam, S. M. (1970c). Variation of population type in *Phragmites communis* Trin. *Annals of Botany* **34**, 147–58.

Haslam, S. M. (1971a). Community regulation in *Phragmites communis* Trin. I. Monodominant stands. *Journal of Ecology* **59**, 65–73.

Haslam, S. M. (1971b). Community regulation in *Phragmites communis* Trin. II. Mixed stands. *Journal of Ecology* **59**, 75–88.

Haslam, S. M. (1971c). The development and establishment of young plants of *Phragmites communis* Trin. *Annals of Botany* **35**, 1059–72.

Haslam, S. M. (1972a). *Phragmites communis* Trin. Biological Flora of the British Isles 128, *Journal of Ecology* **60**, 585–610.

Haslam, S. M. (1972b). *The reed.* Norfolk Reedgrowers Association.

Haslam, S. M. (1973a). Some aspects of the life history and autecology of *Phragmites communis* Trin. *Polskie Archiwum Hydrobiologii* **20**, 79–100.

Haslam, S. M. (1973b). The management of British wetlands. I. Economic and amenity use. *Journal of Environmental Management* **1**, 303–20.

Haslam, S. M. (1973c). The management of British wetlands. II. Conservation. *Journal of Environmental Management* **1**, 345–61.

Haslam, S. M. (1975). The performance of *Phragmites communis* Trin. in relation to temperature. *Annals of Botany* **39**, 881–8.

Haslam, S. M. (1978). *River plants.* Cambridge University Press, Cambridge.

Haslam, S. M. (1979). Infra-red colour photography and *Phragmites communis* Trin. *Polskie Archiwum Hydrobiologii* **26**, 65–72.

Haslam, S. M. (1982). *Vegetation in British Rivers*. Nature Conservancy Council, London.

Haslam, S. M. (1987). *River plants of Western Europe*. Cambridge University Press, Cambridge.

Haslam, S. M. (1989). Early decay of *Phragmites* thatch: an outline of the problem. *Aquatic Botany* **35**, 129–32.

Haslam, S. M. (1990). *Phragmites* culm strength and thatch breakdown: some difficulties. *Landschaftsentwicklung und Umweltforschung, Technische Universität, Berlin* **71**, 58–77.

Haslam, S. M. (1990). *River pollution: an ecological perspective*. Belhaven Press, London.

Haslam, S. M. (1991). *The historic river*. Cobden of Cambridge Press, Cambridge.

Haslam, S. M. (1994). *Wetland habitat differentiation and sensitivity to chemical pollutants (non-open water wetlands)*. Her Majesty's Inspectorate of Pollution, London.

Haslam, S. M. (1995). *A discussion on the strength (durability) of thatching reed (Phragmites australis) in relation to habitat*. Reed Research Report X, Department of Plant Sciences, University of Cambridge.

Haslam, S. M. (1997). *The river scene: ecology and cultural heritage*. Cambridge University Press, Cambridge.

Haslam, S. M. and Wolseley, P. A. (1981). *River vegetation: its identification, assessment and management*. Cambridge University Press, Cambridge.

Hassell, J. (1994). *The implications of global climate change for biodiversity*. Royal Society for the Protection of Birds, Sandy.

Hauber, D. P., White, D. A., Powers, S. P. and De Francesch, F. R. (1991). Isozyme variation and correspondence with unusual infrared reflectance patterns in *Phragmites australis* (Poaceae). *Plant Systematics and Evolution* **178**, 1–8.

Hawke, L. J. and José, P. V. (1996). *Reedbed management*. Royal Society for the Protection of Birds, Sandy.

Hayati, A. A. and Proctor, M. C. F. (1991). Limiting nutrients in acid-mire vegetation: peat and plant analyses and experiments on plant responses to added nutrients. *Journal of Ecology* **79**, 75–95.

Haycock, N. E. (1994). Watershed management for the conservation of water quality in England. *Proceedings of the International Symposium on Agricultural Water Quality Management, November 1994*. Seoul National University, South Korea. NICEM, SNU, South Korea.

Haycock, N. E. and Burt, T. P. (1993). Role of floodplain sediments in reducing the nitrate concentration of subsurface run-off: a case study in the Cotswolds, U.K. *Hydrology Processes* **7**, 287–93.

Haycock, N. E. and Muscutt, A. D. (1994). Landscape management strategies for the control of diffuse pollution. *Landscape and Urban Planning* **31**, 313–21.

Haycock, N. E. and Pinay, G. (1993). Nitrate retention in grass and poplar vegetated riparian buffer strips during the winter. *Journal of Environmental Quality* **22**, 273–8.

Haycock, N. E., Pinay, G. and Walker, C. (1993). Nitrogen retention in river corridors: European perspective. *Ambio* **22**, 340–6.

Helm, P. J. (1963). *Alfred the Great*. Robert Hale, London.

Henshilwood, D., Lacey, P. and Roworth, P. (1998). Reedbed management. *Enact*. Special Supplement.

Higham, N. (1986). *The northern counties to AD 1000*. Longman, London.

Hik, D. S. and Jefferies, R. L. (1990). Increases in the net above-ground primary production of a salt-marsh forage grass: a test of the predictions of the herbivore-optimisation model. *Journal of Ecology* **78**, 180–95.

Hik, D. S., Sadul, H. A. and Jefferies, F. L. (1991). Effects of the timing of multiple grazings by geese on net above-ground primary production of swards of *Puccinella phryganodes*. *Journal of Ecology* **79**, 715–30.

Hofmann, K. (1991). The role of plants in subsurface flow constructed wetlands. In C. Ernier and B. Guterstam (eds), *Ecological engineering for wastewater treatment*. Bokskogen, Gothenburg, pp. 248–59.

Holdgate, M. W. (1984). Concluding remarks. *Philosophical Transactions of the Royal Society of London* B, **305**, 569–77.

Holmes, E. H., Urban, N. R. and Eisenreich, S. J. (1990). Aluminium geochemisry in polluted waters. *Biogeochemistry* **9**, 247–76.

Holmes, P. R., Boyce, D. C. and Reed, D. K. (1991a). *The Welsh peatland invertebrate survey: Merioneth.* Countryside Commission for Wales.

Holmes, P. R., Boyce, D. C. and Reed, D. K. (1991b). *The Welsh peatland invertebrate survey: Monmouth.* Countryside Commission for Wales.

Hook, D. D. and others (eds) (1988a). *The ecology and management of wetlands*, Vol. 1: *Ecology of wetlands.* Croom Helm, London.

Hook, D. D. and others (1988b). *The ecology and management of wetlands*, Vol. 2: *Management, use and value of wetlands.* Croom Helm, London.

Howard, P. H., Boethling, R. S., Jarvis, W. F., Meylan, W. M. and Michalenko, E. M. (1991). *Handbook of environmental degradation rates.* Lewis Publishers, Chelsea, Michigan.

Hughes, F. M. R., Harris, T., Richards, K., Pantou, G., El Hames, A., Barsoum, N., Gird, J., Peiry, J.-L. and Foussadier, R. (1997). Woody riparian species' response to different soil moisture conditions: laboratory experiments on *Alnus incana. Journal of Global Ecology and Biography letters* **6**, 8247–56

Hughes, J. M. R. and Heathwaite, A. L. (eds) (1995). *Hydrology and Hydrochemistry of British wetlands.* John Wiley & Sons, Chichester.

Ismay, J. (1978). The fens of North Wales: an invertebrate survey 1976. NCC Wales, North Region. Internal Report.

Jackson, B. D. (1928). *A glossary of botanical terms.* Duckworth, London.

Jennings, J. N. and Lambert, J. M. (1951). Alluvial stratigraphy and vegetational succession in the region of the Bure Valley Broads. *Journal of Ecology* **39**, 106–19.

Joyce, C. B. and Wade, P. M. (1998). *European wet grasslands.* Wiley, Chichester.

Keddy, P. A. (2000). *Wetland ecology: principles and conservation.* Cambridge University Press, Cambridge.

Kingsley, C. (1866). *Hereward the Wake.* Eversley, London.

Kirby, J. J. H. and Rayner, A. D. M. (1988). Disturbance, decomposition and patchiness in thatch. *Proceedings of the Royal Society of Edinburgh* **94B**, 245–53.

Kirby, J. J. H. and Rayner, A. D. M. (1989). Aspects of the decomposition, mechanical strength and anatomy of water reed (*Phragmites australis*) used in thatching. University of Bath.

Kirby, P. (1992). *Habitat management for invertebrates: a practical handbook.* Royal Society for the Protection of Birds, Sandy.

Koerselman, W., Bakker, S. A. and Blom, M. (1990) Nitrogen, phosphorus and potassium budgets for two small fens surrounded by heavily fertilised pastures. *Journal of Ecology*, **78**, 428–42.

Kostecki, P. T. and Calabrese, E. J. (1989). *Petroleum contaminated soils. Remediation techniques, environmental fate, risk assessment.* Lewis Publishers, Chelsea, Michigan.

Kostecki, P. T. and Calabrese, J. J. (1991). *Hydrocarbon contaminated soils and groundwater. Analysis, fate, environmental and public health effects, remediation*, Vol. 1. Lewis Publishers, Chelsea, Michigan.

Kühl, K. and Neuhaus, D. (1993). The genetic variability of *Phragmites australis* investigated by Random Amplified Polymorphic DNA. *Limnologie Actuall* **5**, 2–17.

Lakatos, G. and Biró, P. (1991). Study on chemical composition of reed-periphyton in Lake Balaton. *BFB-Bericht* **77**, 157–69.

Lambert, J. M. (1951). Alluvial stratigraphy and vegetational succession in the region of the Bure Valley Broads III. *Journal of Ecology* **39**, 149–70.

Lambert, J. N. and Jennings, J. N. (1951). Alluvial stratigraphy and vegetational succession in the region of the Bure Valley Broads II. *Journal of Ecology* **39**, 120–48.

Leicester Polytechnic (1986). *Status and ecology of the warty newt (Triturus cristatus).* Final Report. CSD Report **703**, Nature Conservancy Council, Peterborough.

Letts, J. B. (1993). *Smoke-blackened thatch (SBT): a new source of late medieval plant remains from southern England.* Oxford Environmental Archaeology Unit.

Letts, J. B. (1999). *Smoke blackened thatch. A unique source of late mediaeval plant remains from Southern England.* English Heritage and University of Reading.

Lindsay, R. (1995). *Bogs: the ecology, classification and conservation of ombrotrophic mires.* Scottish Natural Heritage, Perth.

Linsell, S. (1990). *Hickling Broad and its wildlife. The story of a famous wetland nature reserve.* Dalton, Lavenham.

Löfroth, M. (1991). Våtmarkerna och deras betydelse. *Naturvårdsverket Rapport* **3824**.

Lugo, A. E., Brinson, M. and Brown, S. (eds) (1990). *Ecosystems of the world. 15 forested wetlands.* Elsevier, Amsterdam.

Lund, M., Davis, J. and Murray, F. (1991). The fate of lead from duck-shooting and road run-off in three Western Australian wetlands. *Australian Journal of Marine and Freshwater Research* **42**, 139–49.

McVean, D. N. and Ratcliffe, D. A. (1962). *Plant communities of the Scottish Highlands.* Nature Conservancy Monograph **1**. HMSO, London.

Meade, R. and Blackstock, T. H. (1988). The impact of drainage on the distribution of rich-fen plant communities in two Anglesey basins. *Wetlands* **8**, 159–77.

Merritt, A. (1994). *Industrial wetlands.* Wildfowl & Wetlands Trust, Slimbridge, Gloucester.

Mesleard, F., Tan Ham, L., Boy, V., van Wijick, C. and Grillas, P. (1993). Competition between an introduced and an indigenous species: the case of *Paspalum paspalodes* (Michx) Schribner and *Aeluropus littoralis* (Gouan) in the Camargue (southern France). *Oecologia* **94**, 204–10.

Miller, S. H. and Skertchly, S. B. J. (1878). *The fenland, past and present.* Leach & Sons, Wisbech.

Mitsch, W. (ed.) (1994). *Global wetlands: old world and new.* Elsevier, Amsterdam.

Mitsch, W. J. and Gosselink, J. G. (1993). *Wetlands*, 2nd edition. Van Nostrand Reinhold, New York.

Moen, A. (ed.) (1995). Regional variation and conservation of mire ecosystems. *Gunneria* **70**.

Moir, J. and Letts, J. (1999). Thatch. Thatching in England 1790–1940. *English Heritage Research Transactions* **5**, 1–218.

Moore, P. D. and Bellamy, D. J. (1974). *Peat lands.* Springer-Verlag, New York.

Moshiri, G. A. (ed.) (1993). *Constructed wetlands for water quality improvement.* Lewis Publishers, Boca Raton, Florida.

Mountford, J. O. and Sheail, J. (1987). *The Pembrokeshire valleys: a baseline for recording future changes in plant life.* Interim report. Vol. 1, text. CSD Report **748**, Nature Conservancy Council, Peterborough.

Neori, A., Reddy, K. R., Cisková-Koncolova, H. and Agami, M. (2000). Bioactive chemicals and biological-biochemical activities and their functions in rhizospheres of wetland plants. *The Botanical Review* **66**, 350–78.

Newberry, E. (no date). *Let's hear it for the Ridgeway.* Countryside Commission.

Newbould, C. and Mountford, O. (1997). *Water level requirements of wetland plants and animals.* English Nature, Freshwater Series vol. 5.

Norden, J. (1610). *The surveiors dialogue, very profitable for all men to peruse.* Montagu, London (1738 edition).

Novák, M. and Wieder, R. K. (1992). Inorganic and organic sulphur profiles in nine *Sphagnum* peat bogs in the United States and Czechoslovakia. *Water, Air and Soil Pollution* **65**, 353–69.

Novotny, V. and Olem, H. (1994). *Water quality: prevention, identification and management of diffuse pollution.* Van Nostrand Reinhold, New York.

Odum, H. T. (2000). *Heavy metals into the environment – using wetlands for their removal.* Lewis, Florida.

Olson, R. K. (1993). *Created and natural wetlands for controlling nonpoint source pollution.* Smoley, CRC, Boca Raton, Florida.

Ostendorp, W. (1995). Effect of management on the mechanical stability of lakeside reeds in Lake Constance – Untersee. *Acta Oecologia* **16**, 277–94.

Oxford Department of External Studies (1981). *The evolution of marshland landscapes*, Oxford.

Page, S. E. and Rieley, J. O. (1992). Eutrophication and rehabilitation of Wybunbury Moss National Nature Reserve, Cheshire. In E. M. Bragg, P. D. Hulme, H. A. P. Ingram and R. A. Robertson (eds), *Peatlands ecosystems and man: an impact assessment*. British Ecological Society, International Peat Society.

Palmer, S. C. F. and Evans, P. R. (1991). *Lead in an intertidal system (Lindisfarne)*. English Nature, JNCC Report **24**, Peterborough.

Parmenter, J. (1995). *The Broadland fen resource survey*. Broads Authority and English Nature, Norwich.

Patten, B. C. (ed.) (1990). *Wetlands and shallow continental water bodies*, Vol. 1: *Natural and human relationships*. SPB Academic Publishing, The Hague.

Patten, B. C. (ed.) (1994). *Wetland and shallow continental water bodies*, 2: *Case studies*. SPB Academic Publishing, Amsterdam.

Pavaglio, F. L., Bunck, C. M. and Heinz, G. H. (1992). Selenium and boron in aquatic birds from central California.

Pearsall, W. H. (1917). The aquatic and marsh vegetation of Esthwaite Water, I. *Journal of Ecology*, 180–202.

Pearsall, W. H. (1920). The aquatic vegetation of the English Lakes. *Journal of Ecology* **8**, 163–99.

Pellegrin, D. and Hauber, D. P. (1999). Isozyme variation among populations of the clonal species, *Phragmites australis* (Cav.) Trin. ex Steudel. *Aquatic Botany* **63**, 241–59.

Piccolo, A. (1994). Interactions between organic pollutants and humic substances in the environment. In N. Senesi and T. M. Miamo (eds), *Humic substances in the global environment and implications on human health*, Elsevier Science, the Netherlands.

Piccolo, A., Celano, G. and Conte, P. (1996). Interactions between herbicides and humic substances. *Pesticide outlook* April 1996, 21–4.

Pinay, G. and Decamps, H. (1988). The role of riparian woods in regulating nitrogen fluxes between the alluvial aquifer and surface water: a conceptual model. *Regulated rivers: research and management* **2**, 507–16.

Pinay, G., Haycock, N. E., Ruffinoni, C. and Holmes, R. M. (1994). The role of denitrification in nitrogen retention in river corridors. *Proceedings of the International Conference on Wetland Management and Restoration, September 1992*. Intecol IV, Ohio, USA.

Potter, T. W. (1981). Marshland and drainage in the classical world. In *Evolution of marshland landscapes*. Oxford University Department of External Studies, pp. 1–19.

Prach, K. (1992a). Vegetation, microtopography and water table in the Luznice River floodplain, South Bohemia, Czechoslovakia. *Preslia* **64**, 357–67. Prague.

Prach, K. (1992b). Vegetation changes in a wet meadow complex, South Bohemia, Czech Republic. *Folia Geobotanica Phytotaxonomica* **28**, 9–18. Prague.

Prach, K. (1993). Vegetational changes in a wet meadow complex, South Bohemia, Czech Republic. *Folia Geobotanica Phytotaxonomica* **28**, 1–13. Prague.

Prach, K. and Rauch (1992). On filter effects of ecotones. *Ekologia (CSFR)* **11**, 293–8.

Proctor, M. (1989). Notes on mire on Dartmoor. *Reports and Transactions of the Devon Association for the Advancement of Science* **121**, 129–51.

Purseglove, J. (1989). *Taming the flood*. Oxford University Press, Oxford.

Reddy, K. R. and Smith, W. H. (eds) (1987). *Aquatic plants for water treatment and resource recovery*. Magnolia Publishers, Orlando, Florida.

Rieley, J. and Page, S. (1990). *Ecology of plant communities*. Longman, London.

Rimes, C. (1992a). *Freshwater acidification of SSSIs in Great Britain*, I: *Overview*. English Nature, Peterborough.

Rimes, C. (1992b). *Freshwater acidification of SSSIs in Great Britain*, IV: *Wales*. English Nature. Countryside Council for Wales, Bangor.

Robb, D. A. and Pierpoint, W. S. (eds) (1983). *Metals and micronutrients: uptake and utilisation by plants*. Academic Press, London.

Robe, W. E. and Griffiths, H. (1998). Adaptations for an amphibious life: changes in leaf morphology, growth rate, carbon and nitrogen investment, and reproduction during adjustment to emersion by the freshwater macrophyte *Littorella uniflora*. *New Phytologist* **140**, 9–23.

Robe, W. E. and Griffiths, H. (2000). Physiological and photosynthetic plasticity in the amphibious, freshwater plant, *Littorella uniflora*, during the transition from aquatic to dry terrestrial environments. *Plants, Cell and Environment* **23**, 1041–54.

Rodwell, J. S. (ed.) (1991a). *Woodlands and scrub*. Cambridge University Press, Cambridge.

Rodwell, J. S. (ed.) (1991b). *Mires and heaths*. Cambridge University Press, Cambridge.

Rodwell, J. S. (ed.) (1992). *Grasslands and montaine communities*. Cambridge University Press, Cambridge.

Rodwell, J. S. (ed.) (1995). *Aquatic communities, swamps and tall herb fens*. Cambridge University Press, Cambridge.

Royal Society for the Protection of Birds (1998). *Butterbump. News and advice about bitterns and reedbeds*. Sandy.

Royal Society for the Protection of Birds (1983). *Land drainage and birds in England and Wales: an interim report*. Sandy.

Royal Society for the Protection of Birds (no date). *Reedbed management for bitterns*. Sandy.

Royal Society for the Protection of Birds (no date). *Techniques of hydrological management at coastal lagoons and lowland wet grasslands on RSPB reserves*. Sandy.

Royal Society for the Protection of Birds (no date). *Wet grasslands – what future?* Sandy.

Royal Society for the Protection of Birds (no date). *Peat and peatlands*. Sandy.

Rubec, C. D. A. and Overend, R. P. (eds) (1987). *Proceedings symposium 1987. Wetlands/peatlands*. Edmonton, Alberta, Canada.

Rudescu, L., Niculescu, C. and Chivu, I. P. (1965). *Monografia stufului den delta Dunarii*. Academiei Republicii Socialiste, Romania.

Schot, P. (1991). *Solute transport by groundwater flow to wetland ecosystems*. University of Utrecht, Utrecht.

Scottish Natural Heritage (no date). *Boglands*. Perth.

Seddon, B. (1972). Aquatic plants as limnological indicators. *Freshwater Biology* **2**, 101–30.

Seidel, K. (1956). Unsere Flechtbinden. *Hydrobiologische anstall der Max-Planck-Gesellschaft druck: Enil Patzschke*, Nustadt b Coburg.

Seidel, K. (1966). Reinigung von gewässum durch höheren pflanzen. *Naturwissenschaften* **12**, 289–97.

Sellar, W. C. and Yeatman, R. J. (1930). *1066 and all that*. Methuen & Co., London.

Shaw, S. C. and Wheeler, B. D. (1990). *Comparative survey of habitat conditions and management characteristics of herbaceous poor fen types*. CSD Report **1157**, English Nature, Peterborough.

Sheppard, D. (1988). *An entomological survey of selected sites in the West Midlands, 4: Clayhanger SSSI*. England Field Survey Unit, Project No. **65**, Nature Conservancy Council, Peterborough.

Shimp, J. F., Tracy, J. C., Davis, L. C., Lee, E., Huang, W. and Erickson, L. E. (1993). Beneficial effects of plants in the remediation of soil and groundwater contaminated with organic materials. *Critical Review of Environment, Science and Technology* **23**, 41–73.

Singer, C. E. and Havill, D. C. (1985). Manganese as an ecological factor in salt marshes. *Vegetatio* **62**, 287–92.

Sketch, C. and Bareham, S. (1993). *Terrestrial SSSIs at risk from soil acidification in Wales*. Countryside Council for Wales, Bangor.

Smith, A. G. and Morgan, L. A. (1989). A succession to ombrotrophic bog in the Gwent Levels and its demise: a Welsh parallel to the peats of the Somerset Levels. *New Phytologist* **112**, 145–67.

Smith, C. M. (1992). Riparian afforestation effects on water yields and water quality in pasture catchments. *Journal of Environmental Quality* **21**, 237–45.

Stroud, D., *et al.* (1988). *A survey of moorland birds on the Isle of Lewis in 1987*. CSD Report **776**, Nature Conservancy Council, Peterborough.

Sukopp, H. and Markstein, B. (1989). Changes of the reed beds along the Berlin Havel 1962–87. *Aquatic Botany* **35**, 27–40.

Suzuki, T., Kurihara, Y. and Morigana, K. (1989). Distribution of heavy metals in a reedmarsh on a river bank in Japan. *Aquatic Botany* **35**, 121–8.

Symoens, J. J. (1988). *Vegetation of inland waters*. Kluwer Academic Publishers, Dordrecht.

Tansley, A. G. (1911). *Types of British vegetation*. Cambridge University Press, Cambridge.

Tansley, A. G. (1949). *The British Islands and their vegetation*. Cambridge University Press, Cambridge.

Thatch and thatching (1999). English Heritage.

Tithe maps (*circa* 1840).

Torstensson, L. (1990). Occurrence of pesticides in Swedish surface waters and groundwaters. *Proceedings EWRS 8th Symposium Aquatic Weeds 1990*, 215–20.

Trimmer, K. (1866). *Flora of Norfolk*. London.

US Geological Survey (1989). Humic substances in the Suwannee River, Georgia: interactions, properties and proposed structures. *Open-File Report, 87–557.*

van Dam, H. (1988). Acidification of three moorland pools in The Netherlands by acid precipitation and extreme drought periods over seven decades. *Freshwater Biology* **20**, 157–76.

Verhoeven, J. T. A. (ed.) (1992). *Fens and bogs in The Netherlands, history, nutrient dynamics and conservation*. Kluwer, Dordrecht.

Vos, C. C. and Opdam, P. (eds) (1993). *Landscape ecology of a stressed environment*. Chapman & Hall, London.

Vymazal, J. (2001) (eds). *Transformations of nutrients in natural and constructed wetlands.* Backhuys, Leiden.

Vymazal, J., Brix, H., Cooper, P. F., Green, M. B. and Haberl, R. (eds) (1998). *Constructed wetlands for wastewater treatment in Europe*. Backhuys, Leiden.

Ward, D. (1991). *River banks and their bird communities*. Hatfield Polytechnic, Occasional papers in environmental studies. *Riverbank Conservation* **11**.

Ward, D. (1994). The management of lowland wet grassland for breeding waders. *British Wildlife* **6(2)**, 89–98.

Wassen, M. (1990). *Water flow as a major landscape ecological factor in fen development*. University of Utrecht, Utrecht.

Wassen, M. J., Barendregt, A. and Schot, P. P. (1989). Groundwater chemistry and vegetation of gradients from rich fen to poor fen in the Naardermeer (The Netherlands). *Vegetation* **79**, 117–32.

Wassen, M. J., Barendregt, A., Palczunski, A. De Smidt, J. T. and De Mars, H. (1990). The relationship between fen vegetation gradients, groundwater flow and flooding on an undrained valley mire at Biebrza, Poland. *Journal of Ecology* **78**, 1106–22.

Watt, A. S. (1947). Pattern and process in the plant community. *Journal of Ecology* **35**, 1–22.

Watt, A. S. (1955). Bracken versus heather: a study in plant sociology. *Journal of Ecology* **43**, 490–56.

Wells, S. (1978). Interim report on the fungi of Woodwalton Fen National Nature Reserve. Rep. NC171K, Nature Conservancy Council, Peterborough.

Westlake, D. F., Kvet, J. and Sczepaniski, A. (eds) (1998). *The production ecology of wetlands*. Cambridge University Press, Cambridge.

Wheeler, B. D. (1980a). Plant communities of rich-fen systems in England and Wales, II: Communities of calcareous mires. *Journal of Ecology* **68**, 405–20.

Wheeler, B. D. (1980b). Plant communities of rich-fen systems in England and Wales, III: Fen meadow, fen grassland, fen woodland and contact communities. *Journal of Ecology* **68**, 761–85.

Wheeler, B. D. and Shaw, S. C. (1987). Comparative survey of habitat conditions and management characteristics of herbaceous rich fen types. *Contract Surveys* **6**. Nature Conservancy Council, Peterborough.

Wheeler, B. D. and Shaw, S. C. (1991). Above-ground crop mass and species richness of the principal types of herbaceous rich-fen vegetation of lowland England and Wales. *Journal of Ecology* **79**, 285–301.

Wheeler, B. D. and Shaw, S. C. (1992). Biological indicators of the dehydration and changes to East Anglian fens past and present. *English Nature Research Reports* **20**.

Wheeler, B. D., Shaw, S. C., Fojt, W. J. and Robertson, R. A. (1995). *Restoration of temperate wetlands.* John Wiley & Sons, Chichester.

Whitbread, A. and Curson, S. (1992). *Wildlife drying up.* Sussex Wildlife Trust.

White, G. (1789). *The natural history of Selborne.* Oxford University Press, Oxford (1993 edition).

Wilcox, D. A., Shedlock, R. I. and Hendrickson, W. H. (1986). Hydrology, water chemistry and ecological relations in the raised mound of Cowles Bog. *Journal of Ecology* **74**, 1103–17.

Williams, M. (1970). *The draining of the Somerset Levels.* Cambridge University Press, Cambridge.

Willing, M. J. (1972). *A molluscan survey of wetlands in the Rother valley, West Sussex.* JNCC Report **76**, English Nature, Peterborough.

Wilson, K. A. and Fitter, A. H. (1984). The role of phosphorus in vegetational differentiation in a small valley mire. *Journal of Ecology* **72**, 463–73.

Wirdum, G. van (1991). *Vegetation and hydrology of floating rich fens.* Datawyse, Maastricht.

Two useful newsletters are *Aquaphyte* (for the wetter aspects) from the Aquatic (and Invasive) Plants Information and Retrieval Service, University of Florida, Gainesville, USA, and that from the International Mire Conservation Group (for peats and mires), Botanical Institute, Greifswold, Germany.

Basic Figures

Basic figures are drawn as landscape sections. On the left, hills with bog and moor, sloping down to nutrient-poor flood plain (with a raised bog). On the right, lowland hill (chalk unless marked otherwise) sloping down to richer flood plain (fen unless marked otherwise). On the right below (where inserted) are nutrient-poor sands (e.g. New Forest) with valley bog and carr by stream.

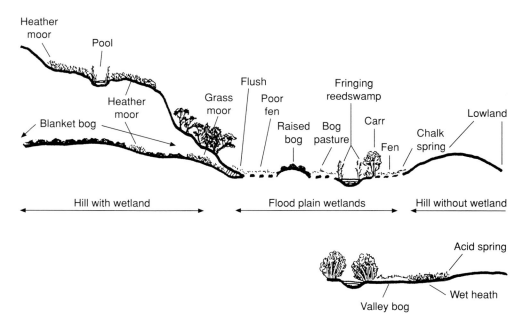

Basic Figure 1 General pattern of interlinking wetlands.

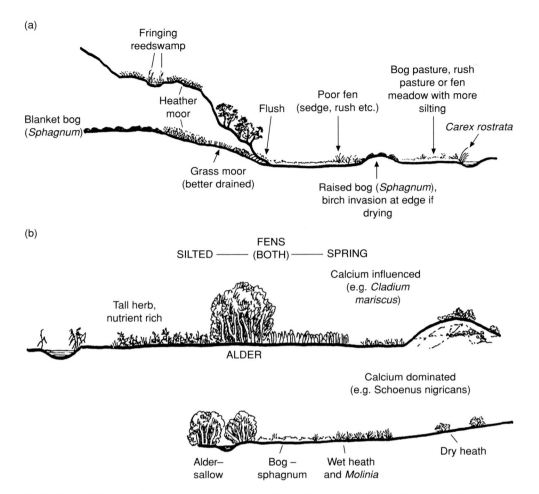

(a)

Fringing
reedswamp

Blanket bog
(*Sphagnum*)

Heather
moor

Grass moor
(better drained)

Flush

Poor fen
(sedge, rush etc.)

Bog pasture, rush
pasture or fen
meadow with more
silting

Carex rostrata

Raised bog (*Sphagnum*),
birch invasion at edge if
drying

(b)

FENS
SILTED ——— (BOTH) ——— SPRING

Calcium influenced
(e.g. *Cladium
mariscus*)

Tall herb,
nutrient rich

ALDER

Calcium dominated
(e.g. Schoenus nigricans)

Alder–
sallow

Bog –
sphagnum

Wet heath
and *Molinia*

Dry heath

Basic Figure 2 Vegetation, general pattern in 'traditional' landscape. (a) Highland. (b) Chalk-lowland and, below, New Forest. A raised bog could have developed, as in (a). Fringing reedswamp is, for example, *Glyceria maxima, Sparganium erectum, Phragmites australis.*

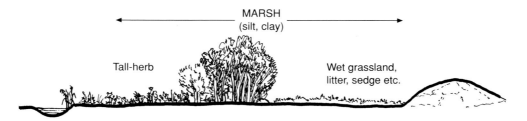

Basic Figure 3 Vegetation, general pattern in 'traditional' landscape, flood plain silt and clay.

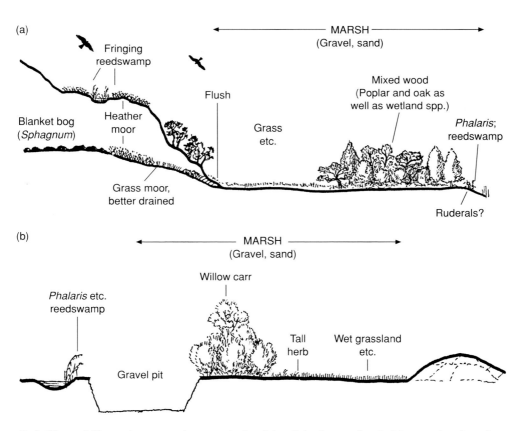

Basic Figure 4 Vegetation, general pattern in 'traditional' landscape, flood plain gravel and sand. Flood plain easily drained, so water rises and falls quickly, and no peat can develop.

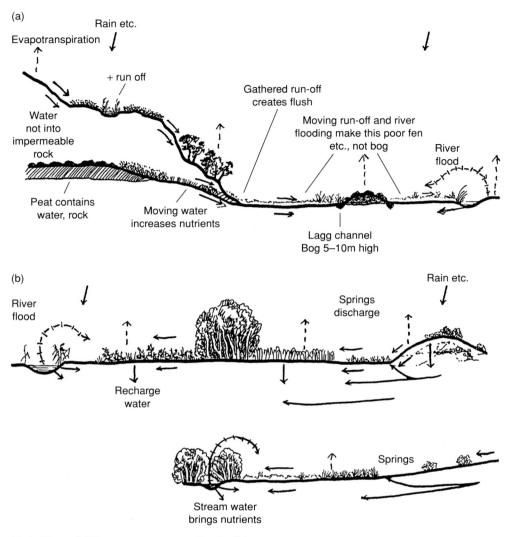

(a)

Evapotranspiration

Rain etc.

\+ run off

Water
not into
impermeable
rock

Gathered run-off
creates flush

Moving run-off and river
flooding make this poor fen
etc., not bog

River
flood

Peat contains
water, rock

Moving water
increases nutrients

Lagg channel
Bog 5–10m high

(b)

River
flood

Springs
discharge

Rain etc.

Recharge
water

Springs

Stream water
brings nutrients

Basic Figure 5 Water movement, aquifers in (b).

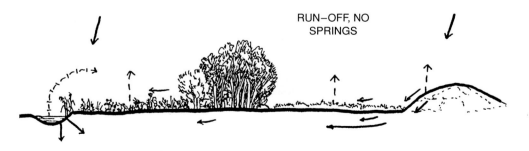

RUN–OFF, NO
SPRINGS

Basic Figure 6 Water movement, non-aquifer.

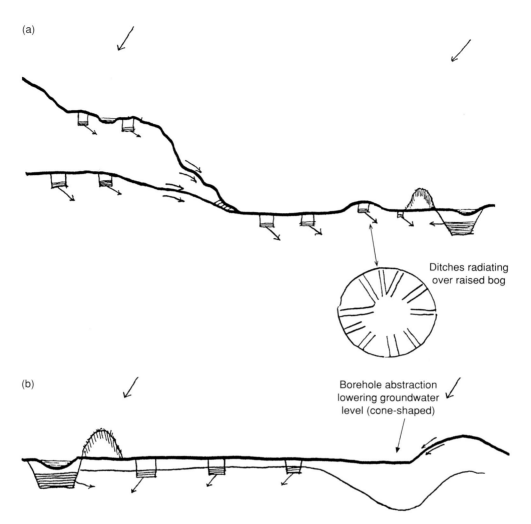

(a)

Ditches radiating
over raised bog

(b)

Borehole abstraction
lowering groundwater
level (cone-shaped)

Basic Figure 7 Drained, embanked. Further drainage erodes and removes bog peat, and in the flood plain, lowers water level far enough for ditches to be redundant.

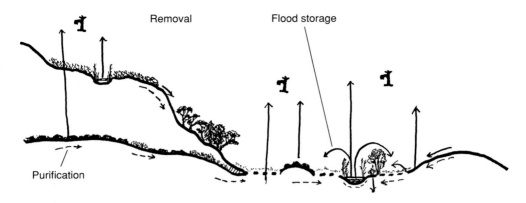

Basic Figure 8 Water resources.

(a)

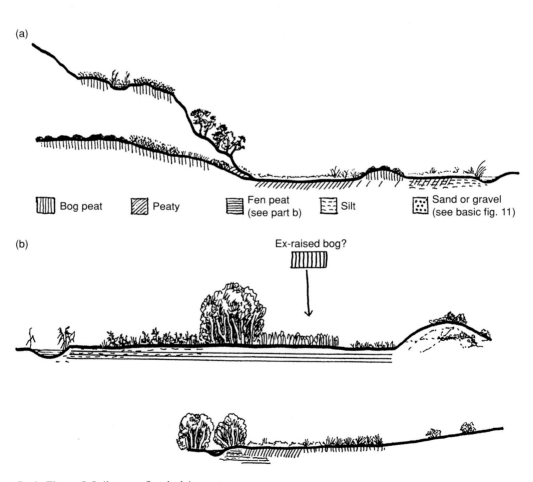

▦ Bog peat	▨ Peaty	▤ Fen peat (see part b)	⊞ Silt	⣿ Sand or gravel (see basic fig. 11)

(b)

Ex-raised bog?

Basic Figure 9 Soil types, flood plain peaty.

(a)

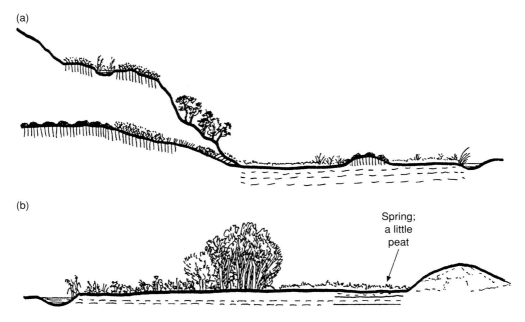

(b)

Spring;
a little
peat

Basic Figure 10 Soil types, flood plain silty. Symbols as in 9.

(a)

No raised bog

Gravel
extraction

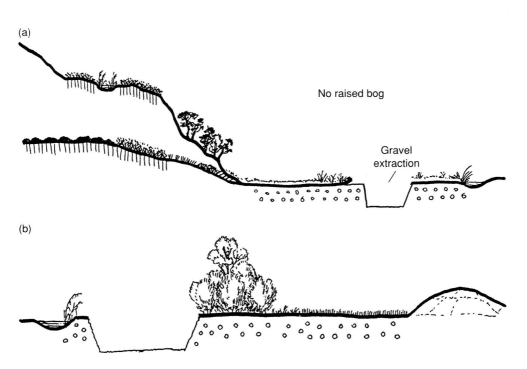

(b)

Basic Figure 11 Soil types, flood plain gravel and sand, symbols as in 9.

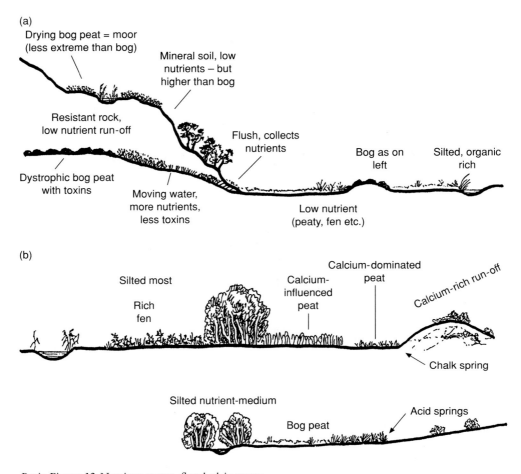

Basic Figure 12 Nutrient status, flood plain peaty.

Basic Figure 13 Nutrient status, flood plain not peaty.

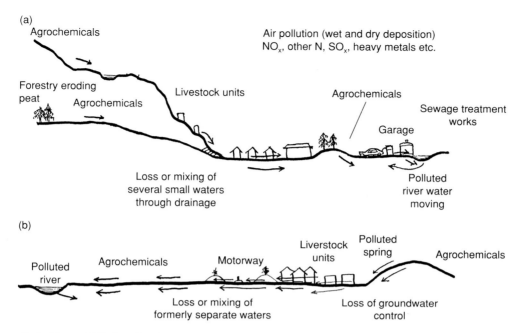

(a)

Agrochemicals

Air pollution (wet and dry deposition)
NO$_x$, other N, SO$_x$, heavy metals etc.

Forestry eroding
peat

Agrochemicals

Livestock units

Agrochemicals

Sewage treatment
works

Garage

Loss or mixing of
several small waters
through drainage

Polluted
river water
moving

(b)

Polluted
river

Agrochemicals

Motorway

Liverstock
units

Polluted
spring

Agrochemicals

Loss or mixing of
formerly separate waters

Loss of groundwater
control

Basic Figure 14 Polluted. The types and amounts vary with the site.

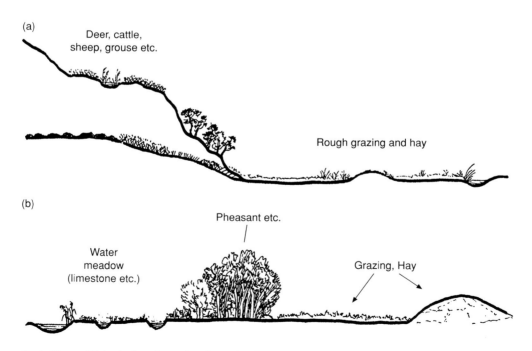

(a)

Deer, cattle,
sheep, grouse etc.

Rough grazing and hay

(b)

Pheasant etc.

Water
meadow
(limestone etc.)

Grazing, Hay

Basic Figure 15 Agricultural use, low-intensity. Intensity is not a uniform trend or progression, it
has varied greatly over the centuries, though never before as intensive as now.

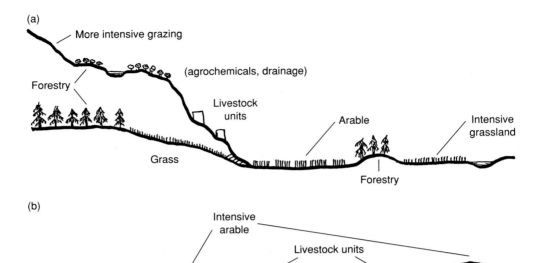

(a)

More intensive grazing

Forestry

(agrochemicals, drainage)

Livestock units

Arable

Intensive grassland

Grass

Forestry

(b)

Intensive arable

Livestock units

Basic Figure 16 Agricultural use, high-intensity.

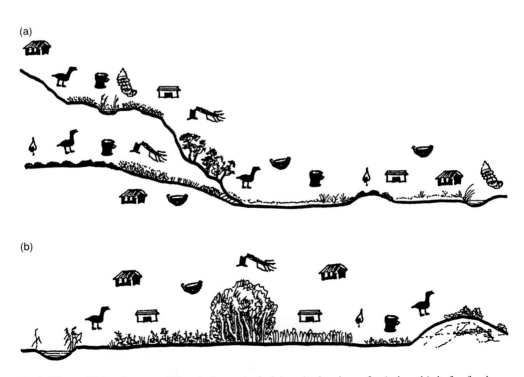

(a)

(b)

Basic Figure 17 Products, traditional: domestic fuel (peat), thatch, craft, timber, birds for food and sport, domestic water, fishing, general farm and house use. Almost all species had a use for man or beast: encouraging diversity.

(a)

(b)

Basic Figure 18 Products, recent: peat for horticulture and electricity generation, birds for
conservation (visitors and funds), other animals, especially those attracting visitors
and funds, mains water, gravel, etc. extraction, education and research.

Basic Figure 19 Recreation: walking, picnicking, sitting, bird watching, nature trails, boating,
angling, painting, bicycling, riding, driving, refreshments, etc.

Basic Figure 20 Historical and archaeological. Pollen and plant remains trapped in growing peat reflect past vegetation and human and other impact. Artefacts show past settlements, burials, etc. Vegetation is (or may be) dependent on centuries of past use.

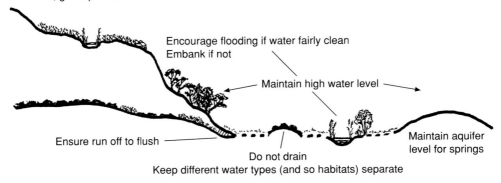

Do not drain
Care for paths
Avoid drainage from paths/roads
If erosion, get expert advice

Encourage flooding if water fairly clean
Embank if not

Maintain high water level

Ensure run off to flush

Do not drain
Keep different water types (and so habitats) separate

Maintain aquifer level for springs

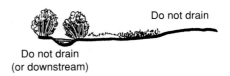

Do not drain

Do not drain
(or downstream)

Basic Figure 21 Conservation 1, water.

All require management, whether annually (graze or mow) or less often (e.g. carr). The first priority, usually, is to preserve existing types and diversity, then to return to an earlier status of habitat or vegetation, to increase revenue etc. Before change, consider ramifications. Pilot study needed. Seldom change habitat for a single species. Grazing should be light and mowing relevant to crop. Control recreation damage and avoid disturbance to breeding birds and other sensitive species. Any planting should be of local species, local strains, in habitats proper to those species.

Basic Figure 22 Conservation 2, other.

Basic Charts

Basic Chart A Principal water patterns in wetlands

Principal water source	Water level	Fluctuation (vertical)	Rising	Descending	Lateral movement
RAIN (precipitation)	Stable, around ground level[a]	Little or some	—	Little	Little, away from higher ground
RUN-OFF	Varies with rain	Great	From subsurface run-off	Yes	Away from higher ground
GROUNDWATER	Stable (or seasonably variable)	Little	Yes	—	Away from spring source
RIVER WATER	Varies with rain, catchment size and type, and spring inflow	Very great	—	Yes	Away from river

[a] Blanket and raised bog.

Basic Chart B Proportion of different water sources in habitats

Principal water source	Wet habitats formed by different water sources				
	Bog	Poor fen	Calcium-dominated fen	Rich fen[a]	Marsh[a]
RAIN (precipitation)	Rain only (raised and blanket bog); (with other sources) valley bog	Part	Minor	Minor	Minor
RUN-OFF	Valley bog[b] part	Major	Minor (locally, main)	Major	Major
GROUNDWATER	—	Minor/major[b]	Main	Minor	Minor
RIVER WATER	—	Part[b]	—	Major	Major

[a] Mostly developed with river flooding originally, usually separated from this now.
[b] Water nutrient-poor.

Basic Chart C Habitat features formed by different water sources

Principal water source	Nutrient status (unpolluted)	Principal wetland types	Soil type formed	Water level when formed
RAIN	Very low	Bog (raised and blanket)	Bog peat	Above main water level. Within or above bog's own water supply from rain. Perched
RUN-OFF[a]	Low to high, depending on catchment surface	Bog, valley Poor fen (Calcium-dominated fen) Rich fen Marsh	Bog peat Poor fen peat Rich fen peat Organic-rich silt Organic-poor silt	Formed under water. Drying causes loss (oxidation, erosion) so for peat to accumulate, drying time must be little or nil
GROUNDWATER	Depends on aquifer type, includes calcium-dominated	Calcium-dominated fen (Poor fen, marsh)	Calcium-dominated fen peat (nutrient-poor peat, silty)	
RIVER WATER	Fairly low to high, depending on catchment	Poor fen Rich fen Marsh	Mineral- to organic-rich	Deposited under water. May be dried rest of year (i.e. storm flooding only), but is then organic-poor. May be flooded all year, or anything in between

[a] Including subsurface run-off and shallow groundwater derived from near-catchment.

Basic Chart D Water regimes of different wetland types

Principal habitats (wet)	Water sources (main)	Water movement[a,b]	Water fluctuation[a]	Water level (usual)[a]
BOG	Rain (run-off)	Little	Little	At surface (for growth)
POOR FEN	(Any except solely rain)	Varies little to considerable	Some	Constant flood to constant damp (wet, for growth)
CALCIUM-DOMINATED FEN	Groundwater (or lime-dominated run-off)	Some	Little with perennial springs, more with run-off	Constantly (or seasonally) near or above surface (wet, for growth)
RICH FEN	Water carrying little or no sediment; now mixed waters	Fairly variable	Little to considerable	Constant flood to (now) constant dry (wet, for growth)
MARSH	Originally river water carrying sediment. Now mixed waters	Fairly to very variable	Little to much	Constant flood to (now) constant dry; may grow with storm flood alone

[a] Excluding severe drought and flood.
[b] Water moving through, in or over the wetland.

Basic Chart E **Soil characters in relation to sediment deposition**

Sediment deposition	Soil formed	Sediment source[a]	Nutrient status	Driest habitat for formation
HIGH	Mineral	River (maybe run-off)	Silt, very high to fairly low; higher than most of the catchment; coarser, lower than most of the catchment	Flooded enough for deposition; can be dry, wet or flooded rest of year
MEDIUM	Organic-rich mineral	1. River 2. Run-off with much sediment[b]		Flooded for deposition; wet or under water most or all of year
LOW	Peat (fen) or peaty	1. Run-off 2. River flood which has dropped most of its sediment	Depends on catchment – rock type, subsoil, soil, land use	Under water most or all of year
NEGLIGIBLE	Peat (bog or fen)	(Air)	Fen, low to medium Bog, poor	Fen, under water most or all of year Bog, able to retain rain

[a] Excluding from air.
[b] Rare in past. Now from intensive arable, construction, forestry works.

Basic Chart F **Some characteristics associated with different soil types**

Soil type	Wetland name	Water source	Habitat size	Occurrence
BOG PEAT	Blanket bog	Rain[a]	Very large to small	North and west highland; gentle slopes and flat (micro- relief can include steep)
	Raised bog	Rain[a]	Large to small	On flat ground – flood plains, fens, blanket bogs, etc., e.g. The Fenland, Ouse–Trent basin
	Valley bog	Nutrient-poor run-off	Medium to small	Nutrient-poor run-off, south to New Forest; in gently sloping valleys
FEN PEAT	Poor fen	Nutrient-poor streams, run-off, springs	Medium to small	Flood plains of nutrient-poor streams (and lakes, during development), flooded most or all of the year, mostly north and west
	Calcium-dominated fen	Calcium-dominated spring water or run-off	Not large, to small. Restricted to area controlled by this water	Most described on East Anglia chalk, also frequent, e.g. Oxon and Anglesey and occurring to north-west Scotland
	Rich fen	Run-off plus river[b,c] or river[b,c]	Large to small	Flood plains without (much) sediment deposition, and (during development) flooded most or all of the year; filling in of lakes; the peat Fens; (part) Somerset levels, etc.
MARSH (MINERAL OR PART-MINERAL SOIL)	Marsh	River[b] or river[b] plus run-off	Large to small, longitudinal along rivers; includes tiny flushes on nutrient-poor places; also flood meadows and water meadows and other penned water systems	Flood plains throughout the country, from under 10 m to tens of miles wide. The silt Fens and levels, Welsh rush (boggy) pasture, valleys of lowland brooks, etc.

[a] All receive precipitation. Rain noted only where other sources are (effectively) absent.
[b] Developed under flooding. River flooding is often prevented now by drainage and/or embanking.
[c] Sediment deposited is very little, so site is usually far from main river current.
Note: intermediates occur.

Basic Chart G Nutrient sources (simplified): processes and results

Soil	Nutrient status	Source	Conditions in which this nutrient status is maintained
LOW NUTRIENTS	1) Small nutrient input 2) Nutrient-low input	Rain; run-off from nutrient-low land; streams from, e.g. bog (some); groundwater from and in nutrient-poor land	Under water or saturated
	3) Depressed nutrient availability within the soil	Chemical reactions in peat (less in mineral soil)	Under water or saturated
	4) This *Sphagnum* etc., with the low nutrients, makes a harsh habitat	*Sphagnum*	Wet *Sphagnum* peat
HIGH NUTRIENTS	1) Large nutrient input 2) Nutrient-high input	Run-off from nutrient-high land; river water or groundwater of high status	(Various)
	3) Mineralisation, high nutrient availability	Chemical reactions in peat and, lesser, in mineral soil	Dry, and (more) dry and disturbed
	4) Added nutrients	Nutrient pollution (intentional or inadvertent)	Human additions

Basic Chart H Formation and loss of peat

1. Lake to bog

Loss   as in (2) and (3)

silt (accumulated at bottom of lake) subsoil or rock	fen peat (accumulated below water level)	bog peat (accumulated above ground level)

⟶

2. Bog only

Loss   drained and eroded; removed for fuel, horticulture (past and present)

soil or rock	bog peat (blanket or raised)

⟶

3. Fen only

Loss   drained and wasted (loss from oxidation, mineralisation, erosion) removal for fuel (more past than present)

silt subsoil or rock	fen peat

⟶

4. Cycles with varying sea/ground levels (theoretical)

subsoil or rock	bog fen silt	fen silt	bog fen silt

⟶

full cycle | cycle truncated by sea incursion before bog stage reached | sea incursion, marine clay | full cycle

5. Western Fens

clay   ↑   /// fen ///   bog

⟶

remaining peat, poor condition   arable wastage   most removed before 1900

lost

6. Patterns in profile

key
- ▨ raised bog
- ■ fen peat
- ▢ silt
- ▢ water

(a)

(b)

(c) river

(d) shallow reedswamp

(e) deeper raft at edge (leads to (b) and (f))

(f) rafts either develop from (e) or from rooted plants lifting with rise in water level (lead to (b))

Basic Chart I Formation and loss of peat (Britain): rates of formation vary with climate

	Peat formed		Human destruction	
	from	*to*	*before 1900*[a]	*after 1900*
RAISED BOG	End Ice Age	Now (local)	Much	Much
BLANKET BOG	At least from 8000 BC. (From end Ice Age?)	Now	Little	Considerable
FEN	End Ice Age	Now (local)	Much	Much

[a] The *rate* of peat loss is greater after 1900.

Basic Chart J Soil characteristics and agriculture

Soil type	Soil composition	Nutrient status	Grazing status	Capable of arable farming?	Soil loss with agriculture
BOG PEAT	*Sphagnum* peat mostly; also e.g. brown mosses (*Hypnum*, etc.) *Eriophorum* spp. small shrub spp. *Betula pubescens*	Nutrient-deficient; with *Sphagnum* effect	Poor; if with some drainage and agrochemicals, then better grazing or forestry may result	If peat lost and subsoil adequate	Complete if eroded; low if kept wet and grazed
FEN PEAT	*Phragmites* peat the commonest; also e.g. *Cladium mariscus, Carex* spp., *Schoenus nigricans*, fen carr (wood, etc.), open water, organic mud	Poor fen is nutrient-low; calcium-dominated fen is calcium-rich, other nutrients low; rich fen is nutrient-rich; disturbance and drying increase nutrients, dramatically so in calcium-dominated fen	Poor to good; increased by drainage and agrochemicals	With drainage (and agrochemicals)	Much wastage and loss
MARSH MINERAL OR PART-MINERAL SOIL	Deposited silt, and coarser sediment; often with organic material	Most are high or middle nutrients, sometimes low if from boggy or deficient catchments	Poor to good; varies with habitat and treatment	Varies with habitat and treatment	Wastage as in land soils (more if organic content high)

Note: intermediates occur.

Basic Chart K The effect of water level fluctuations on vegetation type

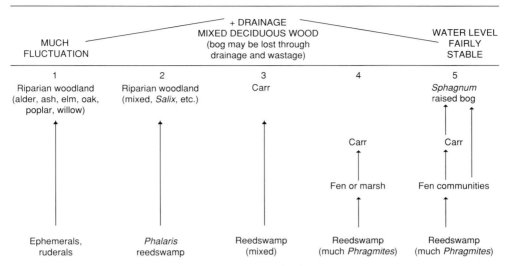

1 Hardly present in Britain now (2000). (Reservoir shores come close.)
2 Often small and fragmented in Britain. No peat. (Some gravel pits.)
3 Frequent. Peat improbable. (Some gravel pits.)
4 Remnants common. Past history of sites is stored in the peat.
5 Remnants common, active succession by bog of carr, (not now recorded), of fen vegetation (rare, and often incomplete). Past history of sites is stored in the peat.

Basic Chart L Patterns of bog development

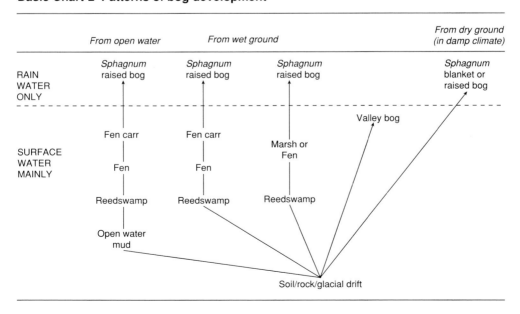

Basic Chart M Vegetation in relation to soil and water level (simplified)

Wetland type	Wet	Medium[a]	Dry[b] (wetland!)
BOG	*Sphagnum* (cotton-grass, *Eriophorum angustifolium*, etc.)	Small shrub (e.g. *Calluna vulgaris, Erica tetralix*) cotton-grass, *Eriophorum vaginatum*	As medium, or with birch and pine woods. With increased nitrogen metabolism, *Molinia*-heath
POOR FEN	Reedswamp, e.g. *Carex rostrata*	Short sedge/rush fen	Carr (sallow, willow, alder, ash, birch) short sedge, rush or grass fen
CALCIUM-DOMINATED FEN	Reedswamp, e.g. *Schoenus nigricans, Cladium mariscus*	*Cladium mariscus,* (*Cladium-*) *Schoenus-Juncus subnodulosus* Short sedge/rush fen	Carr (alder, birch, buckthorn, etc., birch nearer springs, alder further off. Trees colonise tussocks) Tall-herb communities; *Cladium*, tall-sedge vegetation. Short communities nutrient-poor or -rich
RICH FEN	Reedswamp, e.g. *Phragmites australis, Glyceria maxima, Typha* spp.	Reedbed, large-sedge, short grass, etc., fen, tall herb	Carr (sallow, willow, alder, ash, etc.), tall-herb, short grass, etc.
MARSH	Reedswamp, e.g. *Phragmites australis, Glyceria maxima, Typha* spp.	Reedbed, large-sedge, short grass, etc., fen, tall herb	Carr (as last, with more willow and poplar, oak), tall-herb, short grass, etc.

[a] Tree invasion not possible in wetter places; above that, controlled by management (grazing, mowing, burning) as much as water level.
[b] With further draining, mixed deciduous wood develops. On richer soils; oak, poplar, ash, elm increase.

Basic Chart N Vegetation patterns in relation to wetness, nutrient status and **human impact** *(representative and incomplete)*

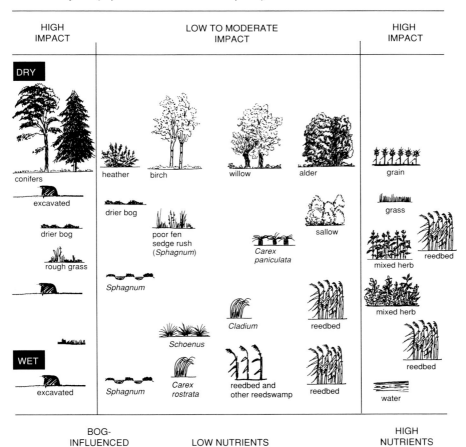

HIGH
IMPACT

LOW TO MODERATE
IMPACT

HIGH
IMPACT

DRY

conifers

excavated

drier bog

rough grass

heather   birch

drier bog

poor fen
sedge rush
(*Sphagnum*)

willow   alder

sallow

*Carex
paniculata*

grain

grass

mixed herb   reedbed

mixed herb

Sphagnum

Cladium   reedbed

reedbed

Schoenus

WET

excavated

Sphagnum

*Carex
rostrata*

reedbed and
other reedswamp   reedbed

water

BOG-
INFLUENCED

LOW NUTRIENTS

HIGH
NUTRIENTS

Basic Chart O Woodland in large flood plains, River Rhône

(modified from Petts, in Cosgrove and Petts 1993)

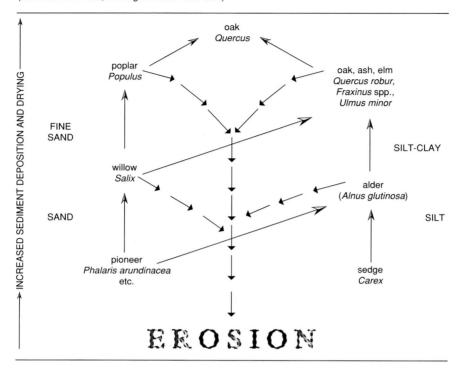

INCREASED SEDIMENT DEPOSITION AND DRYING

oak
Quercus

poplar
Populus

oak, ash, elm
Quercus robur,
Fraxinus spp.,
Ulmus minor

FINE
SAND

SILT-CLAY

willow
Salix

alder
(Alnus glutinosa)

SAND

SILT

pioneer
Phalaris arundinacea
etc.

sedge
Carex

 EROSION

Basic Chart P When herbaceous vegetation occurs instead of carr

1 TOO WET	Seedlings of wetland trees cannot tolerate flooding. Saplings of trees of the drier habitat cannot tolerate prolonged flooding. Woody species can avoid flooding by colonising tussocks
2 DEVELOPMENT TO RAISED BOG	Carr and fen wood die as bog develops (birch and pine may colonise dry bog)
3 WRONG NUTRIENT OR WATER STATUS	Site selects appropriate woody species e.g. *Salix cinerea* is not frequent in calcium-dominated peat. It is often in near-stagnant conditions; *Alnus glutinosa*, occurs with more lateral water movement; *Betula pubescens* low nutrient and calcium-dominant, drier; *Fraxinus excelsior* drier
4 FEW 'WINDOWS OF OPPORTUNITY'	Woody seedlings usually need, at the time of germination, bare soil, damp but not flooded, little frosting, etc. Seeds must not be eaten first. Such conditions may be frequent or very rare
5 MANAGEMENT	Regular mowing, grazing, burning, etc. and activities such as peat-cutting prevent tree development
6 DRAINING AND MANAGEMENT	(a) Drying often leads to mineralised invertebrate-rich peat and (in fen and marsh) tall-herb vegetation. Invasion of woody plants by seed is then difficult (see 3 and 4 above). Planted saplings do well. (If carr is present before drying, it usually persists) (b) Management for intensive grazing or arable prevents tree development

Basic Chart Q Management for some types of fen and marsh herbaceous vegetation

PHRAGMITES Commercial reedbed	Winter-cut, usually annually. (Nutrient-medium to nutrient-rich)
CLADIUM	Summer-cut every three to four years (calcium-influenced or -dominated). Commercial sedge bed
TALL GRASS Former marsh hay	Summer-cut, once or twice a year (nutrient-rich, wet)
MEDIUM GRASS-DOMINATED Community	Summer-cut, about twice in nutrient-rich habitats, once in poorer ones. Or with light grazing
WET GRASSLAND Short grass	Grazed and/or mown more intensively than the two preceding
RUSH-PASTURE	Grazed (less often mown), in nutrient-low places or ones (unintentionally) managed for *Juncus* spp.
SHORT SEDGE OR RUSH Communities	Summer grazing or mowing. Variation in the patterns of these, together with variation in soil and water, leads to a wide variety of communities. The higher the nutrient status, the more the treatment required to keep vegetation short (but less so than for wet grassland)
TALL-HERB Community	Develops from any of the others, if abandoned, and if nutrient-poor also dried to release nutrients through mineralisation, etc.

These communities are created by management. This is either prolonging into drier places communities typical of wetter ones, or creating large areas of herbaceous vegetation of types otherwise restricted to local areas of, e.g., shallow soil and disturbance. (Management here includes abandonment.) In the natural succession, these habitats would be covered by carr.

Basic Chart R Drying habitats for agriculture

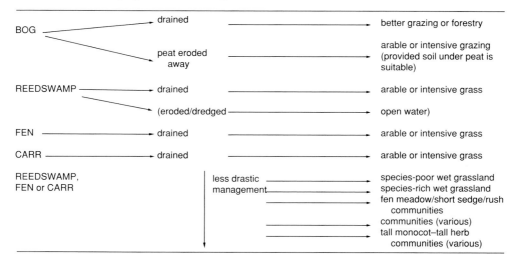

Index

Species referred to only in tables are not included in the index. Page numbers for plant name cross references are inserted once, mainly under the Latin entry. **Bold** page numbers indicate chapter or section number.

abandonment (*see also* neglect) 28, 41, 98, 102, 161, Basic Chart Q
abstraction, water *xi*, 19, 23, 52, 96, 123, 126, 141, 144–52, 155, 158–9, 161, 163–5, 227, 230–1, 234, 237–8, 240–2, 246
acid rain 154, 160, 231
acidity 45, 154, 162, 167, 176, 232
acrotelm 79, 124, 234
advancing margin, vigour 156, 194, 215, 217, 220, 222, 224–5
Aeluropus littoralis 221
aerobic, aeration 80, 98, 119, 120, 126, 134, 136, 168, 172, 208, 224, 235
agriculture 24, 27–9, 32, 47, 79, 110, 113–14, 123–4, 148, 154, 234, 237–9, 245, 249, Basic Charts J, R
agrochemicals 177, 231, 232, 242
Agrostis stolonifera 55, 68, 127, 193–4, 219, 225, 239
alder – see *Alnus glutinosa*
alder buckthorn – see *Frangula alnus*
Alfred the Great, King 10, 252
alluvial, alluvium 2, 7, 19, 24, 26, 35, 44–5, 49, 50, 61, 65, 72–5, 97, 227, 246, 251, 253, 255
Alnus glutinosa, alder 5, 12, 13, 25, 29–30, 46–8, 50, 54–5, 58, 66–8, 60, 61, 63, 73, 84, 86, 89, 90, 92, 98, 107–8, 102, 104, 107, 111, 126–7, 136–7, 136, 157, 159, 163–4, 161, 169, 175, 203–4, 209, 214, 224, 232, 239, 240–1, Basic Charts K, M, N, O, P
aluminium 154, 157–8, 175, 231, 253
ammonia (*see also* nitrogen, nitrate) 154, 232
amphibia, amphibian **118**, 125
Anagallis tenella 79, 132, 141, 156, 163, 214, 239
Angelica sylvestris 55, 67, 72, 99, 205
Anglesey 50, 141–2, 250, 254, Basic Charts F

Anglo-Saxon 2, 5, 10, 28–9, 32, 178
animals **103**
annual changes 105, 132, 138, 147, 154, 160, 180, 189, 196, 199
Anthriscus sylvestris 158
aquifer 24, 44–5, 60, 122–4, 142–3, 145–7, 157–8, 175, 232, Basic Figures 6
arable (*see also* crops) 5–7, 19, 26, 28–9, 34, 39–40, 47, 49, 57, 88, 98, 110, 113–14, 116, 118, 123, 125, 150, 175–6, 196, 203, 227–8, 236, 239–41, 246, Basic Charts E, H, J, P, R
archaeological 88, 246, Basic Figures 20
Arrhenatherum elatius 99, 133
Arthur, King 10
ash – see *Fraxinus excelsior*
Ashby Warren 241
Askham Bog 231
Athelney 10–11
auroch 83, 115
Austen, Jane 8
Azotobacter 169

bacteria 53, 118–19, 151–2, 168, 171, 173, 197
Badley Moor Fen 139–40
Barton Broad 31, 241
Barton Fen 239
Basic Charts 121 (*All at end of book. Numbers next to each entry here relate to referrals in the text*)
  Basic Chart A 43, 93, 153
  Basic Chart B 44, 93, 153
  Basic Chart C 45, 93, 153
  Basic Chart D 45, 93, 153
  Basic Chart E 43, 45–6, 93, 153
  Basic Chart F 43, 50, 93, 153
  Basic Chart G 45, 93, 153
  Basic Chart H 47, 93, 153

Basic Chart I 47, 93, 153
Basic Chart J 47, 93, 153
Basic Chart K 47, 50, 86, 89, 93, 98, 125, 153
Basic Chart L 47, 93, 98, 125
Basic Chart M 49, 98, 125
Basic Chart N 36, 49, 98, 125
Basic Chart O 49, 98, 125
Basic Chart P 50, 98, 108
Basic Chart Q 98, 108
Basic Chart R 98
Basic Figures 153, 121 *(All at end of book. Numbers next to each entry here relate to referrals in the text)*
Basic Figure 1: 6, 16, 19, 46, 50, 76–7, 86, 93, 121, 152, 226, 246
Basic Figure 2: 6, 16, 18–19, 43, 50, 76–7, 92–3, 121, 152, 157, 226, 246
Basic Figure 3: 6, 16, 19, 50, 76–7, 83, 92–3, 121, 152, 226
Basic Figure 4: 6, 19, 50, 76, 86, 92–3, 121, 152, 226
Basic Figure 5: 18–19, 50, 86, 93, 121–2, 226, 246
Basic Figure 6: 50, 93, 121, 226, 246
Basic Figure 7: 17, 50, 52, 93, 121–2, 141, 226, 228, 246
Basic Figure 8: 7, 18, 23, 50, 93, 121, 123, 246
Basic Figure 9: 50, 93, 121
Basic Figure 10: 50, 93, 121, 152
Basic Figure 11: 50, 91, 93, 121, 152
Basic Figure 12: 50, 52, 93, 121, 150, 152
Basic Figure 13: 50, 52, 93, 121, 150, 152
Basic Figure 14: 18, 50, 52, 93, 121, 153, 228
Basic Figure 15: 7, 23, 50, 93
Basic Figure 16: 7, 17, 23, 50, 52, 93, 227–8
Basic Figure 17: 23, 50, 93
Basic Figure 18: 23, 50, 93, 93, 227
Basic Figure 19: 23, 50, 93, 235
Basic Figure 20: 50, 93, 235, 246
Basic Figure 21: 93
Basic Figure 22: 93, 246
beaver 26, 83, 115–16
Beccles Marshes 164, 240
beetle 250
behaviour, plant 70, 77, 131, 132–3, 174, 178, 211, 223, 225
Beowulf 9
Betula spp., birch 54, 30, 32, 48, 54–5, 66, 68, 71, 73, 79, 87, 96, 98, 102, 107, 109, 119, 127, 137, 152, 153, 161–2, 214–15, 231, 235, 238–9, Basic Charts J, M, N, P
bicarbonate 145, 157, 159–60

biodiversity 23, 40, 41, **52**, 53–4, 56, 243–4
birch – see *Betula* spp.
birds *(see also* fowl, waterfowl) 1, 9, 12–14, 27, 30, 38–9, 41, 53–4, 56, 75, 80, 92, 104, 106–7, **110**–16, 125, 128–9, 166, 178, 198, 228, 230, 233, 235–6, 241–2, 244–7
bittern 111, 129
black bog rush – see *Schoenus nigricans*
Blackmore, R, D 12
blanket bog *(see also* bog) 6–7, 16–17, 23, 28, 31–2, 43, 46, 50–1, 56, 59, 66, 74, 78, 84, 93, 112–13, 119, 150, 153, 226–7, 234, 246, Basic Charts B, C, F, I
Blo Norton Fen 162
bog myrtle – see *Myrica gale*
bog pasture 72
bog peat *(see also* peat) 5, 24, 29, 31, 33, 43–4, 46–7, 50, 52, 79, 82–4, 231, Basic Charts C, F, H, J
bog 77
    blanket 6–7, 16, 17, 23, 28, 31, 32, 43, 46, 50–1, 56, 59, 66, 74, 78, 79–80, 81–2, 83, 84, 93, 112, 113, 119, 150, 153, 226–7, 234, 246, Basic Figures, Basic Charts
    raised 6–7, 16–17, 19, 28, 30, 32–4, 50, 52–8, 74, 75–6, 79–80, 83, 84, 96, 112, 126, 148, 153, 156, 162, 227, 231, 234, Basic Figures
    valley 6–7, 17, 44, 50, 52, 84, 100, 116, 122, 153, 157, 217, 233, 241, Basic Figures
brackish *(see also* salt) 2, 7, 32, 63, 146–9, 159–61, 163, 179, 194–5, 219, 221, 225
bramble – see *Rubus fruticosus*
Braughing 188
Briza media 96, 163
Broadland, Broad 4, 8, 14–15, 17, 21, 23, 29–34, 46, 49–50, 56, 68, 96, 98–100, 104, 111, 117–18, 133, 161–4, 232–3, **236**, 238–42
brook *(see also* river, stream, watercourse, water) 102, 145, 148, 175
bryophytes *(see also* mosses, Sphagnum) 4, 65, 82, 214, 217
buckthorn, alder – see *Frangula alnus*
buds, *Phragmites* 181, 183–7, 189, 195, 197, 206, 218
buffer strips, zones 25, 65, 167, **172**–6, 245
building, stage 20, 24, 47–8, 73, 82, 87, 96, 211–12, 214, 218, 220
Buitengoor Fen, Mol 140–1, 143, 156–7
bulrush – see *Scirpus lacustris (see also Typha* spp.)
Bure, river 83, 96

burn, burning (*see also* fire) 9, 10, 23, 32, 37, 69, 71, 75, 79, 98, 108, 165, 187, 202, 211, 213–15, 223–4, 231, 235, 238–9
bur-reed – see *Sparganium erectum*
butterfly 103, 106–7, 241
butterwort – *Pinguicula vulgaris*

caespitose (*see* tussock species) 204, 216
Calamagrostis spp. 39, 72, 98, 215, 225, 238
calcium (*see also* Fen: calcium-dominated, calcium-influenced, calcium-poor, calcium-rich, and chalk, lime, limestone, water quality) 4–5, 29, 38, 45, 69, 96, 137, 141, 145–6, 148–9, 151–2, 156–9, 160, 163–4, 175, 203, 231
Calliergon spp. 54, 66
Calluna vulgaris, heather 3–4, 9, 25, 48, 54–5, 62, 66, 71, 78, 81, 83–4, 110, 112, 119, 127, 235, 246
Camargue, France, The **221**
Car Dyke 28
Carex appropinquata 135–6, 161, 239
   C. davalliana 70, 74, 97, 157, 159, 160
   C. elata 93, 132, 137, 161, 220
   C. paniculata 55, 58, 92, 126, 129, 133–4, 136, 139, 161, 163, 204, **205**, 206, 208, 211–12, 216, 225, Basic Charts N, M
   C. riparia 55, 98, 163
   C. rostrata 13–14, 48, 55, 64, 66, 68, 71–2, 80, 83, 92–3, 116, 153, 159, 220, 239
   C. spp., other 14, 50, 55–6, 58, 61–2, 64–9, 71–2, 74, 80, 83, 86, 94, 96–7, 104, 127, 131, 133, 141, 145, 156, 158, 160, 163, 218, 238–40, Basic Charts J, O
Carices (*see also Carex*) 41, 98
Carabus spp. 110
carr 5, 7, 29–30, 32, 34, 36, 38–9, 42, 46, 48, 58, 60–1, 63, 68, 84, 89, 90, 92, 94, 97–9, 100, 102, 109, 111–14, 116–17, 126, 130, 134, 136, 153, 157, 161–4, 174–5, 179, 182, 196, 199, 203, **204**, 209, 214–15, 220, 224, 232, 239–1, 246–7, Basic Figures, Basic Charts B, C, J, K, L, M, P, Q, R
catchment 6, 19–20, 24, 29, 43–5, 60, 138, 153, 159, 165, 167, Basic Charts A, C, E
Catfield 141
Catfield Fen 161
cation 77, 96, 158
catotelm 79, 124, 234
cattle 3, 12, 14, 30, 40–1, 50, 52, 103, 115–17, 141–2, 158, 246
Cavenham 133–6, 138–9, 141, 161, 190, 191–2, 237
chalk (*see also* calcium, limestone, rock type) *xi*, 6, 19, 45, 50, 57, 61, 122, 134, 136–7, 139, 141, 145, 153, 157, 161, 163–4,

Basic Figures, Basic Figures 2, Basic Charts F
Chara 211
chloride (*see also* salt) 145–6, 149, 157, 160
Cicuta virosa 159–60
Cirsium arvense 77, 133, 158
Cirsium dissectum 54–5, 67, 96, 127
Cladium mariscoides, saw or fen sedge (America) 37, 48, **222**
Cladium mariscus, fen sedge 37, 48, 55, 66–7, 71, 76, 92–3, 96, 104, 126–7, 129, 136–7, 159, 161–2, 213, **215**–16, 219–20, **222**, Basic Charts J, M
Cladonia spp. 78, 82, 234
Clashnessie 191
classification 60–1, 64
clay (*see also* rock type) 20, 29–30, 32, 34, 46, 50, 60, 86–7, 110, 124–5, 134, 139, 145, 147, 153, 166, 171, 234, 241, Basic Figures 3, Basic Charts H, O
clone 94, 179, 184–5, 191, 193, 197, 199
Clostridium 119, 169
Clyde, river 233
Combined ills, **223**
community 15, 28, 39, 43, 48–50, 52, 54, 56, 59–62, 64, 66, 68–9, 75–6, 94, 96, 98–9, 104, 110, 117, 130, 132, 145–6, 149, 155, 160, 165–7, 175, 195–6, 203–6, 208, 211–16, 223–4, 227, 231, 235–6, 238, 241, 245–7
competition 38, 129, 132, 168–9, 181, 183, 187, 191–3, **194**, 195–7, 202–3, 220, 222, 225
conifer 73, 152, 155, 227, 231, 241
conservation (and *see* Principles of) *xi*, 23, 28, 30, 31, 33–4, 39–41, 45, 70, 75, 90, 96, 103, 109, 124, 165, 202, 233, 236, 238, 243–7
constructed wetlands (*see also* wetlands) 25, 50, 138, 146, 157, 167–8, 170–2, 178, 236, 240
construction 7, 26, 123, 160, 167, 169, 228, 240, 242, Basic Chart E
contaminant, contamination (*see also* toxin, pollution) 25, 44, 148, 152, 172, 231
CORINE 70–1, 74–6
Cors Erddreiniog 141
Cors Gogh 141
cotton grass – see *Eriophorum* spp.
Cowles Bog 148
coypu 115, 118, 236, 241
crafts 25, 27
Crestwick 163
crops (*see also* arable) 5, 9, 23, 26, 29, 31, 33, 36, 38, 48, 111, 118, 134, 170, 172–3, 175–6, 234, 236, 245
Crowland 9–10

curlew 112–14
cut, cutting 26, 28–9, 31–4, 36–8, 46, 79, 90, 95, 98, 100, 108, 121, 129–30, 134, 136, 138–9, 141, 163, 187, 195, 197–8, 202, 205, 211, 214–15, 224, 230–1, 234, 236, 238, 240, 247, Basic Charts P, Q

damage 19, 23–4, 33, 56, 111, 123, 154, 164, 166, 195, 197–8, 204, 214, 229–30, 233, 235–8, 241–2, 244, 246
Dane, Danish 5, 10
dangers and threats **226–41**
Danube, river 21, 27, 47, 73, 101, 131, 179
Dartmoor 14, 26, 82–3, 86, 56, 236
Davalliana (sedge fens) 72, 74
de Weeribben 146
Decoy Carr, Acle 240
deer 23, 26, 40–1, 104, 115–17, 166, 202, 204, 235, 246
definition(s) 165
Defoe, Daniel 12, 38, 40, 47, 117
degenerate stage 214
deposition 20, 29, 47, 86, 153, 154, Basic Charts E
   air 150, 159–60, 229
   acid 154, 231
   etc. 20, 24, 44, 150, 155, 220, Basic Charts E, F, O
Deterioration (of waterfowl and wet grassland) **241**
dieback, of reedswamp, etc. **241**
diffuse source pollution 228
Dilham Broad 239
discharge (of wetland or aquifer) 24, 122, 161, 163, 173, 175, 231–2
disease 15, 224, 242
distribution 17, 37, 50, 69–70, 81, 84, 116, 129, 145–6, 169, 204, 215, 228, 232
disturbance 44–5, 56, 66, 71, 96, 98, 104, 113, 117, 155, 173, 196–7, 203, 215, 224, 230, 235–6, 242, 246–7
ditch 49, 52, 57, 135–6, 144, 148, 157–8, 164, 175, 177, 205, 232, 234
diversity 23–6, 41, 52–3, 56, 68, 72, 75, 83, 87, 96, 98, 100, 103, 108–11, 117, 125, 134, 139, 145–6, 175–6, 226, 231, 233, 241, 246
dog 115
dome 16, 80, 141
Domesday Book 29–30, 32, 87, 133
Doyle, A, Conan 140
drainage, draining (*see also* drought, drying, water loss) *xi*, 4–7, 10–11, 16–21, 23–32, 34–6, 38, 41–3, 46–8, 50, 56, 61, 63–5, 69, 71, 73, 75, 79, 84, 86–7, 88, 91–2, 99–100, 106, 108–10, 113–14,

116, 118, 121–3, **124**, 125–6, 132–5, 137, 139, 140–2, 145–9, 150–2, 155–6, 158–62, 164–6, 171–2, 176–7, 218, 226–35, 237–42, 245–6
Drenthe plateau 144
Drenthe valleys 144, 149, 158
Drosera spp., sundews 12, 46, 62, 71, 74, 79, 81, 96, 109, 131, 137, 156, 214, 217, 239
drought (*see also* drainage, draining, drying, water loss) 44, 78, 80, 130, 159, 197, 204, 238, Basic Charts D
drying (*see also* drainage, draining, drought, water loss) 9, 20, 24, 29, 32, 36, 38, 40, 44–5, 48–9, 50, 52, 58, 65–6, 68, 71, 73, 75, 79, 84–5, 96, 98–9, 102, 109, 113, 116–18, 120, 123, 126, 129–30, 134, 141–2, 145–6, 149, 155, 159, 161–2, 164–5, 175, 187, 189, 191, 197, 199, 203, 212, 214–15, 220, 223, 229–30, 232, 235–6, 238–42, 246
Dryopteris cristatus 162, 240
duckweed – see *Lemna* spp.
dunlin 114, 233
dyke 8, 28, 40, 162, 240
dystrophic 4, 44, 92, 116, 153, 227

East Anglia *xi*, 4, 7, 19, 29, 50, 61, 124, 133, 142, 157, 159, 200, 216, 230, **236**
East Ruston Common, mown fen 237–9
edge effects 104, 106
education (*see also* study) *xi*, 11, 27, Basic Figures 18
effluent 25
Eichornia crassipes 170
Eleocharis palustris 55, 67, 68, 92, 102, 127, 168, 220
elk 83, 115
Ellenberg's moisture value 131
elm – see *Ulmus* spp.
Epilobium hirsutum, great willowherb 49, 65, 72, 102, 131, 133, 202–3, 205, 210, 224–5, 232, 238
Epipactis palustris 60, 96, 159, 214, 239
Equisetum spp. 13, 55, 67, 92, 131, 159, 213–14, 220
Erica tetralix, heath 8, 54–5, 61–2, 66, 71, 81, 127, 141, 145, 234–5, Basic Charts M
Eriophorum spp., cotton grass 4, 14–15, 30, 34, 42, 54–5, 61–2, 64, 66, 68–9, 71–2, 74, 78, 94, 96, 102, 127, 156, 164, 218, 231, 234–5, 239, Basic Charts J, M
Eriswell 132, 136–41, 148–9, 161, 213, 216, 218–19

erosion 24, 42, 48, 50, 78–80, 86, 88, 123, 125, 129, 132, 151, 155, 173, 178, 195–6, 224, 231, 233–4, 242, 246

Essex marshes 12, 22

establishment (of plants) 50, 125–6, 199, 202–4, 208, 224

Esthwaite Fen 60, 97

Etheldreda, St 9

Eupatorium cannabinum 55, 67, 72, 127, 137, 157, 217

eutrophic 63–4, 71, 73, 77, 93, 146, 155, 223

evapotranspiration 15–16, 18, 134, 136, 138–9, 140, 143, 234–5

Everglades, Florida, The **222**

Exmoor 12–13

extraction 31–2, 36, 48, 125, 228, 234–5, 237–8, 240, 242, 246, Basic Figures 18
    water 31, 48, 228, Basic Figures 18
    peat 31–2, 125, 228, 234–5, 237, 246
    gravel 36, 240, 242, Basic Figures 18
    sand 238

exudate, root 124, 129, 147, 168, 170, 172, 181, 183, 195–6, 206, 210

fables 2, **8**

Fallopia japonica 91

fen peat (*see also* peat) 5, 17, 21, 24, 26, 29, 31–3, 36, 46–7, 50, 84, 86, 93–4, 96, 136, 153, 155, 158–9, 233, Basic Charts C, F, H, J

fen sedge – see *Cladium mariscoides* (America) and *Cladium mariscus*

fen(s) (*see also* calcium, nutrient status) **93**
    calcium-dominated 5, 7, 17, 19, 31, 36–7, 44–6, 48–50, 52, 56, 59, 68, 77, 92–3, 96, 102, 122–3, 131, 137, 140–2, 148, 153, 155–6, 161–3, 192–4, 203–4, 211–12, 214–15, 220, 228, 238, 240, 244, Basic Charts B, C, P
    calcium-influenced 45, 66, 96, 141–2, 162, 215–16, Basic Charts Q
    calcium-rich 4–5, 7, 19, 37, 44, 50, 71, 83, 92–3, 96–8, 126, 135, 145, 148, 153, 157, 159, 160, 163, 230–1, 241, Basic Charts J
    calcium-poor 161

Fenland, The 4–5, 7–10, 14, 21–2, 29–31, 33, 46, 50, 87, 164, 235–6

Ferric, ferrous iron 168, 129, 145

fertility 38, 47, 52, 131, 151, 167

Festuca ovina 158

Filipendula ulmaria, meadowsweet 39, 54–5, 65, 67, 72, 91, 96, 99, 127, 133, 135–6, 139, 164, 202, 210

fire (*see also* burn, burning) 213

fish 28, 59, **110**, 150

Flixton Decoy area 240

flood meadow (*see also* meadow) 6, 22, 61, Basic Charts F

flood plain 21, 61, 86, 88–9, 110, Basic Figures 4

flood, flooding 12, 20, 24, 44–5, 61, 73, 83, 86, 111, 115, 197, Basic Charts B, C, F, P

Florida Everglades – *see* Everglades

floristics 69

flush (fens) 59

forest (*see also* New Forest) 7, 31–2, 48, 50–1, 73–4, 90, 116, 120, 150, 237, 239

forestry 235

fowl (*see also* birds, waterfowl) 14, 19, 25, 27, 28–9, 38, 57, 151

fox 115

France 7, 17, 74, 110, 221

Frangula alnus, alder buckthorn, 5, 60, 63, 98, 107, 126, Basic Charts M

Fraxinus excelsior, ash 30, 34, 47–8, 54, 67–8, 73, 107, 127, 154, 170, Basic Charts K, M, P

frog 118

frost, effects of 186–7, 190, 192–3, 195, 197, 202, 210

fuel 22, 25, 28, 31–4, 38, 41, 47, 151, 160, Basic Figures 17, Basic Charts H

Fungi **118**

Galium aparine, goosegrass 49, 69, 76, 102, 131, 133, 136, 209–11, 225
    G. palustre 52, 54–5, 61, 67–8, 94, 96, 127, 131, 202
    G. uliginosum 135–6

Garboldisham Fen 162

Glyceria maxima, reedgrass (see also *Phalaris arundinacea*) 39, 45, 55, 65, 67, 71, 74, 77, 93, 98, 145, 164, 220, 224–5, 240, Basic Figures 2, Basic Charts M

glyphosate 91, 168–9

godwit 114

goose, geese 22, 40–1, 115–16, 166, 223

goosegrass – see *Galium aparine (G, palustre, G, uliginosum)*

grass, grassland (*see also* pasture, meadow, wet grassland) 3, 5–6, 9, 13–14, 19, 22–4, 26, 29, 34–6, 38–42, 48–9, 54, 56–7, 61–4, 66, 68–9, 71–6, 84–6, 88, 90–1, 96–9, **102**–3, 105–6, 109–18, 123, 126–8, 141, 153, 156, 167, 170, 172–3, 175–6, 203, 215, 219, 221–2, 227–8, 231, 241, 245–6

grass-of-Parnassus – see *Parnassia palustris*

gravel pit 42, 52, 228

grazing 6, 8, 12, 19, 23, 26, 36, 40–1, 43, 48, 50, 56, 68–9, 83–4, 90–1, 98, 100, 102, 106, 108–9, 111–12, 114–17,

136, 141–2, 164–5, 179, 181, 187, 189, 191, 196, 198, 202, 204, 214, 217–18, 223–4, 230–1, 233, 235, 238–41, 246, Basic Charts J, M, P, Q, R
Great Ouse, river 37, 87, 226
great willowherb – see *Epilobium hirsutum*
groundwater (*see also* water) 4, 15, 18–19, 23–4, 43–4, 52, 59–61, 86, 88–9, 96, 109, 122–4, 137–8, 140–50, 152, 155–61, 163, 165, 227, 230–2
growth of plants 3, 5, 42, 46, 48, 68, 94, 106, 110, 169, 180, 183–4, 186–8, 191–5, 197, 199, 210–12, 214, 218, 221–5
  bog 44, 47, 80, 115–16
  wetlands 2, 6–7, 48, 53, 56, 83, 121, 126, 165, 172, 212, 228, 235, 243, 245
  other 20, 45, 47, 79, 157, 168, 234
Guthlac, St 9–10

habitat *xi*, 1, 5–7, 15, 17–20, 23–6, 36, 38–9, 41, 45, 47–8, 50–6, 62, 64, 66–70, 75–8, 80, 82, 84, 91–4, 96–8, 102–6, 108–15, 117–18, 121–2, 125–6, 129–32, 135, 137, 139, 144, 146, 148, 150, 152–3, 155–8, 160, 164–9, 171, 173, 175–6, 179, 182–4, 187, 191–2, 194–5, 197, 203–5, 212, 214–17, 219–20, 222–6, 228–31, 233–4, 236, 241–2, 245, 247
Halvergate Marshes 23, 30, 241
Hardness ratio, Seddon's 195, 198–9, 202
harrier 111, 129
harvest, harvesting 31, 36–8, 90, 113, 155, 187, 197–8, 227
Hatfield Moor, Chase 34
heath – see *Erica tetralix*
heather – see *Calluna vulgaris*
hedgehog 115
Heracleum montegazzianum 91
heritage 26–7, 75, 243–5, 247
heron 111–12
Hickling 15, 162, 200
historical 233–4, 228, 238, Basic Figures 20
Holcus lanatus 55, 68, 127, 158
hollow 3, 47, 60, 80, 81, 137, 144, 172
Holme Fen 5, 17, 29, 33, 124, 162
horses (*see also* ponies) 3, 39, 40–1, 116–17
Horsey-Breydon marsh 163
Hoveton Marsh 240
humic, humus 46, 77, 161, 168, 175–6, 150, 172
humid 72–3, 75–6
hummock 12, 57–8, 63, 80–2
Hydrocotyle vulgaris 61, 80, 132
hydrogeological 59–60
hydrology 76, 124, 147–8, 160, 165, 167, 176
hydromorphic 59

Ice Age 7, 17, 31, 42, 47, 83, 106, 115, 246, Basic Charts I
Icklingham 133–5, 138–9, 161, 190–2, 205–6, 237
Ijsselmeer 146
impact *xi*, 6–7, 17, 23, 28, 38, 42, 45, 47–9, 52, 74–5, 83, 85, 94, 98, 106, 116, 121, 125, 130, 139, 143, 153, 155, 164–5, 167, 177, 215, 218, 224, 235, 241
Impatiens glandulifera 91
indicator species 103, 130, **132**
indicators 10, 76, 132, 155, 241
insect 103–4, 113
intensive (farming, agriculture) 24, 113, 117, 123, 167, 227, 231, 238
International Biological Programme (IBP) 2, 64–5
invasion 38, 42, 48, 95–6, 102, 108, 116, 126, 132, 134, 137, 161–2, 179, 199, 204–5, 208, **220**, 224, 236, 238–40, 246, Basic Charts M, P
invertebrate (*see also* insect, beetle, spider, millipede, etc.) 53, 103, **104**, 105–9, 114, 134, 198, 223, 233
ion 77, 129, 145, 171
Iris pseudacorus 13, 54–5, 67, 127, 168
iron, ferrous, ferric 129, 145, 148, 154, 157, 159, 160, 168, 170, 175, 232–3

Juncus, rush 1, 3, 4, 7, 12, 13–14, 25, 29, 33, 39, 48, 69, 71, 74, 84–5, 104, 109, 141, 160–1, 195, 238, 240, Basic Charts M, N, Q, R
  J. effusus 13, 54–5, 61, 68–9, 80, 84, 96, 99, 127
  J. subnodulosus 39, 54, 66–8, 96, 99, 127, 134–6, 141, 159, 161, 195, 205–8, 216–18, 225, 232, 240
  J. spp., other 14, 62, 69, 72–4, 127, 160–1, 163, 168, 238, 240

Keeper, The **222**
Kingsley, Charles 13–15, 28

lagg channel 16, 50, 65, 71, 76
lake 3, 6, 9, 17, 19, 30, 31–2, 34, 36, 46, 52, 57, 59–60, 72–3, 77, 79, 86, 93, 96–7, 121, 148, 151, 160, 175, 220, 221, 236, 239
lapwing 112–14
Lathyrus palustris 164
lead 233
leaf-hoppers 105
Lemna spp., duckweed 14, 170
Lewis 104, 113

limestone (*see also* chalk) 17, 19, 44–5, 50, 52, 82–3, 97, 118, 122–3, 130, 141–2, 153, 156, 159, 164, 216
Liparis loeselii 96, 239–40
litter 27, 38, 41, 98, 100, 104, 108, 110, 120, 134, 163, 172, 179, 185, 193, 199, 202, 205, 214, 216–17, 220, 224, 236, 238
Littorella uniflora 232
livestock 27, 39–40, 111–12, 115–17, 160, 171, 202, 204, 218, 230, 233, 235–6, 246
lodging 224–5
long-term changes 146
Lopham Fen 140, 159, 214, 237
losses 47, 134, 138–9, 226–42
of wetlands 23, 28, 84, 124, 134, 227, 234, 236
species 38, 110, 235, 238, 240–1
etc. 28, 34, 38–9, 87, 110, 114–15, 117–18, 120, 123, 125, 136, 141, 152, 155, 159, 164–6, 169, 173, 175, 197, 223, 226, 230, 233–5, 240–2, 245, 247, Basic Charts C, H, I, J
Luznice, river 161, 163, **222**
Lychnis flos-cuculi 49, 73, 96, 102

Magnocaricetum 62, 66, 71, 98
malaria 8, 12
mallard 111–12, 129
mammals 56, **115**–16, 125, 228
management *xi*, 7–8, 24, 26, 31, 36, 38, 40, 49–50, 52, 56, 99, 102–3, 108–9, 117, 126, 129–30, 134, 136, 145, 152, 161–5, 173, 179–81, 187, 191, 196–7, 220, 225, 231, 235–6, 238–47
marsh 1–9, 11–12, 14–15, 17, 19, 20, 24, 26, 28–9, 31, 36–41, 44–5, 47–50, 52, 57, 59, 61, 65–7, 71–2, 74, 77, 84, 86–9, 98, 100, 102, 111, 114–15, 117, 129, 146, 155, 158, 161, 167, 172, 187, 195, 205, 220, 221, **226**–7, 237–40, 245
marsh hay 26, 38–9, 41, 48–9, 88, 100, 102, 245, Basic Charts Q
Martham Broad 163
mature, stage 214, 218
meadow (*see also* grassland), flood, water 3–4, 6, 29, 40, 56, 66–8, 88, 103, 110, 112, 127–8, 133, 141, 145, 153, 158–9, 163–4, 167, 172, 179, 239, 241, 245, 247
meadow pipit 112, 241
meadowsweet – see *Filipendula ulmaria*
medicines 26, 178
Mediterranean 7, 28, 71–3, 76, 196, 222
medium fen 48–9, 66, 92–3, 96–8, 157, 174
medium marsh 66–7
Menyanthes trifoliata 72, 80, 131, 239–40

mercury 233
Mere 3, 60
mesotrophic 45, 63, 145, 148
mesotrophic grassland 66, 75, 102
metal 119, 154, 164–5, 169, 171
microbes (*see also* micro-organisms) 166, 168–9
micro-organisms **118**, 152
migrant 38, 75, 104, 111, 113
migration 30, 54
migratory 113–14, 233
Mildenhall 133, 161
millipede 104
mineral 4, 6, 17–18, 35, 45–7, 52, 61, 65, 77, 80, 82–3, 85–7, 93–4, 96, 116, 121–3, 144, 150, 153, 155–6, 158, 160, 168, 171, 176, 220, 234, Basic Charts B, C, E, F, G, J
mineralisation 49, 102, 142–3, 148, 153, 155, 158–9, 162, 172, 215, 234, Basic Charts G, H, Q
mining 79
mink 115, 117–18
mire 1, 3–4, 11, 13–14, 54, 61, 64, 66–9, 71, 73–6, 118, 127, 234, 243
mixed herb Basic Charts N
moisture, soil 1–2, 119, 120, 131, 136
mole 115, 178
Molinia caerulea 41, 48, 54–5, 61–2, 66–9, 71–4, 96, 98, 109, 127, 129, 141, 156, 213–14, **215**, 225, 234
moorland 3, 5, 12, 84–5, 114, 156, 160
morass 3, 13–14
mosaic 100, 106, 109, 145, 147, 220, 244, 246
mosses (*see also* bryophytes, Sphagnum) 4–5, 71–2, 74, 77–9, 82, 86, 132, 166, 210–11, 213–14, **215**, 231, Basic Charts J
mouse 115
mow, mowing 3, 39, 41, 48, 65, 71, 75, 98–9, 104, 106, 108, 114, 136, 142, 159–60, 164, 171, 175, 216, 222, 224, 230, 233, 236, 239–40, 246, Basic Charts M, P, Q
mud 3, 11, 14, 17, 32, 72, 78, 80, 109, 113–14, 118, 133, 221, 233, 236, 238, Basic Charts J, L
mycorrhiza 79, 119, 169
Myrica gale, bog myrtle 61, 71, 73, 79–80, 96, 109, 169, **222**–3

Nardus stricta 69, 73, 231
Narthecium ossifragum 54–5, 61–2, 66, 71, 80, 119, 127, 156
National Vegetation Classification (NVC) 64, 68, 70, 74–6, 92–3, 102
neglect (*see also* abandonment) 162, 236, 238

Netherlands, The 7, 19, 34, 45, 48, 50, 79, 114, 121, 144–7, 158, 227, 231–2, 241
nettle – see *Urtica dioica*
New Forest, The 51, 79, 83, 100, 116, 122, 153–4, 157, 218, Basic Figures, Basic Figures 2, Basic Charts F
newt 118
Nitella 203
nitrogen, nitrate (*see also* ammonia) 19, 46, 49, 79, 119, 124, 151–2, 154–60, 164, 167, 169, 170–2, 175–7, 215, 229, 231–2, 249, 252–3, 255–6, Basic Charts M
nutrient status (*see also* fen, bog, mire, marsh) 44–5, 50, 61, 66–7, 96, 113, 122–3, 129, 132, 148, 155–7, 159, 161, 164–5, 192, 196, 215, 220, 225, 228, 230, Basic Figures 12, 13, Basic Charts C, E, G, J, N, Q
  poor (low) 4, 97–8, 160
  medium (intermediate) 4, 97, 121, 145, 160, 225, 230
  rich (high) 4, 97, 121, 145, 160, 225, 230
  calcium-influenced 66, Basic Charts Q
  calcium-dominated 5, 7, 17, 19, 31, 36, 37, 44–5, 46, 48–9, 50, 52, 59, 68, 77, 92–3, 96, 102, 122–3, 131, 137, 140–1, 148, 153, 155–6, 159, 162–3, 192–4, 211–12, 215, 228, 238, 240, 244, Basic Charts C, P
nutrients (*see also* water quality, pollution, calcium, trophic status) 4–5, 15, 19, 38–9, 44–6, 53, 56, 67–9, 77, 79, 83, 92–4, 96, 98, 102, 111, 115, 122, 124, 129, 132, 135, 145, 148–9, 152–3, 156, 159, 161, 165–7, 169, 171, 175, 187, 189, 191, 194–6, 203, 215, 220, 224, 230–7, 233, 237, 241

oak – see *Quercus* spp.
oligotrophic 4, 44, 52, 63, 69, 71, 92, 109, 137, 145, 153, 241
organic matter 25, 79, 150, 158, 169, 171
Ormsby Broad 239
osier (*see also* *Salix*) 102, 107, 126, 245
otter 26, 115
Oulton 164
Oxon, Oxfordshire 29, 50, 215, Basic Charts F
oyster catcher 113–14

Parnassia palustris, grass-of-Parnassus 61, 71, 74, 77, 96, 129, 137, 214, 217, 239
Parvocaricetum 62, 66, 74
Paspalum paspalodes 221
pasture (*see also* grassland) 3, 7, 23, 29, 33, 40, 68–9, 72, 99, 100, 102, 110, 123, 127, 134, 150, 167, 232, 238, 240, 245, 247

peat 3–7, 13–14, 16–17, 19, 20–2, 25–6, 28–9, 30–4, 36, 42–8, 50–3, 56, 58, 61, 64–5, 69, 74–5, 77–80, 82–8, 93–4, 96, 98, 108, 110, 112, 116, 120, 122–5, 128, 133–4, 136–7, 140–2, 144–5, 147–8, 150, 152–3, 156–9, 161–3, 165, 168, 176, 178, 192–3, 195, 203, 205, 209, 210–12, 214–15, 220, 228, 230–8, 241, 246–7
  bog 3–5, 24, 29, 31–3, 43–4, 46–7, 50, 52, 63, 79, 82–4, 129, 151, 231, Basic Figures 7, Basic Charts C, F, H, J, E
  fen 5, 17, 21, 24, 26, 29, 31–9, 36, 46–7, 50, 84, 86, 93–4, 96, 136, 153, 155, 158–9, 233, Basic Charts C, E, F, H, J
  loss 34, 155, 233, Basic Charts I
  cutting 29, 31–3, 34, 79, 98, 100, 108, 134, 136, 139, 141, 163, 215, 238, 240, Basic Charts P
penned water system (*see also* water meadow) 3, 6, 29, Basic Charts F
Pennines 12, 61, 82–5, 154, 231
Periphyton 169
pesticides 25, 39, 115, 117, 151, 155, 167, 172–3, 175, 228, 233
Phalaris arundinacea, reedgrass (see also *Glyceria maxima*) 39, 48, 55, 58, 67–8, 71–2, 77, 93, 96, 130, 131, 145, **222**, 223, 240, Basic Charts O
phosphorus, phosphate 8, 19, 45, 124, 137, 148–9, 151, 155–60, 167, 170–2, 175, 193, 223, 231–2
Phragmites australis, reed 5, 36, 42, 48, 52, 54–5, 61, 65–8, 71, 76–7, 92, 96, 108, 113, 126–7, 134, 136–7, 145, 159, 161–2, 164, **178**, 197, 203, 205, 220–1, 231, 238–9, 240–9
  community 48, 99, 132, 196, 203, 208
  peat 30, 32, 36, 42, 46, 94, 134–6, 148, 163, 178, 195, 238
  seasonal cycle 184
  behaviour 70, 77, 178
physiognomy 62, 68–9
phytosociology 68, 70, 75–6
picturesque 8–9
Pilgrim's Progress 11
pine – see *Pinus* spp.
pingo 60
Pinguicula vulgaris, butterwort 46, 54, 66, 74, 96, 127, 131, 239
Pinus spp., pine 30, 32, 35, 48, 71, 73, 79, 88, 102, 107, 119, 153, 156, 214, Basic Charts M, P
pioneer, stage 214
Po, river 7, 23
pollards 90, 108

pollution (*see also* contamination, contaminant, toxin, water quality) *xi*, 10, 18–20, 46, 49, 52, 56, 71, 79, 92, 96, 99, 104, 117–18, 151, 154–5, 157, 159, 160, 164–7, 169, 172, 176–7, 222–4, 227–33, 235, 238, 240, 242, 244, 246–7

Polygonum hydropiper 80

Polytrichum spp. 69, 83

pond 8, 41, 118, 167, 171–2, 236, 240

ponies (*see also* horses) 40, 116

pool 3, 15, 57–8, 66, 71, 76, 80, 82, 109, 127, 160, 246

pool, bog 15, 57, 66, 76, 82, 127

poplar – see *Populus* spp.

population type, *Phragmites* 189

Populus spp., poplar 26, 50, 73, 90, 107, 169, 170, 175, Basic Charts O

post–glacial, regimes 16, 32

Potamogeton pectinatus 231

Potamogeton polygonifolius 71, 87, 156

Potentilla erecta 62, 66–9, 74, 127, 137, 213–14

Potentilla palustris 61, 66, 72, 83, 94, 127, 132, 239–40

precipitation (*see also* rain) 4, 15, 79, 150, 152

Principles of Conservation **246**

processes 1–2, 19, 22–3, **42–4**, 64, 77, 103, 150, 152, 156, 158, 167–8, 171, 175–6, 197, 227

products 23, 31, 151, 158, 171, 227, 231, 236, Basic Figures 17, 18

purification, purify 25, 87, **166**, 169, 174, 177, 246

quagmire 3, 12–14, 46, 74, 83, 96, 121, 226

quaking (*see also* quagmire) 46–7, 65, 72, 74, 83, 148, 239

Quercus spp., oak 12, 29, 30, 48, 50, 73, 90, 98, 102, 107, 150, Basic Charts K, M, O

rabbit 40, 115

rain (*see also* precipitation) 4, 6–7, 15–17, 19, 43–4, 46, 58, 71, 77–9, 83, 88, 93, 96, 98, 118, 121–4, 134, 139, 141–4, 147–9, 152–4, 157–60, 163, 167, 171, 176–7, 198, 231, 234, Basic Charts A, B, C, D, E, F, G, L

raised bog (*see also* bog) 6–7, 16–17, 19, 28, 30, 32–4, 45–8, 50, 52, 74–80, 83–4, 96, 112, 126, 148, 153, 156, 162, 227, 231, 234, Basic Figures 2, Basic Charts A, F, H, I, K, L, P

Ranunculus lingua 164, 214, 239

Ranworth Broad 30, 200, 240

rat 115–16

recharge (of aquifer or wetland) 24, 122–3, 160, 161, 163, 232

recreation 22, 23, 27, 30, 124, 220, 235–6, 242, 246, Basic Figures 19

Redgrave Fen 140–1, 144, 159, 213–15, 237

redshank 112, 114, 233

reed bunting 111–12, 114, 231

reed – see *Phragmites australis* (community peat, seasonal cycle, behaviour)

reedbed 26, 36, 57, 66–7, 93, 97, 108, 126, 129, 141, 180, 182, 187, 192, 194–6, 198–9, 201–2, 227–8, 240, 242

reedbug 187

reedgrass – see *Phalaris arundinacea* and *Glyceria maxima*

Reedham Marshes 240

reedmace – see *Typha* spp.

reedswamp 5, 7, 9, 17, 24, 29–30, 32, 36, 42, 46, 48, 50, 53, 57, 60–2, 66–7, 70, 76, 84, 86, 90, **92**, 94, 97–8, 100, 106, 109, 113, 116–18, 121–2, 125, 129–30, 132, 148, 151, 153, 174, 178, 191, 194, 196, 199, 218, 220–1, 225, 231, 241–2, Basic Figures 2, Basic Charts H, K, M, N, R

reptiles **92**

resilience 106, 182, 196–7, 218, 224

resistant rock (*see* rock type) 20, 52

Rhacomitrium lanuginosum 62, 71

Rhine, river 73, 121, 148, 159, 231–2

Rhizobium 169

rhizomes 118, 159, 178, 181–4, 186–7, 189, 191–5, 197, 199, 203, 205–6, 208, 211–12, 215–16, 218, 222, 224–5

rich fen 5, 7, 17, 36–7, 45, 48–9, 52–3, 67, 71, 96, 98, 145, 157, 164, 203, 232, 239

rich marsh 7

ridge (in bog, in landscape) 78, 148, 158

riparian 71, 173

    strip 30, 65

    woodland 48, 73

river (*see also* stream, brook, watercourse, water) 3–4, 6–7, 14–15, 17–21, 24, 31, 34, 44, 50, 52, 59, 60, 77, 83, 86–9, 91–2, 104, 110, 114, 117–18, 121–3, 125, 133–4, 139, 141–2, 144, 147–8, 152–3, 155, 160, 161, 163–4, 167, 172, 174–7, 211, 220, 227–9, 231, 240

riverine 7, 61, 72, 114, 125

rock type 17, 20, 82, 109, 150, Basic Charts E

    resistant 17, 20, 52

    chalk (*see also* limestone) *xi*, 6, 19, 45, 50, 57, 137, 139, 153, 161, Basic Figures, Basic Figures 2, Basic Charts E, F, H

clay 20, 29–30, 32, 34, 46, 50, 60, 86–7, 110, 124–5, 134, 139, 145, 147, 153, 161, 171, 234, 241, Basic Figures 3, Basic Charts H, O
    sandstone 17, 19
roden 53
Romney Marsh 12, 15, 28–9, 40, 57, 87
roots 9, 77, 79–80, 129, 132, 152, 159, 165, 166, 168–9, 170–2, 174, 179, 183, 186, 193, 196–7, 208, 211, 220, 232
Rosa canina 107, 129
Rossshire 62
Roydon Fen 162
Rubus fruticosus, bramble 157, 238
ruff 113–14, 241
run-off 4, 15, 17–20, 24, 27, 29, 30, 42, 43, 44, 50, 52, 60, 71, 80, 83–4, 92–3, 96, 98, 116, 121–5, 132, 134, 137, 139, 141–6, 150–3, 155, 8, 161, 164, 167, 171–2, 174–7, 227, 229–32, 234–6
rush pasture 7, 68, 102, 127, 167, 247, Basic Charts Q
rush – see *Juncus*

saline, salinity 173
Salix cinerea, *Salix* spp., sallow 54–5, 66–8, 97–8, 126–7, 136–7, 169, 202, **204**, 208
Salix spp., willow, sallow, osier 13, 26, 48, 50, 54, 63, 66–8, 73, 83, 86, 89, 90–1, 107–8, 111–12, 127, 169, 175, 202–3, 240, Basic Charts K, O, P
sallow – see *Salix cinerea*, *Salix* spp.
salt (*see also* brackish, chloride, saline) 2, 32, 79, 154, 161–4, 181, 189, 193–6, 221
sandstone 17, 19
saw sedge – see *Cladium mariscoides*
Schoenus nigricans, black bog–rush 31, 43, 54, 64, 66, 68, 71–2, 74, 96, 98, 104, 127, 129, 131, 136–7, 149, 159, 161–2, 211, 213, **215**–16, 220, 224, 231, 239, 249, Basic Charts J, M
Scirpus fluitans 232
Scirpus lacustris, bulrush 3, 55, 67, 71, 93, 127, 153, 167–8, 220–1, 241
Scotland 24, 64, 75, 78, 83, 108, 113, 129, 155, 246
scrub 23, 54, 56, 59, 63, 73, 91, 106, 112, 114, 141, 162, 232, 237, 239
seasonal changes 106
Seddon's Hardness Ratio – *see* Hardness Ratio
Sedge fen 22, 32, 94, 97, 153, 241, 247
sedge swamp (*see also* swamp) 57, 66–7, 92, 95
sediment 6–8, 18–20, 24, 29, 44–5, 47–8, 86, 88, 171–3, 220, Basic Charts D, E, F, J, O

seepage 19–20, 106, 109, 121–2, 132, 134–6, 138, 141, 143, 145–8, 155, 159–62, 176, 205, 234, 239, 240
selenium 233
settlement 157, 164, 234–5
shade 52, 56, 104, 117, 129, 130, 187, 194–6, 199, 202, 204–5, 209, 214–15, 217–20, 222, 224–6
sheaf 38, 41, 100
sheep 12, 14, 30, 40–1, 46, 91, 103, 115–17, 235, 246
short grass 113, 219, Basic Charts M, Q
short herb **98**, 102, 111, 126
short herb community 98, 111
silt 4–7, 12, 20, 26, 28–9, 32, 40, 44–6, 49–50, 52, 77, 88, 93, 97, 110, 121, 122, 128–9, 132, 135, 145, 150–1, 153, 155, 164, 182, 220, Basic Figures 3, Basic Charts C, E, F, H, J, O
skylark 112, 241
slough 3, 11–13, 47, 226
snail 104
snipe 112, 114
sociology, plant 62–3
sodium 144, 149
soil 1–7, 14, 17, 19–20, 25–6, 35–6, 39–40, 42–7, 49, 52, 65, 77, 80, 82–3, 86–7, 90, 96, 102, 110–11, 113–14, 119, 120, 121–3, **124**–6, 128–9, 131–2, 134–6, 143–4, 146, 150–3, 155–8, 165, 167–9, 171–7, 181, 183–4, 187, 192–4, 196–7, 202, 204–5, 214–15, 218, 220, 224–5, 228, 230, 232, 234, 242
Somerset Levels 10, 12, 27–8, 40, 50, 79, 83, 87, 106, 235
Sparganium erectum, bur-reed 13, 48, 55, 67, 92–3, 220, Basic Figures 2
Sphagnum community 55, 61, 66, 68, 76, 94, 96, 145, 235
    peat Basic Charts G
    bog 13, 86
    reedlands (*see also* bryophytes, mosses) 94, 141, 159
spider 108
spongefly 108
spring 19, 20, 37, 44, 52, 59, 63, 122–3, 131–2, 136–40, 142–4, 147–8, 153, 155, 157–9, 161–2, 164, 176–7, 211, 216, 227, 232
starling 111–13
stoat 115
Stratiotes aloides 159
stream (*see also* river, brook, watercourse, water) 3, 18, 23–4, 30–1, 43, 61, 73, 83–4, 108, 115, 117–18, 122, 129, 136–8, 145–6, 153, 158, 162, 167, 174–6, 227, 246

strip (*see also* riparian, buffer) 30, 91, 118, 151, 173–6, 229

structure, of vegetation 35, 40, 102–4, 106, 108–9, 110–11, 113, 116, 128, 163, 166

study (*see also* education) 23, 50, 57–9, 70, 121, 124, 130, 146, 178, 191, 199, 201, 203, 225, 246

subsurface water 16–18, 43, 144, 172, 174

succession 48, 69–70, 86, 94, 97, 126, 208, 215, 217, 220, 241

Succisa pratensis 94, 158

sulphate, sulphur 119, 144–5, 149, 154, 157, 160, 168, 172

sundews – see *Drosera* spp.

surface water 2, 4–6, 18, 27, 48, 59, 60, 61, 71, 86, 93, 140, 146–7, 154, 160, 167, 230–1, 238

Surlingham Marsh 240

Sutton Broad Fen 239

swallowtail butterfly 106

swamp (*see also* sedge swamp) 1, 4, 54, 66–8, 75, 108, 127, 153, 167, 239, 241

  tree swamp 42, 73, 120

  sedge swamp 57, 66–7, 92, 95

swan 112

tall grass 63, 106, Basic Charts Q

tall herb 19, 49, 72–4, 91, **98**, 99, 100, 102, 113–14, 126, 134, 136, 139, 141–2, 148

tall herb community 39, 49, 66–7, 72, 74–5, 96, 98, 102, 104, 117, 142, 195, 204, 215, 224, 231, 236, 238

tall monocotyledons 4

Tansley, Sir Arthur *xi*, 4, 30, 60, 61–2, 64, 68, 70, 75–7, 86, 96–7, 116, 221

thatching 22, 25, 27, 36, 126, 134, 178, 187, 191, **197**–9, 202

Thelnetham Old Fen 217

Thorne Waste 34, (moors) 234–5

Thorpe Marshes 240

threats **226**–42

tits 111

toad 118

Torridge, river 142–3

toxin (*see also* contamination, contaminant, pollution) 215, 218

trampling 98, 102, 109, 111, 114, 132, 202, 204, 224, 246

transpiration 15, 172

Triglochin palustris 240

trophic status (*see also* nutrient status, nutrients, pollution, calcium, water quality) 63

tufa 74, 139, 140

turf 3, 8, 22, 26, 29

tussock species, caespitose 204, 216

Typha domingensis 223

Typha jamaicense **222**

Typha spp., bulrush, reedmace, 48, 64–5, 71–2, 76, 92–3, 98, 126, 168, 172, **222**, 224, Basic Charts M

Ulmus spp., elm 48, 73, 90, Basic Charts K, M, O

under drainage 125, 227, 241

Urtica dioica, nettle 23, 39, 47, 49, 54–5, 66–8, 80, 91, 96, 104, 127, 131, 133, 139, 157–8, 164, 203, 210, 215, **222**–3, 231, 238

Utricularia spp. 15, 71, 78, 211, 239

Valeriana spp. 96, 134, 136, 159

valley bog (*see also* bog) 6, 7, 17, 44, 50, 52, 84, 100, 116, 122, 153, 157, 217, 233, 241, Basic Figures, Basic Charts B, F, L

Vecht fens, etc. 148, 159

Vecht plain 147, 158, 164

vigour, plant 156, 194, 215, 217, 220, 222, 224–5

virus 53, 224

warblers 12, 91, 111–12, 129

wastage, of peat 29, 233, Basic Charts H, J, K

water level changes 125, 130, 138–9

water loss (*see also* drainage, draining, drought, drying) 136, 152, 159, 164, 234, 247

water meadow (*see also* meadow, penned water system) 3, 56, 68, 88

water quality (*see also* nutrients, pollution, calcium, trophic status) 7, 19–20, 23, 25, 27, 165, 227–8

water vole (water rat) 116–17

water (*see also* brook, groundwater, river, stream, watercourse)

  rain 16–17, 24, 42–4, 46–8, 50, 59, 78, 83, 94, 140–1, 144–8, 156, 158–60, 162, 165, 196

  level 2, 4, 6, 47, 19–20, 27, 30–1, 36–7, 42–4, 46–7, 52, 65, 67, 77, 79, 81–2, 88, 92, 109, 122–6, 128–31, 133–4, 136–9, 143–4, 146, 153, 160–3, 174–5, 192, 196–7, 202, 204, 209, 211, 214–15, 222, 225, 227, 240, 247

  table 2, 6, 17, 19, 27, 31, 45–6, 52, 69, 79, 109, 112, 123–4, 131, 133, 139, 143, 146, 148, 158, 162, 196, 211, 215, 230

  patterns 20, 49, 50, 52, 59, 109–10, 121–2, 146–7, 165, 172, 220, 230

  regime 19, 31, 37–8, 42–3, 45, 48–50, 52, 67, 77, 102, 110, 125–6, **132**–4, 136, 140, 142, 146, 149, 155–6, 163, 165, 196–7, 205, 224–5

watercourse (*see also* river, stream, brook, water) 73, 229, 231
waterfowl (*see also* birds, fowl) 12, 19, 23, 26, 38–9, 53, 104, 111, 113, 115, 150, 166, 221, 233, 241, 244, 247
waterlogging 32, 36, 50, 69, 125, 175–6, 220
Watt, Dr A, S. 133, 214, 217
weasel 115
weather 11, 109, 111, 125, 146, 160, 165, 185, 191, 210–11, 225
wet grassland 5–6, 19, 23, 35, 39–40, 48, 74, 91, 99, 105, 109, 111–13, 117, 126, 128, 167, 227, 241, 245
wet heathland 109
wet woodland 6
wetlands (*see also* constructed wetlands) *xi*, 1–2, 4, 6–10, 12, 14–21, 22–7, 23, 30–1, 38–9, 41, 44–5, 47–50, 52–4, 56–7, 59–60, 64–5, 75–6, 87, 103–4, 106, 110–11, 113–15, 117–19, 121, 123–5, 129, 133, 142–3, 145–6, 148, 150–2, 155, 158, 161, 163–72, 175–80, 182, 203, 224, 226–7, 229–30, 231–6, 238, 240, 243–6
Whitecast 164

Wicken Fen 5, 23, 34, 47, 53, 56, 98, 100, 104, 111, 162, 232, 236, 241, 244
wigeon 111–12, 233
will o' the wisp 8
willow – see *Salix* spp.
window of opportunity 194, 204, 208–9, 214, 223
withy 48
wood, woodland (*see also* riparian) 5–7, 13, 25–6, 28–30, 32, 48, 50, 53–4, 56, 58, 63, 66–8, 73, 87–8, 96, 98, 100, **102**, 104–6, 108, 110–14, 116, 120, 126–7, 152–3, 162, 175–6, 231, 235–6, 238, 240–1, 244
Woodbastwick 163
woodlice 104
woodpeckers 111
World Charter for Nature **22**, 245
Wynbunbury Moss 232

Yare, river 96, 240

zonation 69–70, 97, 220
zones (*see also* buffer strips) 25, 65, 172, 176, 245